SLEEP AND NEUROLOGIC DISEASE

SLEEP AND NEUROLOGIC DISEASE

Edited by

MITCHELL G. MIGLIS
Neurology Department
Stanford University School of Medicine
Stanford, CA, USA

ACADEMIC PRESS
An imprint of Elsevier
elsevier.com

Academic Press is an imprint of Elsevier
125 London Wall, London EC2Y 5AS, United Kingdom
525 B Street, Suite 1800, San Diego, CA 92101-4495, United States
50 Hampshire Street, 5th Floor, Cambridge, MA 02139, United States
The Boulevard, Langford Lane, Kidlington, Oxford OX5 1GB, United Kingdom

Notices
Knowledge and best practice in this field are constantly changing. As new research and experience broaden
our understanding, changes in research methods, professional practices, or medical treatment may become
necessary.

Practitioners and researchers must always rely on their own experience and knowledge in evaluating and
using any information, methods, compounds, or experiments described herein. In using such information or
methods they should be mindful of their own safety and the safety of others, including parties for whom they
have a professional responsibility.

To the fullest extent of the law, neither the Publisher nor the authors, contributors, or editors, assume any
liability for any injury and/or damage to persons or property as a matter of products liability, negligence
or otherwise, or from any use or operation of any methods, products, instructions, or ideas contained in the
material herein.

Library of Congress Cataloging-in-Publication Data
A catalog record for this book is available from the Library of Congress

British Library Cataloguing-in-Publication Data
A catalogue record for this book is available from the British Library

ISBN: 978-0-12-804074-4

For information on all Academic Press publications
visit our website at https://www.elsevier.com/

Working together
to grow libraries in
developing countries

www.elsevier.com • www.bookaid.org

Publisher: Mara Conner
Acquisition Editor: Melanie Tucker
Sr. Editorial Project Manager: Kristi Anderson
Production Project Manager: Karen East and Kirsty Halterman
Designer: Matthew Limbert

Typeset by Thomson Digital

Contents

12. Sleep and the Autonomic Nervous System

M.G. MIGLIS

List of Contributors

L. Ashbrook The Stanford Center for Sleep Sciences and Medicine, Redwood City, CA, USA

T.J. Braley University of Michigan School of Medicine; University of Michigan Multiple Sclerosis Center and Sleep Disorders Center, Ann Arbor, MI, USA

J. Cheung The Stanford Center for Sleep Sciences and Medicine, Redwood City, CA, USA

R.P. Cowan Department of Neurology and Neurological Sciences, Palo Alto, CA, USA

J.W. Day Department of Neurology and Neurological Sciences, Palo Alto, CA, USA

E.H. During The Stanford Center for Sleep Sciences and Medicine; Division of Sleep Medicine, Stanford University, Redwood City, CA, USA

R.S. Fisher Department of Neurology and Neurological Sciences, Palo Alto, CA, USA

C. Guilleminault The Stanford Center for Sleep Sciences and Medicine, Redwood City, CA, USA

M. Kawai The Stanford Center for Sleep Sciences and Medicine, Redwood City, CA, USA

D.J. Kimbrough Harvard Medical School; Partners Multiple Sclerosis Center at Brigham & Women's Hospital, Boston, MA, USA

G.J. Meskill Comprehensive Sleep Medicine Associates, Houston, TX, USA

M.G. Miglis Neurology Department, Stanford University School of Medicine, Stanford, CA, USA

E. Mignot The Stanford Center for Sleep Sciences and Medicine, Redwood City, CA, USA

R.P. Najjar Department of Psychiatry and Behavioral Sciences, Stanford University, Stanford; Mental Illness Research, Education and Clinical Center, Veterans Affairs Palo Alto Health Care System, Palo Alto, CA, USA; Department of Visual Neurosciences, Singapore Eye Research Institute (SERI), The Academia, Singapore

M. O'Hare Department of Neurology and Neurological Sciences, Palo Alto, CA, USA

B.R. Peters Pulmonary and Sleep Associates of Marin, Novato, CA, USA

B. Razavi Department of Neurology and Neurological Sciences, Palo Alto, CA, USA

C.M. Ruoff The Stanford Center for Sleep Sciences and Medicine, Redwood City, CA, USA

S. Sakamuri Department of Neurology and Neurological Sciences, Palo Alto, CA, USA

L. Schneider The Stanford Center for Sleep Sciences and Medicine, Redwood City, CA, USA

S.J. Sha Department of Neurology and Neurological Sciences, Palo Alto, CA, USA

K. Yaffe University of California, San Francisco, CA, USA

J.M. Zeitzer Department of Psychiatry and Behavioral Sciences, Stanford University, Stanford; Mental Illness Research, Education and Clinical Center, Veterans Affairs Palo Alto Health Care System, Palo Alto, CA, USA

Foreword

This is a very exciting time for the field of sleep neurology. Through knowledge gained in sleep neuroanatomy and neurophysiology, the vast and complex network of systems that regulate our states of being is just starting to be unraveled. As new discoveries challenge empirically established conventions, no field of the neurosciences is more primed for transformation than the field of sleep medicine. Sleep disorders offer a lens through which we can better understand the impact of healthy sleep on neurological function. This neurologic "black box" of sleep is just starting to be deciphered, revealing new insights into our understanding of neurological diseases.

Since the initial descriptions of rapid eye movements (REM) and their correlation with dream mentation just over half a century ago, explorations into the functions of various sleep states have rapidly expanded. Since this time, we have come to realize that monitoring of sleep physiology can provide key insights into neurologic dysfunction. Whether it's the loss of REM atonia that heralds the rostral spread of α-synuclein in neurodegenerative disease, the role of sleep and memory consolidation, or the association of sleep disruption, and fatigue in many of the central nervous system disorders that result in network inefficiency (e.g., multiple sclerosis, stroke, etc.), there are many things that sleep can teach us about neurological function.

Moreover, when the functional reserve of the brain deteriorates, sleep's powerful influence on health becomes even more apparent, as is most evident when sleep/wake disruptions in our elderly inpatient populations induce delirium. With stroke incidence peaking in the early morning hours and certain seizure subtypes occurring predominantly at night, the role of our body's internal clock becomes readily apparent. There are countless examples of how addressing the quality and quantity of sleep can impact the lives of our patients, such as the morning headaches that are precipitated by a number of sleep-related factors, white matter disease burden correlating with obstructive sleep apnea, and impaired amyloid clearance during states of sleep deprivation and the implications this may have for those with cognitive impairment.

And, don't forget that sleep disorders are neurologic diseases! Once housed squarely in the realm of psychiatry, narcolepsy type 1 is now known to be related to a loss of hypocretin neurons in the hypothalamus, correlating well with symptoms experienced by Ma-2-associated encephalitis and NMO patients who bear hypothalamic pathology. Additionally, the sensory integration issues related to restless legs syndrome are starting to be elucidated as we search to explain the efficacy, but inevitable augmentation, of dopamine agonist medications. Even sleep apnea, the bread and butter of all sleep specialists, is under a variety of elaborate neurologic control mechanisms, an association which becomes readily apparent when we look at how prevalent sleep apnea is in the acute stroke period.

There is still so much to learn, and the explanations for why we sleep are speculative at best. Each state of being (from wake to NREM to REM) serves different functions, and it is time to return our focus to the brain as we move forward in further developing our understanding of sleep. We must incorporate more than just respiratory analysis from our polysomnograms and implement more sophisticated analyses for disease characterization, thus striving for greater granularity. With a more nuanced understanding, sleep has the potential to provide the neurologic community with more robust treatment options and even opportunities for preventive medicine. Most importantly, now that we recognize the impact that poor sleep can have on neurologic disease, simply being aware of this association is of the utmost importance for better treatment, and thus better quality of life, for all of our patients.

Logan Schneider

Preface

A few years ago I was invited to speak at our department's grand rounds on a topic of my choice. I knew immediately that I wanted to present something on the subject of sleep and neurology, but on what specifically? Neurology had always been my first interest, as it is for most neurologists, but once I began my clinical rotations in medical school I quickly became interested in the field of sleep medicine. Sleep seemed to serve such important neurological functions, vital even; and yet the physiology was incompletely understood, the treatments were limited, and the discipline seemed to generate more questions than answers.

Despite the fact that we now have a greater understanding of some of these concepts, sleep still remains a topic that we understand on a relatively superficial level. How much sleep do we need? Why do we dream? Why do we *sleep*? These are questions that my patients ask me almost daily, and they remain questions that I struggle to answer, for the most part because we still do not know the answers. While frustrating in some regards, this is also part of the excitement. Of all the neurological subspecialties, sleep medicine to me possesses the greatest potential for future growth, both at the basic science level and in the clinic.

I began to think about some of these issues when I was preparing my presentation, and instead of focusing on one area of sleep and neurologic disease, as I had initially decided to, I chose to highlight several: stroke, epilepsy, headache, neuromuscular disease, neurodegenerative disease…the list quickly grew and I easily ran out of time and space, and ended up cutting a great deal. I realized at this point that there was so much to cover on this topic that a 45-min talk could, at best, only scratch the surface. The subject could easily span an entire textbook. Hence the genesis of *Sleep and Neurologic Disease*.

Our goal in the creation of this book was to provide a resource for practicing neurologists and sleep medicine physicians. In my experience, I have yet to encounter a patient in the neurology clinic who does not also have some problem with their sleep. This may be an overstatement in some regards, and partly reflective of a greater problem in our society, but it is undeniable that nearly all neurological diseases can and do result in distinct patterns of sleep disruption. Conversely, many sleep disorders also impact neurological disease. While we still have more questions than answers, our objectives are to strengthen the dialogue between sleep physicians and neurologists, to realize that sleep medicine is more than just treatment of sleep apnea, and that sleep is truly a neurological function. Patients that do not sleep well manifest clear signs and symptoms of neurological impairment.

I am indebted to the prior generation of sleep physicians and sleep scientists who helped to establish sleep as not only a viable but a necessary field of medicine, and am privileged to count some of them as colleagues. I am indebted to all of the authors of this text, who put in the time and effort to help promote this very important field. We hope that it can offer something for your practice, even in some small way, that you can use and take back to your patients. We can all attest to the power of a good night's sleep, and being able to provide this to our patients can be a very powerful thing.

Mitchell G. Miglis

Anatomy and Physiology of Normal Sleep

L. Schneider

The Stanford Center for Sleep Sciences and Medicine,
Redwood City, CA, USA

INTRODUCTION

The human brain exists in three primary states: wake, sleep with rapid eye movements (REM), and sleep without rapid eye movements (NREM). While sleep clearly subserves an essential role, so much remains unexplained about this physiologic state. Since the discovery of REM by Dement and Kleitman in the 1950s,[1] there has been much interest in the psychiatric and neurologic communities to develop a better understanding of the physiologic underpinnings and neurobehavioral correlates of sleep. Even though each of the states of sleep serve a vital role in the maintained function of all animals, as demonstrated by the physical deterioration and eventual death that some animals experience with sleep deprivation,[2-4] there remain vast

Sleep and Neurologic Disease. **http://dx.doi.org/10.1016/B978-0-12-804074-4.00001-7**

deficits in the fundamental understanding of sleep's purpose. As Allan Rechtschaffen aptly noted, "If sleep doesn't serve some vital function, it is the biggest mistake evolution ever made."[5]

Sleep is a globally coordinated, but locally propagated phenomenon. Despite incredible progress in scientific understanding of the major brain areas and neurotransmitters involved in sleep and wake, the mechanisms by which transitions between wake, NREM, and REM occur are still somewhat elusive. Even with distinct neuroanatomical regions showing clear changes in their firing patterns or neurotransmitter levels in correlation with different sleep states, the neurophysiologic monitoring of sleep indicates that sleep happens in a progressive fashion, without discrete or complete transitions between stages. In fact, activity-dependent accumulation of tumor necrosis factor alpha (TNFα), a sleep-inducing "somnogen," can promote sleep-like activity in localized cortical neuronal assemblies,[6] and sleep spindles are noted more profusely over the motor strip contralateral to motor learning tasks.[7] This suggests that sleep initiation is a property of local neuronal networks that are dependent upon prior activity specific to that network.

INITIAL DISCOVERIES OF SLEEP CIRCUITRY

As far back as the early 20th century, a basic understanding of the importance of the brain in generating sleep and wake was promoted by the neurologist Baron Constantine von Economo. Based on observations of postmortem central nervous system (CNS) lesions in patients of the encephalitis lethargica epidemic, von Economo found that those patients suffering from excessive sleepiness often had lesions at the junction of the posterior hypothalamus and midbrain, whereas those patients suffering from insomnia had lesions localized more anteriorly in the hypothalamus and the basal forebrain.[8]

However, it was not until 1935 that the first evidence of an arousal circuit was revealed when Bremer noted that transection of the brainstem at the pontomesencephalic junction (as compared to the spinomedullary junction) would produce coma in anesthetized cats.[9] Over a decade later, support for an arousal system originating in the brainstem was furthered by the work of Moruzzi and Magoun, after they demonstrated the ability to induce EEG desynchronization from slow-wave activity by stimulating the rostral pontine reticular formation in anesthetized cats.[10] Hence, the concept of the ascending reticular activating system (ARAS) was born; however, the question as to the nature of the anatomical pathways and neuronal populations that defined the ARAS remained a mystery.

Decades later, evidence of a bipartite arousal system originating from distinct neuronal populations emerged: the first, a cholinergic system, originating in the pedunculopontine (PPT) and laterodorsal (LDT) tegmental nuclei and projecting to the thalamic midline and intralaminar nuclei; the other, a monoaminergic system, bypassing the thalamus to directly activate neurons in the hypothalamus, basal forebrain, and cortex. The cholinergic neurons projecting to the thalamus serve to prevent burst firing of thalamic neurons, thereby allowing for sensory transmission to the cortex.[11] The existence of the thalamo-cortico-thalamic system is supported by the fact that thalamic relay neuronal firing patterns correlate with cortical EEG.[12] However, persistent low-amplitude, mixed-frequency EEG patterns characteristic of arousal and REM sleep can be noted despite lesions of the LDT/PPT or thalamus,[13–15] suggesting that the role of the thalamo-cortical relay is not to serve as a source of cortical arousal, but rather as a means of providing content *to* the aroused cortex.[16] This can best be illustrated by the transient lapses in conscious processing of external sensory stimuli at sleep onset[17] and the insensate nature of sleepwalking[18]—even resulting in one patient waking to severe frostbite of the feet after a somnambulistic event (Mahowald M. Personal communication, 2015). Conversely, the monoaminergic system bypasses the thalamus, projecting from brainstem nuclei directly to the lateral hypothalamic area (LHA), basal forebrain (BF), and cortex.[19,20] These neuronal populations generally demonstrate diminishing firing rates as the brain progresses from wake to NREM to REM. Conceptualizing the duality of the arousal system might best be illustrated by a comparison between REM sleep (where, despite the absence of monoaminergic tone, the cortex is still able to process sensory stimuli from the thalamocortical network) and delirium (in which a monoaminergically aroused cortex is no longer effectively processing sensory inputs due to cholinergic suppression).[21,22]

Also inherent in von Economo's initial observations was the concept of active promotion of the state of sleep. In the years following his initial observations, confirmation of the importance of more rostral brain structures in facilitating sleep was shown to be preserved across species. Insomnia-inducing lesions were initially reproduced surgically through basal forebrain and preoptic area ablation in rats by Nauta,[23] and subsequently reproduced in felines via the preoptic lesioning experiments performed by McGinty and Sterman.[24]

NEUROANATOMY AND NEUROTRANSMITTERS

Wake-Promoting Neurotransmitter Systems

Acetylcholine (ACh)

The primary locations of cholinergic neurons are in the LDT/PPT and the BF. The LDT is a heterogeneous region, lateral to the periaqueductal gray (PAG), which extends rostrally from the PPT (Fig. 1.1). These brainstem nuclei project primarily to the thalamus (the dorsal path of the bipartite arousal system), lateral hypothalamus, and basal forebrain. However, it is the release of acetylcholine into the thalamus, that is, the primary contributor to the cortical activation during wake and REM sleep.[25,26] The basal forebrain cholinergic population is comprised of the medial septum, magnocellular preoptic nucleus, diagonal band of Broca, and substantia innominanta, which are located in the region surrounding the rostral end of the hypothalamus (Fig. 1.1). Similar to the brainstem nuclei, these cholinergic neurons are primarily active during wakefulness and REM sleep, and cortical acetylcholine levels are noted to be elevated during these two states, while there is negligible release noted during NREM sleep[27] (Table 1.1). In concert with GABAergic inhibition of cortical interneurons, increased levels of cortical and hippocampal acetylcholine have been shown to result in faster EEG activity.[28,29]

Pharmacologic manipulations of the cholinergic system have led to a greater understanding of the biological pathways that contribute to REM. Muscarinic receptor subtypes, located in the pons, mediate the induction of REM sleep, as has been demonstrated in both rats and dogs.[30–32] Injections of cholinergic agents ranging from acetylcholine and nicotine to muscarinic receptor agonists and acetylcholinesterase inhibitors result in desynchronized EEG activity and can precipitate REM.[33–36] Conversely, the duration of REM is reduced and cortical slow-wave activity predominates following administration of muscarinic antagonists such as scopolamine and atropine, predominantly through their actions at the M2 receptor subtype.[37–39]

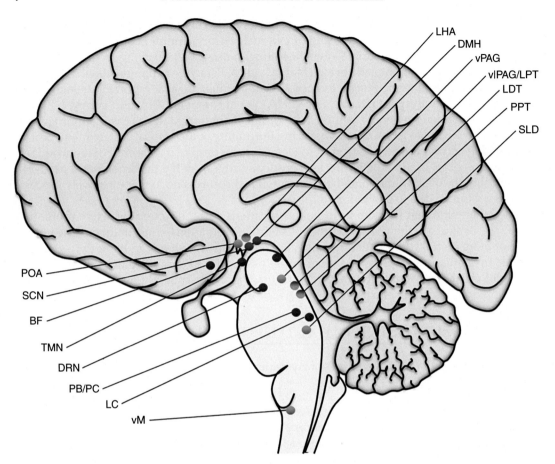

FIGURE 1.1　**General location of the neuroanatomic structures critical to wake/sleep control.** The colors of the marker indicate the predominant role played by the structure: red for arousal, blue for sleep, green for REM, purple for circadian regulation, and multicolored markers indicating multistate activity. Abbreviations: *BF*, basal forebrain; *DMH*, dorsomedial hypothalamic nucleus; *DRN*, dorsal raphe nucleus; *LC*, locus ceruleus; *LDT*, laterodorsal tegmental nucleus; *LHA*, lateral hypothalamic area; *PB/PC*, parabrachial nucleus/preceruleus; *POA*, preoptic area (containing ventrolateral and median preoptic nuclei); *PPT*, pedunculopontine tegmental nucleus; *SCN*, suprachiasmatic nucleus; *SLD*, sublaterodorsal nucleus; *TMN*, tuberomammillary nucleus; *vlPAG/LPT*, ventrolateral periaqueductal gray/lateral pontine tegmentum; *vM*, ventral medulla; *vPAG*, ventral periaqueductal gray.

Norepinephrine (NE)

Of all the noradrenergic brainstem nuclei, the locus ceruleus (LC) has the greatest influence in wake/sleep regulation. As with most of the monoaminergic system, the noradrenergic projections of the LC that promote wakefulness do so along the ventral division of the arousal system, heading from the floor of the fourth ventricle to the forebrain (Fig. 1.1). Firing rates of these neurons and extracellular NE levels are greatest during wake, and progressively drop off in NREM sleep, becoming almost quiescent during REM sleep[40–42] (Table 1.1).

TABLE 1.1 Characterization of the Firing Patterns of the Primary Sleep-Wake Regulatory Systems and Neurotransmitters

System	Primary neurotransmitters	Wake	NREMS	REMS
vPAG, LC, TMN, DRN	Monoamines (MA)	++	+	−
LDT/PPT	Acetylcholine (ACh)	++	−	++
LHA	Hypocretin (Hcrt)	++	−	−
	MCH	−	+	++
POA	GABA and galanin	−	++	++

++, Indicates high activity; +, indicates moderate activity; −, indicates little or no activity; vPAG, ventral periaqueductal gray; LC, locus ceruleus; TMN, tuberomammillary nucleus; DRN, dorsal raphe nucleus; LDT/PPT, laterodorsal tegmental/pedunculopontine tegmental nuclei; LHA, lateral hypothalamic area; POA, preoptic area; MCH, melanin concentrating hormone; GABA, gamma-aminobutyric acid.

The noradrenergic system appears to contribute to multiple aspects of wakefulness through activation of autonomic arousal and selective attention. In fact, LC neuronal firing rates are notably increased during periods of stress and exposure to salient stimuli.[40,42,43] Excessive activity of this system may underlie anxiety-associated insomnia, given the benefits of α1 antagonists such as prazosin in posttraumatic stress disorder (PTSD) patients with nightmares and insomnia.[44] Additionally, antagonists directed at the presynaptic autoinhibition through α2 receptors result in a net increase in adrenergic tone and heightened states of arousal, correlating with increased LC activity.[45] Furthermore, direct noradrenergic α1- and β-receptor stimulation of the medial septal and preoptic area of the basal forebrain promotes both behavioral and EEG measures of wakefulness.[46,47] In contrast, inhibition of the locus ceruleus, either through α2 agonism with clonidine, or α1 and β antagonism with prazosin or timolol, results in an increase in the physiologic and behavioral characteristics of NREM sleep.[48,49]

Dopamine (DA)

Dopaminergic projections are diffuse and thus integral to many neurological functions, such as motor control, learning, reward, and wakefulness. The dopaminergic systems are generally divided into four major pathways: mesolimbic, mesocortical, nigrostriatal, and tuberoinfundibular. However, neurons located in the substantia nigra (SN) and ventral tegmentum do not demonstrate firing pattern variability in response to sleep/wake changes, as they do in response to movement and reward.[50–53] More recently, dopaminergic neurons originating in the ventral periaqueductal gray (vPAG), which have reciprocal connections with the sleep-wake circuitry and lie in close approximation to the serotonergic raphe nuclei, have been shown to influence wake activity[54] (Fig. 1.1). Nonetheless, the factors influencing firing in this neuronal population have not been elucidated, although a connection to motivated physical activity as a means of volitional override of sleep onset seems most likely.

The evidence for dopamine's roll in potently promoting wakefulness is best demonstrated through the primary mechanism of stimulant medications. Amphetamines, methylphenidate, and related compounds act to prevent reuptake through dopamine transporter (DAT) blockade, but also disrupt vesicular packaging, thereby promoting dopamine release. However, these indiscriminant effects result in overactivation of reward pathways and, at higher doses, sympathetic side effects due to the added blockade of the vesicular monoamine transporter (VMAT). A more specific DAT blockade, achieved with agents like modafinil, confirms the central role of dopamine in promoting alertness in the absence of strong reward or autonomic activation.[55–58] The sedating effects of the D2 receptor agonists used in the treatment of Parkinson disease (PD) and restless legs syndrome (RLS) lend further support for the direct influence of dopamine on wakefulness.[59,60] The D2 receptor's short isoform functions as an autoinhibitor, which provides the most likely explanation for the soporific consequence of these medications.[61]

Histamine (His)

The sole source of histamine in the human brain is the tuberomammillary nucleus (TMN). Located at the base of the posterior hypothalamus, adjacent to the paired mammillary bodies, it projects to the basal forebrain and caudally to the brainstem sleep-wake circuitry (Fig. 1.1). Histamine activity, either through direct His administration or through H1 receptor agonism, augments cortical activation and EEG desynchrony.[62,63] As with the other monoamine neurotransmitter systems, the histaminergic neurons fire most readily during wake, with gradual decrements of firing in NREM and even less in REM sleep[64,65] (Table 1.1). While histamine can augment motivated behaviors such as grooming and feeding as well as psychomotor performance, it may also play a critical role in the initiation of arousal in situations mandating vigilance or at the start of the wake period, which has been posited to underlie the "sleep drunkenness" seen in some patients with idiopathic hypersomnia.[66–68]

Perhaps the most notable examples of the histamine system's impact on wakefulness are the side effects of first generation antihistamine allergy medications (e.g., diphenhydramine). The CNS penetration of these histamine antagonists has been shown to produce sleepiness in adults, without clear changes in sleep architecture.[69] Unlike most neurotransmitters in the sleep-wake system, histamine acts not through synaptic transmission but via volume transmission. Targeted histamine receptor subtype 1 manipulations in animals have, however, prompted increases in both NREM and REM sleep.[70] While the exact mechanism of histamine's action remains to be elucidated, optogenetic experiments have confirmed the wake-promoting mechanisms of histamine through multisynaptic, reciprocal connections between the TMN and ventrolateral preoptic nucleus (VLPO).[71] Toward this end, drug development has recently focused on inverse agonists at the recently identified H3 receptor subtype, which is a presynaptic autoinhibitory receptor that may regulate wakefulness not only through regulation of histamine release and biosynthesis but also through inhibiting release of all of the other neurotransmitters essential to sleep and wake.[72,73]

Serotonin (5-HT)

Of the many serotonergic raphe nuclei that line the midline of the brainstem, the dorsal raphe nucleus (DRN) is the main neuronal population responsible for sleep-wake control (Fig. 1.1). As with the other monoaminergic systems, multiple cerebral and brainstem structures implicated in the sleep-wake circuitry receive inputs from the DRN, including the preoptic area, basal forebrain, and hypothalamus. Also, consistent with their wake-promoting behavior, serotonergic neurons tend to fire most frequently during wake, less so during NREM sleep, and have the lowest firing rate during REM sleep[74,75] (Table 1.1).

The abundance of serotonin receptor isoforms and their ubiquity make clarification of the role of serotonin in wakefulness a challenging process. Nonetheless, evidence for serotonin's involvement in sleep comes indirectly from the clinical manifestations of serotonergic medications. Selective serotonin reuptake inhibitors (SSRIs) and other antidepressants with serotonergic activity are known to suppress REM sleep by decreasing REM density and prolonging REM latency.[76] In addition, these medications may also augment REM-sleep behavior disorder (RBD), RLS, and periodic limb movements of sleep (PLMS),[77] and have been used therapeutically to prevent cataplexy, which can be thought of as the intrusion of REM atonia into the waking state.[78] More specific investigations have identified that the 1A, 1B, 2, and 3 receptor subtypes are able to promote wakefulness through their activation.[79–83] Using this novel pathway for drug development, research has recently focused on the development of a 5-HT 2 receptor antagonist as a potential treatment for insomnia.[80,82,84–86]

Hypocretin/Orexin (HCRT)

The relatively recent description of hypocretin (also known as orexin) deficiency in the sleep disorder narcolepsy type I has defined it as a critical moderator in the maintenance of wakefulness.[87–89] A relatively modest number of cells in the lateral hypothalamus are the sole source of hypocretin in the brain (Fig. 1.1). Despite the limited number of hypocretinergic neurons, their sprawling projections reach all major arousal regions, with the greatest number found in the LC and TMN.[90,91] Hypocretin neurons fire during wakefulness, with heightened activity reflecting behaviors such as grooming, feeding, and exploring[92–96] (Table 1.1). The neurons are silent during NREM and REM sleep, and optogenetic activation of the neurons in a sleeping rodent can precipitate an abrupt awakening from sleep.[97,98]

While hypocretin cell loss is the hallmark of narcolepsy type I, other neurological conditions have been associated with lower levels of cerebrospinal fluid (CSF) hypocretin. Parkinson disease, multiple systems atrophy (MSA), myotonic dystrophy type-II, Neiman-Pick, and traumatic brain injury (TBI) have all been associated hypocretin deficiency.[99–103] While the multiple roles of hypocretin are still being elucidated, it is known that it not only serves as a sleep-wake gatekeeper (noting that type I narcoleptics have normal daily sleep amounts, despite the instability in the sleep/wake state),[104] but also likely plays an important role in behavioral regulation. Through influences of somatic humoral (e.g., ghrelin and leptin) and metabolic (glucose) factors on the hypocretin system, and its direct influence on the mesolimbic reward pathways, the association between waking behaviors and the waking state is most evident in the actions of this neurotransmitter.[105–110] Playing upon this unique feature, medications targeting both of the hypocretin receptor isoforms (HCRTR1 and HCRTR2) have been developed to treat insomnia.[111] Such pharmacologic suppression of hypocretin activity has also been noted to decrease drug-seeking behaviors.[105,106,108,109]

Sleep-Promoting Neurotransmitter and Signaling Systems

Gamma-Aminobutyric Acid (GABA)

The active promotion of sleep is reflected in the global increase in inhibitory tone in the sleep-wake system. The primary source of this inhibition resides in the ventrolateral and median preoptic nuclei (VLPO and MnPO, respectively)[112,113] (Fig. 1.1). Lesions in this region result in dramatically reduced and fragmented sleep, similar to the findings of von Economo's patients with lesions at the junction of the hypothalamus and basal forebrain.[24,114] The start of MnPO neuronal firing immediately preceding NREM sleep suggests that this region may play a role in sleep initiation.[115,116] In contrast, the VLPO neurons are implicated in the maintenance of sleep because they fire most vigorously during NREM sleep, with scant firing during REM, and virtual silence during wakefulness.[116,117] Both the VLPO and MnPO project to the primary arousal system nuclei (LDT/PPT, LC, DR, TMN, and HCRT neurons), where they promote sleep through both GABA and galanin.[112,113] This reciprocal inhibition between the wake- and sleep-promoting systems allows for smooth transitions between states. NREM-active, GABAergic neurons are also located in the basal forebrain and lateral hypothalamus, though their role in sleep facilitation is less clear.[118–120]

GABA has also been noted to play an important role in the production of REM sleep. A cluster of cells ventral to the locus ceruleus, known as the sublaterodorsal nucleus (SLD), plays critical roles in promoting some of the hallmark features of REM sleep.[121] As part of the mechanisms that result in the atonia of REM sleep, direct and indirect pathways from this cell group contribute to the release of GABA on the spinal cord ventral horn cells and glutamate on the ventromedial medulla (vM), respectively.[122] It has also been shown that REM sleep EEG signatures can be promoted through stimulation of the SLD, and REM sleep can be reduced through targeted ablation.[121–126] Toward this end, optic stimulation of vM neurons promotes NREM-to-REM (but not wake-to-REM) transitions and dramatic increases in REM-sleep duration, presumably through GABAergic projections onto inhibitory, REM-suppressing vlPAG neurons.[127] Nonetheless, the role of GABA in REM sleep seems to be more of an indirect, multisynaptic process of disinhibition of REM-active neurons and silencing of REM-inactive neurons, which will be discussed in more detail later.

The circadian regulation of sleep is also heavily influenced by GABA signaling. Differential coupling of the dorsal and ventral cell groups in the SCN via variations in the phasic and tonic firing patterns of GABAergic neurons are critical to aligning the master clock to the season-dependent variations in day length.[128] Additionally, GABAergic signaling from the dorsomedial hypothalamic nucleus (DMH) transmits circadian cues from the SCN to the VLPO in order to allow for environmental regulation of sleep onset.[16]

In the general facilitation of sleep, the primary receptor through which many sedative/hypnotic medications act is the GABA-A receptor subtype.[129,130] This chloride channel has binding sites for a range of sleep-inducing medications: from benzodiazepines, barbiturates, and nonbenzodiazepine receptor agonists (the "Z-drugs") to alcohol to sedatives such as propofol. However, due to the extensive network of GABA receptors and the lack of site specificity of any of these medications, the exact mechanisms by which GABA-A R agonists promote sleep are incompletely characterized. As opposed to the GABA-A ligand-gated ion channel, the GABA-B receptor is a G protein-coupled receptor that is most susceptible to activation by baclofen and gamma-hydroxybutyrate.[131] Similar to GABA-A, however, activation of the GABA-B receptor readily induces somnolence and EEG characteristics of slow wave sleep, although the mechanism remains incompletely understood.[131] Knockout models of the GABA-B receptor have revealed profound disruption of sleep patterns in mice, suggesting a role for this receptor subtype in the circadian organization of sleep.[131] In addition to tolerance and withdrawal, as well as the lack of circuit-specific agonist activity preventing the development of ideal GABA-modulating sleep aids, the dosage ceiling effect imposed by respiratory drive depression and muscle relaxation pose further concerns for the impact on comorbid sleep-disordered breathing.[132]

Acetylcholine (ACh)

A critical constituent of the wake system, acetylcholine also plays a key role in the active promotion of REM sleep. As previously mentioned, subpopulations of the LDT/PPT neuronal group are theorized to be active in both REM sleep and wake, as well as some that are active only during REM sleep[123,133–135] (Table 1.1). Thalamic ACh levels increase during REM sleep, resulting in suppression of spindle activity, while depolarized thalamic neurons are able to transmit single spikes of information to the cortex.[136] Cortical ACh levels also increase, contributing to a suppression of slow-wave activity (SWA) and promoting EEG desynchronization/cortical activation.[27] The combined thalamic transmission and activated cortex suggest that the dream content during REM sleep is the consequence of the cortex's emotional processing of sensory input and may offer particular insight into the violent content of dreams in patients with RBD: they may be dreaming out their acts, rather than acting out their dreams, an interesting theory that has not been firmly established.

Another role for ACh in the promotion of REM is through the facilitation of ponto-geniculo-occipital (PGO) waves. These waves are theorized to be important in the phasic firing patterns induced by the cholinergic subpopulation of the caudal parabrachial body (cPB), thereby facilitating the transition from NREM to REM sleep.[137] While PGO waves have only been inferred in humans through deep brain stimulation and epilepsy monitoring protocols,[137,138] their presence in cats (and, similarly, P-waves in rodents) is a hallmark of the transition to REM sleep.[139] Despite the critical balance between ACh and GABA defining PGO wave/phasic REM-sleep activity, experiments to suppress cholinergic propagation of this phenomenon do not generally result in suppression of REM sleep, suggesting that this is an independent REM sleep-related phenomenon.[140]

Not only does the direct activation of the M2 & M3 receptors in the pontine reticular formation induce REM sleep EEG and behavioral phenomena but the characteristic muscle atonia of REM sleep is also mediated, in part, by acetylcholine.[123,135,136,141] Via projections to the ventromedial medulla (vM), acetylcholine likely aids the SLD neurons in the activation of atonia-producing neurons.[123,141] Nonetheless, the drastic reductions in REM sleep caused by LDT/PPT lesions suggest that this is the primary region responsible for active REM-sleep promotion.[126,142]

Melanin-Concentrating Hormone (MCH)

Melanin-concentrating hormone (MCH) neurons provide a logical juxtaposition to the hypocretin neurons with which they are anatomically associated. GABAergic/MCHergic neurons originating in the hypothalamus parallel the projections of the hypocretin neurons providing an inhibitory activity on all of the same targets in the brainstem arousal nuclei.[143–147] MCH agonists effectively increase REM sleep quantity, while antagonists decrease it.[147,148] Furthermore, based on their firing activity being maximal in REM sleep and absent during wakefulness (Table 1.1), MCH neurons are presumed to serve a role opposite to hypocretin neurons.[118] However, MCH neurons demonstrate a moderate level of firing during NREM sleep and they have been noted to result in decreases in both REM and NREM in knockout models,[149] suggesting that there is still much to learn about the contribution of MCH neurons to sleep state regulation.

Somnogens

Early investigations into the "substance" of sleep began over a century ago. Applying Koch's postulates, Ishimori (1909), followed independently by Legendre and Pieron (1913),

---

BOX 1.1

SLEEP REGULATORY SUBSTANCE CRITERIA.[153]

1. Should promote sleep (or inhibit it, if a waking substance)
2. If the SRS is inhibited, the expected state should decrease
3. Levels in the brain (or receptor sensitivity or abundance) should vary with sleep propensity
4. The SRS should act on sleep regulatory circuits
5. Changes are proportionate with pathologies that are associated with sleep/sleepiness or wake/wakefulness

---

induced sleep in normal dogs through transfusion of cerebrospinal fluid from sleep-deprived dogs.[150,151] Sixty years later, Pappenheimer and others recovered muramyl peptide (Factor S) from goats; however, this was later revealed to be a bacterial contaminant (though it still may have been soporific through induction of interleukin-1β).[152] Since these initial discoveries, much effort has been invested in discovering these nonneurotransmitter somnogens as a means of understanding the pathogenesis of sleep/wake disorders as well as for the development of more targeted therapeutics. Criteria for sleep-regulatory substances (SRSs) have been proposed (Box 1.1),[153] and a limited number of substances qualify for the promotion of NREM sleep (growth hormone-releasing hormone, adenosine, interleukin-1β, tumor necrosis factor alpha, prostaglandin D_2, and nitric oxide), REM sleep (vasoactive intestinal peptide, and prolactin), and wake (corticotrophin-releasing hormone, and ghrelin). Only a few of the well studied, sleep-promoting SRSs will be discussed here.

ADENOSINE/ADENOSINE TRIPHOSPHATE (ATP)

Perhaps the most well known SRS is adenosine. First proposed in 1984 by Radulovacki and coworkers, adenosine remains the best example of an SRS underlying the homeostatic sleep drive.[154] Adenosine follows the expected pattern of a somnogen: increasing as a consequence of high metabolic activity and prolonged wakefulness and falling with recovery sleep.[154–157] The primary receptors for adenosine are in the purine P1 receptor family: the inhibitory A1 receptor, which is ubiquitous throughout the brain; and the excitatory A2a receptor, which is primarily located in the meninges underlying the VLPO. While the role of adenosine as the primary regulator of the homeostatic drive seems most apparent from the efficacy of caffeine, an adenosine receptor antagonist,[156,158–161] A1R and A2aR knockouts do not result in impaired sleep homeostasis.[16] Furthermore, the critical role in adenosine signaling played by support cells, such as astroglia, is underscored by the fact that the expected increases in sleep and delta power following sleep deprivation can be reduced by astrocytic manipulations.[156,157,162] The mechanisms of purinergic sleep regulation are elaborate. The abundance of ATP in vesicles that are coreleased with the majority of neurotransmitters (GABA, ACh, NE, and glutamate) predominantly bind to the P2 family of purine receptors, located on both the postsynaptic membrane and local glia. At the same time, ectonucleotidases convert ATP in the synaptic cleft into adenosine. Glial-based, ATP-induced release of TNFα, interleukin-1β (IL-1β), brain-derived neurotrophic factor (BDNF), and additional ATP results in NFκB-mediated transcription of adenosine 1 receptors (A1R) and glutamate AMPA receptors (AMPAR) in the postsynaptic membrane, thereby scaling the sensitivity of the postsynaptic neurons to the prior use of the synapse.[153] This augmented receptor sensitivity to adenosine as a consequence of the degree of neuronal activity is most supportive of the theory that sleep is a locally initiated phenomenon. This effect is demonstrated in the cellular network activity of cortical columns independently oscillating between sleep-like and wake-like states,[163] and is suggested electrophysiologically by augmented slow-wave activity in the hemisphere contralateral to motor learning tasks.[164,165] Further support for the homeostatic regulation of adenosine is noted in the conversion of adenosine to ATP with sufficient energy availability.[155] Thus, it is likely that it is the balance of adenosine and ATP rather than either metabolite individually that truly modulates purinergic sleep regulation.

The soporific activity of adenosine is not just a central nervous system-mediated process. In 1972, ATP was first proposed to serve a peripheral nonadrenergic/noncholinergic autonomic afferent role.[166] More recent studies have revealed the ability of peripheral intramuscular injection of combinations of metabolites (protons, lactate, and ATP) to induce global fatigue and even a sense of pain and muscle ache.[167] Thus, distortion of the concerted peripheral and central purinergic signaling of adenosine/ATP may be involved in the pathogenesis of systemic exercise intolerance disease (SEID, formerly known as chronic fatigue syndrome/myalgic encephalitis), although this has not been substantiated.

CYTOKINES

The most notable cytokines playing a role in sleep homeostasis are IL-1β and TNFα. Their role in the purinergic homeostat has already been discussed; however, they also have an independent role in the regulation of sleep. IL-1β and TNFα show characteristics of a physiologically normal sleep-regulatory substance: levels of IL-1β and TNFα increase with prolonged wakefulness, reach a maximum around sleep onset, and decline with sleep.[168–171] Additionally, NREM activity is notably increased through physiologic manipulations that increase IL-1β and TNFα, such as high-fat diets or increases in ambient temperature. Furthermore, direct application of IL-1β and TNFα (as well as other cytokines such as linoleic acid and prostaglandin D₂) to the surface of the cortex increases c-Fos activation in the VLPO and enhances delta EEG power during NREM sleep, pointing to activation of sleep-regulatory circuitry.[153] Furthermore, the NREM rebound that characteristically follows sleep deprivation can be blocked through IL-1β and TNFα antagonists, gene knockout animal models, and interfering antibodies.[168] Nonetheless, it is the pathologic manifestations of increased NREM and decreased REM sleep that result from cytokine production in the setting of infection (specifically mediated by lipopolysaccharide and muramyl peptide) that suggest a role for sleep in the recovery process.[172] However, this sleep-wake alteration may come at the cost of inducing the twilight state of delirium (typified by the characteristic encephalopathic slow-wave activity on EEG) in those individuals most susceptible.

PROSTAGLANDIN D₂ (PGD2)

As mention earlier, the application of prostaglandin D₂ results in increased c-Fos expression in the VLPO as well as enhanced EEG delta power, as a result of direct cortical application, highlighting the activation of sleep-regulatory circuits.[153] In fact, the normal production of PGD2 demonstrates the expected diurnal variation—with a maximum during the sleep period[173]—and increases with sleep deprivation.[174] Production of PGD2 is predominantly noted in the basal meninges,[175] and preoptic injections of PGD2 have been shown to activate the VLPO (increasing both NREM and REM sleep), possibly mediated by adenosine's A2aR activity.[176–180] Like most prostaglandins, PGD2 is a byproduct of cyclooxygenase action on lipid membrane fatty acid esters. The soporific role of inflammation may be mediated in part by PGD2, since patients with African sleeping sickness have been noted to have elevated cerebrospinal PGD2 levels.[181]

SLEEP-WAKE CIRCUITRY

In attempting to understand the interactions between the many neurotransmitter systems that exist to promote wakefulness, consider the following: if each of the aforementioned arousal systems can independently promote a state of arousal, then why are there so many? A possibility would be that there needs to be biologically assured redundancy in the system so that a dysfunction in any one system does not impair it entirely (Fig. 1.2). This thought is teleologically intuitive, making the rostral midbrain/posterior hypothalamus one of the only regions that can produce coma from a single lesion. However, an additional interpretation would be that each arousal system contributes a different aspect of arousal and input to the maintenance of wakefulness and the activities/behaviors necessitated therein. Attention is enhanced by norepinephrine (NE) and histamine (His) in the setting of novel or stressful stimuli, while DA seems to be associated with reward-motivated behaviors, based on its connections to the limbic system. Hypocretin's wake-to-sleep gating activity points to a critical role in maintaining wakefulness, particularly in the context of goal-oriented behaviors and locomotion. Nevertheless, the ultimate result of the arousal systems' individual effects is excitement of the thalamus and cortex.

The cortical excitation caused by monoamines and AChincreases the neuronal sensitivity to incoming sensory stimuli,[182,183] and it is through interactions between these subcortical systems and the thalamus and cortex that consciousness and associated EEG activity originate. The desynchronization of the EEG during wake and REM sleep indicates the cortical receptivity to the thalamic transmission of sensory and limbic signals. ACh is also the primary regulator of the thalamic depolarization that clears the way for information passing through the thalamus.[136] During NREM sleep, when cholinergic tone is lowest, the thalamic neurons are hyperpolarized, changing the reciprocal communication with the cortex. With GABAergic activity predominating among cortical projection neurons and the thalamic reticular nucleus, the hallmark slow-wave activity and spindles of NREM sleep can predominate.[184,185] However, despite the loss of consciousness imparted by thalamic lesions (as in the overwhelming majority of patients in a persistent vegetative state),[186] the persistence of the organization of wakefulness and NREM/REM sleep highlights that the thalamus is not a critical constituent of the brain's sleep circuitry.[187–190]

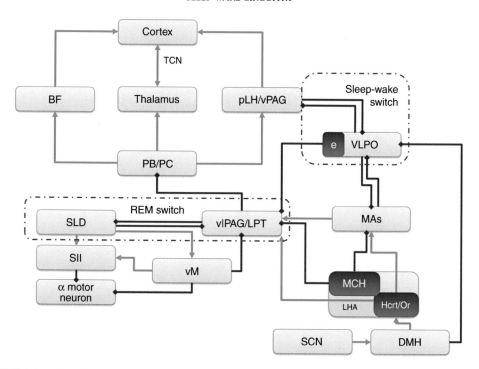

FIGURE 1.2 **Essential components of the sleep-wake circuitry.** Not all connections between areas are indicated, but the fundamental network connectivity is highlighted and separated out in order to show different components of the circuitry in relation to the sleep-wake and REM flip-flop switches, rather than anatomic relationships. Excitatory connections are in green, and inhibitory connections in red. The VLPO plays a major regulatory role in sleep-wake activity via projections to the pLH/vPAG, but also influences REM switching at the vlPAG/LPT. The incorporation of the PB/PC glutamatergic activity into the model emphasizes the critical role this region plays in sleep-wake behavior (via pLH/vPAG), conscious processing (via thalamus), and cortical arousal (via BF). Abbreviations: *BF*, basal forebrain; *DMH*, dorsomedial hypothalamic nucleus, *(e)VLPO*, (extended) ventrolateral preoptic nucleus; *Hcrt/Or*, hypocretin/orexin; *LHA*, lateral hypothalamic area; *MAs*, monoaminergic systems (locus ceruleus, tuberomammillary nucleus, dorsal raphe nucleus, and ventral periaqueductal gray); *MCH*, melanin-concentrating hormone; *TCN*, thalamocortical network; *PB/PC*, parabrachial nucleus/preceruleus; *SCN*, suprachiasmatic nucleus; *SII*, spinal inhibitory interneurons; *SLD*, sublaterodorsal nucleus; *vlPAG/LPT*, ventrolateral periaqueductal gray/lateral pontine tegmentum; *vM*, ventral medulla; *pLH /vPAG*, posterior lateral hypothalamus/ventral periaqeuductal gray.

Sustaining Wakefulness

As mentioned previously, there appear to be two main branches to the ascending arousal system: the dorsal pathway leading from the LDT/PPT to the thalamus, promoting sensory transmission through to the cortex, and the ventral pathway, which projects from the monoaminergic (MA) brainstem nuclei to the LHA, BF, and cortex (Fig. 1.3). This latter branch of the arousal system has firing patterns that are greatest during wakefulness, diminished during NREM sleep, and all but silent during REM sleep. It is through the mutually inhibitory connections to the preoptic area that the primary flip-flop switch controlling wake and sleep theoretically operates[19,20] (Fig. 1.2). The VLPO has primary inhibitory tone that is balanced through reciprocal, counter-inhibitory connections from the TMN, DRN, vPAG, LC, and LHA.

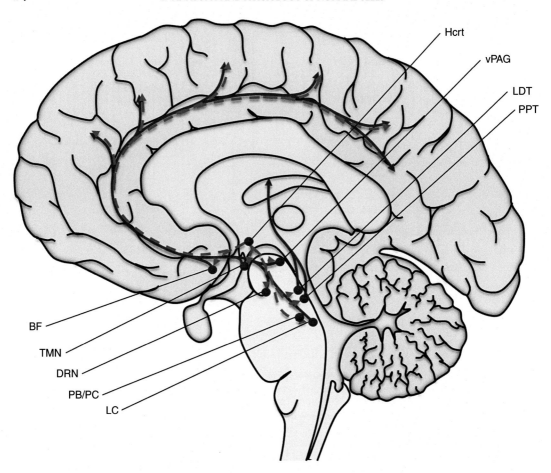

FIGURE 1.3 **The arousal-promoting pathways.** The dorsal, cholinergic pathway is demonstrated to originate in the LDT/PPT and proceed up to the thalamus. The ventral flow of the monoaminergic systems is also illustrated in red. The purple, *dashed lines* indicate projections of the hypocretin system feeding into the monoaminergic system, as well as the BF and cortex. Abbreviations: *BF*, basal forebrain; *DRN*, dorsal raphe nucleus; *Hcrt*, hypocretin; *LC*, locus ceruleus; *LDT*, laterodorsal tegmental nucleus; *PB/PC*, parabrachial nucleus/preceruleus; *PPT*, pedunculopontine tegmental nucleus; *TMN*, tuberomammillary nucleus; *vPAG*, ventral periaqueductal gray.

One of the critical nodes of this wake-promoting circuitry is the LHA, the seat of hypocretin neurons. Hypocretin neurons are wake-active and promote arousal via TMN, LC, DRN, and cortical projections;[191] however, hypocretin neurons do not project directly to the VLPO (which consequently lacks hypocretin receptors), suggesting that they serve more of a wake-state-stabilizing role external to the primary sleep-wake circuit. As evidence of this, hypocretinergic neuronal loss, as in type I narcolepsy, results in frequent state transitions with an overall normal sleep duration.[104] Similarly, lesions to the closely associated MCH neurons, which are noted to be predominantly REM-active, do not result in changes in the amount of wakefulness.[192] Nonetheless, the LHA's reciprocal connections to the monoaminergic system, as well as its basal forebrain and cortical projections, highlight the important role that this region plays in maintaining wakefulness, as noted by the fact that lesions to the LHA region generally result in hypersomnia.[23,104,147,191,193,194]

The BF is composed of the nucleus basalis of Meynert, the magnocellular preoptic nucleus in the substantia innominata, the medial septal nucleus, and the nucleus of the diagonal band of Broca. The BF receives inputs from the MA system as well as the LHA, and appears to serve as a waystation for cortical arousal signals. Support for this hypothesis is provided by the fact that the BF has predominantly cholinergic outputs, as well as the fact that cortical EEG activation is time-locked to burst firing in stimulated BF neurons.[195,196]

A more recently discovered wake-promoting glutamatergic neuronal group, which originates in the parabrachial nucleus (PB), parallels MA system projections. In comparison to the relatively minor decrements to wakefulness noted from lesions to any of the monoaminergic pathways, disruption of this glutamatergic system results in nearly 40% increases in total sleep time,[122] suggesting a more central role in the maintenance of wakefulness. It may be through the primary connections to the posterior lateral hypothalamus (pLH), BF, and thalamus that the primary wake-promoting effects are realized.[122]

Turning Off the Arousal System

It is the balance of activity from both sleep-promoting and wake-promoting regions that constitute the flip-flop switch necessary to transition into sleep (Fig. 1.2). The first indication that sleep was an active process were the insomniac patients with basal forebrain/anterior hypothalamic area lesions who were documented by von Economo.[8] Subsequent studies recapitulated these findings and eventually suggested that this sleep-promoting region is located around the sleep-active GABAergic and galaninergic cell populations in the preoptic area.[8,23,24,197]

The VLPO plays a critical role in regulating sleep-wake behaviors (Fig. 1.4). As mentioned earlier, the mutual inhibition between the VLPO and the monoaminergic and hypothalamic components of the arousal system contribute to the sleep-wake flip-flop switch. Consistent with this belief is the finding that lesions to the VLPO produce profound insomnia and sleep fragmentation.[114] While the firing patterns of the VLPO neurons are greatest throughout sleeping states (most notably during NREM), it is the median preoptic nucleus (MnPO) that may actually flip the switch, since firing rates of this neuronal population precede transitions to NREM sleep.[115,116]

Transitioning to REM

REM sleep is characterized by autonomic instability, skeletal muscle atonia, and the desynchronized cortical EEG patterns that resulted in the descriptions of this state as "active" or "paradoxical" sleep. Building upon the feline transectioning studies performed by Jouvet and a growing body of neuropharmacologic evidence, a flip-flop switch of "reciprocal interaction" was first proposed as the mechanism by which the brainstem regulates the ultradian cycling into REM sleep.[198] Although initially purported to rely upon MA and cholinergic stimulation, similar to that which subserves the bipartite arousal circuitry, subsequent experiments on REM sleep, demonstrating noncholinergic neurons active in the primary REM circuitry (e.g., SLD) and minimal changes in REM sleep as a consequence of selective lesions in either

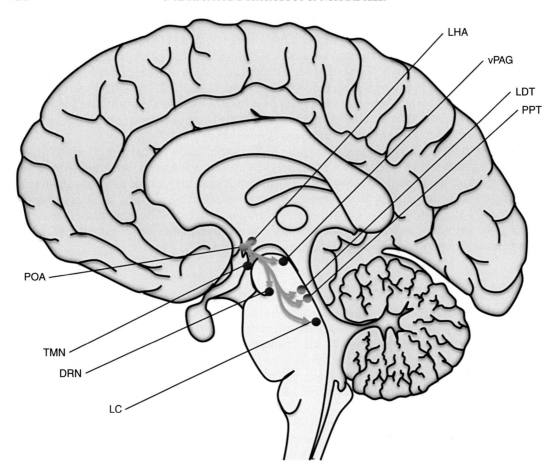

FIGURE 1.4   **The primary pathways subserving sleep promotion originating in the POA (in the ventrolateral and median preoptic nuclei).** The colors of the marker indicate the predominant role played by the structure: red for arousal, blue for sleep, green for REM, and multicolored markers indicating multistate activity. Abbreviations: *DRN*, dorsal raphe nucleus; *LC*, locus ceruleus; *LDT*, laterodorsal tegmental nucleus; *LHA*, lateral hypothalamic area; *POA*, preoptic area (containing ventrolateral and median preoptic nuclei); *PPT*, pedunculopontine tegmental nucleus; *TMN*, tuberomammillary nucleus; *vPAG*, ventral periaqueductal gray.

cholinergic or monoaminergic brainstem nuclei, suggested that the MA and cholinergic systems are neither necessary nor sufficient for the production of REM sleep.[123,199,200]

Again, the VLPO serves a primary role in the promotion of both primary states of sleep, however it is through extended VLPO (eVLPO) projections to the ventrolateral periaqueductal (vlPAG) and the lateral pontine tegmentum (LPT) that it is involved in REM sleep promotion.[122] Through a double-inhibitory mechanism, the VLPO feeds into the REM-promoting flip-flop switch (Fig. 1.2): (1) REM-active, GABAergic neurons in the eVLPO project to and inactivate (2) the REM-inactive, GABAergic neurons in the vlPAG and LPT, which, in turn, have reciprocally inhibitory connections with (3) REM-active GABAergic and glutamatergic neurons in the SLD (Fig. 1.5). In this model, the vlPAG and LPT serve as a waystation for REM

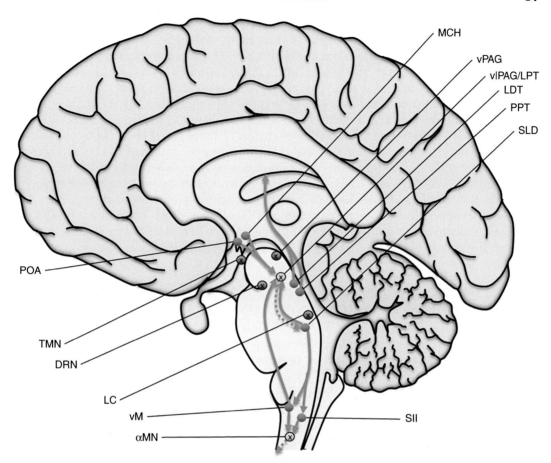

FIGURE 1.5 **The current characterization of the REM sleep architecture.** The colors of the marker indicate the predominant role played by the structure: red for arousal, blue for sleep, and green for REM. The *solid arrows* indicate REM-active limbs of the sleep-wake circuitry. Faded markers with an "X" and dashed pathways indicate REM-inactive neuronal populations and projections, respectively. Abbreviations: *αMN*, alpha motor neuron; *DRN*, dorsal raphe nucleus; *LC*, locus ceruleus; *LDT*, laterodorsal tegmental nucleus; *MCH*, melanin-concentrating hormone; *POA*, preoptic area (containing ventrolateral and median preoptic nuclei); *PPT*, pedunculopontine tegmental nucleus; *SII*, spinal inhibitory interneuron; *SLD*, sublaterodorsal nucleus; *TMN*, tuberomammillary nucleus; *vlPAG/LPT*, ventrolateral periaqueductal gray/lateral pontine tegmentum; *vM*, ventral medulla; *vPAG*, ventral periaqueductal gray.

regulation. The REM-inactive (i.e., REM-suppressing, when active) GABAergics in this region are also stimulated by monoaminergic and hypocretinergic inputs, while MCH and vM neurons provide an inhibitory (REM-promoting) input mediated primarily through GABA.[92,147,201–203]

With the withdrawal of the vlPAG/LPT inhibitory tone, REM phenomena can predominate. The uninhibited PB and precoeruleus (PC) together promote activation of the cortical and hippocampal EEG by way of glutamatergic projections onto corticopetal cholinergic and GABAergic pathways in the BF and medial septum.[16] Furthermore, the atonia of REM is principally controlled by the SLD, lying at the other end of the REM flip-flop switch[122] (Fig. 1.2).

The SLD's GABAergic projections back onto the vlPAG/LPT are essential in the transition to and maintenance of REM. The potent REM promotion of the SLD glutamatergic outputs activating vM GABAergic suppression of the vlPAG/LPT, further emphasizes the role of the SLD in REM sleep. However, it is important to note that the SLD-vM, multisynaptic REM promotion is a phenomenon that is only observed as an NREM-to-REM transition, rather than a wake-to-REM transition,[127] suggesting that the vM's inhibitory activity on the vlPAG/LPT is not sufficient for REM induction and, therefore, is not a primary constituent of the flip-flop switch. Furthermore, glutamatergic outputs of the SLD are essential to both direct and indirect induction of atonia. The pathway mediated by SLD-activated vM cell groups as well as direct synapses on spinal inhibitory interneurons use GABA and glycine to prevent the corporeal manifestations of the central pattern generators and myoclonic activity through the production of atonia during REM sleep.[122] The central role of these pontine and medullary glutamatergic neurons in the generation of REM-sleep atonia is consistent with Braak's hypothesized spread of α-synuclein,[204] confirming that the loss of REM atonia in REM-sleep behavior disorder is a biomarker of synucleinopathies, predating disease onset by more than a decade.[205]

CIRCADIAN AND HOMEOSTATIC REGULATION OF SLEEP

The interaction between a circadian alerting signal (process C) and a homeostatic soporific signal (process S) was first proposed by Borbély and Tobler in 1982, and was dubbed the two-process model of sleep regulation.[206] Forced desynchronization experiments performed by Dijk and Czeisler and subsequent studies in the field have sought to confirm that the circadian signal is alerting in humans.[207–209] However, the circadian signal not only regulates the level of wakefulness throughout the day, it also serves as the body's primary time keeper, aligning all physiologic functions through a network of multisynaptic systems.[210,211] While the master clock, housed in the SCN, is active during the *light* cycle, the VLPO is always active during the *sleep* cycle, regardless of circadian phenotype (e.g., diurnal vs. nocturnal). Thus, it is theorized that the complex connectivity linking the SCN to the main sleep-wake and autoregulatory behavioral activities of the hypothalamus allows for adaptation of rest-activity cycles to the needs of an organism through an integration of environmental influences—such as temperature, feeding, social cues—with the primary zeitgeber (time giver) of light. The influence of the circadian system on sleep is discussed in detail in the next chapter.

CONCLUSIONS

Sleep is an elaborate and complex process. The redundancy and multinodal nature of the neurocircuitry allows sleep to be sculpted to suit the needs of the organism. While strategically placed lesions in only a few of the key neuroanatomical areas are sufficient to cause substantial changes in sleep-wake duration, all of the areas contribute essential modulating effects on the coordination of sleep. The variety of neurotransmitters, neuropeptides, and neuromodulators that feed into the pair of flip-flop switches allow for the integration of external cues in the adaptation of sleep to meet the needs of the organism. Behavioral entrainment is possible through a number of the monoamines as well as hypocretin. The somnogen-based homeostatic system, which ensures that the drive for recovery is scaled to the degree of use, may also serve to promote learning by means of Hebbian plasticity. Finally, environmental alignment is possible through the coupling of the externally activated SCN and the internally active VLPO. We have only just begun to understand the many roles that sleep serves: from energy renewal, to waste removal and damage repair, to information organization and learning consolidation. While sleep's complexity and multisystem redundancies highlight the importance of this physiologic function, it is through damage to the architecture underlying this globally coordinated process that the clinical impact of neurologic diseases can be understood, and the pathophysiology of the diseases themselves explored.

References

1. Dement W, Kleitman N. Cyclic variations in EEG during sleep and their relation to eye movements, body motility, and dreaming. *Electroencephalogr Clin Neurophysiol*. 1957;9(4):673–690. *http://www.ncbi.nlm.nih.gov/pubmed/13480240*.
2. Everson CA, Bergmann BM, Rechtschaffen A. Sleep deprivation in the rat: III. Total sleep deprivation. *Sleep*. 1989;12(1):13–21. *http://www.ncbi.nlm.nih.gov/pubmed/2928622*.
3. Gilliland MA, Bergmann BM, Rechtschaffen A. Sleep deprivation in the rat: VIII. High EEG amplitude sleep deprivation. *Sleep*. 1989;12(1):53–59. *http://www.ncbi.nlm.nih.gov/pubmed/2928626*.
4. Kushida CA, Bergmann BM, Rechtschaffen A. Sleep deprivation in the rat: IV. Paradoxical sleep deprivation. *Sleep*. 1989;12(1):22–30. *http://www.ncbi.nlm.nih.gov/pubmed/2928623*.
5. Eyes Wide Shut: Thoughts on Sleep. *New York Times*. http://www.nytimes.com/2007/10/23/science/23quot.html, 2015.
6. Churchill L, Rector DM, Yasuda K, Fix C, Rojas MJ, Yasuda T, Krueger JM. Tumor necrosis factor alpha: activity dependent expression and promotion of cortical column sleep in rats. *Neuroscience*. 2008;156(1):71–80.
7. Morin A, Doyon J, Dostie V, Barakat M, Hadj Tahar A, Korman M, Benali H, Karni A, Ungerleider LG, Carrier J. Motor sequence learning increases sleep spindles and fast frequencies in post-training sleep. *Sleep*. 2008;31(8):1149–1156. *http://www.pubmedcentral.nih.gov/articlerender.fcgi?artid=2542961&tool=pmcentrez&rendertype=abstract*.
8. Von Economo C. Sleep As a Problem of Localization. *J Nerv Ment Dis*. 1930;71:249–259.
9. Bremer F. Cerveau "isolé" et Physiologie du Sommeil. *C R Soc Biol*. 1935;118:1235–1241.
10. Moruzzi G, Magoun HW. Brain stem reticular formation and activation of the EEG. *Electroencephalogr Clin Neurophysiol*. 1949;1(1–4):455–473.
11. McCormick DA, Bal T. Sleep and arousal: thalamocortical mechanisms. *Annu Rev Neurosci*. 1997;20:185–215.
12. Steriade M, McCormick DA, Sejnowski TJ. Thalamocortical oscillations in the sleeping and aroused brain. *Science*. 1993;262(5134):679–685.
13. Starzl TE, Taylor CW, Magoun HW. Ascending conduction in reticular activating system, with special reference to the diencephalon. *J Neurophysiol*. 1951;14(6):461–477. *http://jn.physiology.org.laneproxy.stanford.edu/content/14/6/461.long*.
14. Villablanca J, Salinas-Zeballos ME. Sleep-wakefulness, EEG and behavioral studies of chronic cats without the thalamus: the "athalamic" cat. *Arch Ital Biol*. 1972;110(3):383–411. *http://www.ncbi.nlm.nih.gov/pubmed/4349191*.
15. Vanderwolf CH, Stewart DJ. Thalamic control of neocortical activation: a critical re-evaluation. *Brain Res Bull*. 1988;20(4):529–538.
16. Fuller PM, Lu J. In: Amlaner C, Fuller PM, eds. *Neurobiology of Sleep*. Westchester: Sleep Research Society; 2015.
17. Aminoff MJ. *Electrodiagnosis in Clinical Neurology*. 6th ed. Philadelphia, PA: Elsevier Health Sciences; 2012. *https://books.google.com/books?id=bPYDhpOiZaMC&pgis=1*.
18. Lopez R, Jaussent I, Dauvilliers Y. Pain in sleepwalking: a clinical enigma. *Sleep*. 2015;38(11):1693–1698.
19. Fuller PM, Gooley JJ, Saper CB. Neurobiology of the sleep-wake cycle: sleep architecture, circadian regulation, and regulatory feedback. *J Biol Rhythms*. 2006;21(6):482–493.

20. Saper CB, Scammell TE, Lu J. Hypothalamic regulation of sleep and circadian rhythms. *Nature*. 2005;437(7063):1257–1263.

21. Flacker JM, Lipsitz LA. Neural mechanisms of delirium: current hypotheses and evolving concepts. *J Gerontol A Biol Sci Med Sci*. 1999;54(6):B239–B246. *http://www.ncbi.nlm.nih.gov/pubmed/10411009*.

22. Trzepacz PT. Is there a final common neural pathway in delirium? Focus on acetylcholine and dopamine. *Semin Clin Neuropsychiatry*. 2000;5(2):132–148.

23. Nauta WJH. Hypothalamic regulation of sleep in rats; an experimental study. *J Neurophysiol*. 1946;9:285–316. *http://www.ncbi.nlm.nih.gov/pubmed/20991815*.

24. McGinty DJ, Sterman MB. Sleep suppression after basal forebrain lesions in the cat. *Science*. 1968;160(3833):1253–1255. *http://www.ncbi.nlm.nih.gov/pubmed/5689683*.

25. Hallanger AE, Levey AI, Lee HJ, Rye DB, Wainer BH. The origins of cholinergic and other subcortical afferents to the thalamus in the rat. *J Comp Neurol*. 1987;262(1):105–124.

26. Satoh K, Fibiger HC. Cholinergic neurons of the laterodorsal tegmental nucleus: efferent and afferent connections. *J Comp Neurol*. 1986;253(3):277–302.

27. Marrosu F, Portas C, Mascia MS, Casu MA, Fà M, Giagheddu M, Imperato A, Gessa GL. Microdialysis measurement of cortical and hippocampal acetylcholine release during sleep-wake cycle in freely moving cats. *Brain Res*. 1995;671(2):329–332. *http://www.ncbi.nlm.nih.gov/pubmed/7743225*.

28. Gritti I, Mainville L, Mancia M, Jones BE. GABAergic and other noncholinergic basal forebrain neurons, together with cholinergic neurons, project to the mesocortex and isocortex in the rat. *J Comp Neurol*. 1997;383(2):163–177. *http://www.ncbi.nlm.nih.gov/pubmed/9182846*.

29. Henny P, Jones BE. Projections from basal forebrain to prefrontal cortex comprise cholinergic, GABAergic and glutamatergic inputs to pyramidal cells or interneurons. *Eur J Neurosci*. 2008;27(3):654–670.

30. Marks GA, Birabil CG. Enhancement of rapid eye movement sleep in the rat by cholinergic and adenosinergic agonists infused into the pontine reticular formation. *Neuroscience*. 1998;86(1):29–37. *http://www.ncbi.nlm.nih.gov/pubmed/9692741*.

31. Reid MS, Tafti M, Nishino S, Siegel JM, Dement WC, Mignot E. Cholinergic regulation of cataplexy in canine narcolepsy in the pontine reticular formation is mediated by M2 muscarinic receptors. *Sleep*. 1994;17(5):424–435. *http://www.ncbi.nlm.nih.gov/pubmed/7991953*.

32. Baghdoyan HA. Location and quantification of muscarinic receptor subtypes in rat pons: implications for REM sleep generation. *Am J Physiol*. 1997;273(3 Pt 2):R896–R904. *http://www.ncbi.nlm.nih.gov/pubmed/9321865*.

33. Marks GA, Birabil CG. Infusion of adenylyl cyclase inhibitor SQ22,536 into the medial pontine reticular formation of rats enhances rapid eye movement sleep. *Neuroscience*. 2000;98(2):311–315. *http://www.ncbi.nlm.nih.gov/pubmed/10854762*.

34. Davila DG, Hurt RD, Offord KP, Harris CD, Shepard JW. Acute effects of transdermal nicotine on sleep architecture, snoring, and sleep-disordered breathing in nonsmokers. *Am J Respir Crit Care Med*. 1994;150(2):469–474.

35. Velazquez-Moctezuma J, Gillin JC, Shiromani PJ. Effect of specific M1, M2 muscarinic receptor agonists on REM sleep generation. *Brain Res*. 1989;503(1):128–131. *http://www.ncbi.nlm.nih.gov/pubmed/2482113*.

36. Yamamoto KI, Domino EF. Cholinergic agonist-antagonist interactions on neocortical and limbic EEG activation. *Int J Neuropharmacol*. 1967;6(5):357–373. *http://www.ncbi.nlm.nih.gov/pubmed/6069723*.

37. Benington JH, Heller HC. Monoaminergic and cholinergic modulation of REM-sleep timing in rats. *Brain Res*. 1995;681(1-2):141–146. *http://www.ncbi.nlm.nih.gov/pubmed/7552271*.

38. Imeri L, Bianchi S, Angeli P, Mancia M. Selective blockade of different brain stem muscarinic receptor subtypes: effects on the sleep-wake cycle. *Brain Res*. 1994;636(1):68–72. *http://www.ncbi.nlm.nih.gov/pubmed/8156412*.

39. Spehlmann R, Norcross K. Cholinergic mechanisms in the production of focal cortical slow waves. *Experientia*. 1982;38(1):109–111. *http://www.ncbi.nlm.nih.gov/pubmed/7056349*.

40. Aston-Jones G, Bloom FE. Activity of norepinephrine-containing locus coeruleus neurons in behaving rats anticipates fluctuations in the sleep-waking cycle. *J Neurosci*. 1981;1(8):876–886. *http://www.ncbi.nlm.nih.gov/pubmed/7346592*.

41. Berridge CW, Abercrombie ED. Relationship between locus coeruleus discharge rates and rates of norepinephrine release within neocortex as assessed by in vivo microdialysis. *Neuroscience*. 1999;93(4):1263–1270. *http://www.ncbi.nlm.nih.gov/pubmed/10501450*.

42. Foote SL, Aston-Jones G, Bloom FE. Impulse activity of locus coeruleus neurons in awake rats and monkeys is a function of sensory stimulation and arousal. *Proc Natl Acad Sci USA*. 1980;77(5):3033–3037. *http://www.pubmedcentral.nih.gov/articlerender.fcgi?artid=349541&tool=pmcentrez&rendertype=abstract*.

43. Dayas CV, Buller KM, Crane JW, Xu Y, Day TA. Stressor categorization: acute physical and psychological stressors elicit distinctive recruitment patterns in the amygdala and in medullary noradrenergic cell groups. *Eur J Neurosci.* 2001;14(7):1143–1152. *http://www.ncbi.nlm.nih.gov/pubmed/11683906.*

44. Byers MG, Allison KM, Wendel CS, Lee JK. Prazosin versus quetiapine for nighttime posttraumatic stress disorder symptoms in veterans: an assessment of long-term comparative effectiveness and safety. *J Clin Psychopharmacol.* 2010;30(3):225–229.

45. De Sarro GB, Ascioti C, Froio F, Libri V, Nisticò G. Evidence that locus coeruleus is the site where clonidine and drugs acting at alpha 1- and alpha 2-adrenoceptors affect sleep and arousal mechanisms. *Br J Pharmacol.* 1987;90(4):675–685. *http://www.pubmedcentral.nih.gov/articlerender.fcgi?artid=1917214&tool=pmcentrez&rendertype=abstract.*

46. Berridge CW, Isaac SO, España RA. Additive wake-promoting actions of medial basal forebrain noradrenergic alpha1- and beta-receptor stimulation. *Behav Neurosci.* 2003;117(2):350–359. *http://www.ncbi.nlm.nih.gov/pubmed/12708531.*

47. Berridge CW, Foote SL. Effects of locus coeruleus activation on electroencephalographic activity in neocortex and hippocampus. *J Neurosci.* 1991;11(10):3135–3145. *http://www.pubmedcentral.nih.gov/articlerender.fcgi?artid=3058938&tool=pmcentrez&rendertype=abstract.*

48. Berridge CW, España RA. Synergistic sedative effects of noradrenergic alpha(1)- and beta-receptor blockade on forebrain electroencephalographic and behavioral indices. *Neuroscience.* 2000;99(3):495–505. *http://www.ncbi.nlm.nih.gov/pubmed/11029541.*

49. Berridge CW, Page ME, Valentino RJ, Foote SL. Effects of locus coeruleus inactivation on electroencephalographic activity in neocortex and hippocampus. *Neuroscience.* 1993;55(2):381–393. *http://www.ncbi.nlm.nih.gov/pubmed/8104319.*

50. Schultz W. Predictive reward signal of dopamine neurons. *J Neurophysiol.* 1998;80(1):1–27. *http://www.ncbi.nlm.nih.gov/pubmed/9658025.*

51. Schultz W. Multiple dopamine functions at different time courses. *Annu Rev Neurosci.* 2007;30:259–288.

52. Trulson ME. Simultaneous recording of substantia nigra neurons and voltammetric release of dopamine in the caudate of behaving cats. *Brain Res Bull.* 1985;15(2):221–223. *http://www.ncbi.nlm.nih.gov/pubmed/4041929.*

53. Trulson ME, Preussler DW. Dopamine-containing ventral tegmental area neurons in freely moving cats: activity during the sleep-waking cycle and effects of stress. *Exp Neurol.* 1984;83(2):367–377.

54. Lu J, Jhou TC, Saper CB. Identification of wake-active dopaminergic neurons in the ventral periaqueductal gray matter. *J Neurosci.* 2006;26(1):193–202.

55. Mignot E, Nishino S, Guilleminault C, Dement WC. Modafinil binds to the dopamine uptake carrier site with low affinity. *Sleep.* 1994;17(5):436–437. *http://www.ncbi.nlm.nih.gov/pubmed/7991954.*

56. Qu W-M, Huang Z-L, Xu X-H, Matsumoto N, Urade Y. Dopaminergic D1 and D2 receptors are essential for the arousal effect of modafinil. *J Neurosci.* 2008;28(34):8462–8469.

57. Volkow ND, Fowler JS, Logan J, Alexoff D, Zhu W, Telang F, Wang G-J, Jayne M, Hooker JM, Wong C, Hubbard B, Carter P, Warner D, King P, Shea C, Xu Y, Muench L, Apelskog-Torres K. Effects of Modafinil on Dopamine and Dopamine Transporters in the Male Human Brain. *JAMA.* 2009;301(11):1148.

58. Wisor JP, Nishino S, Sora I, Uhl GH, Mignot E, Edgar DM. Dopaminergic role in stimulant-induced wakefulness. *J Neurosci.* 2001;21(5):1787–1794. *http://www.ncbi.nlm.nih.gov/pubmed/11222668.*

59. Arnulf I, Leu S, Oudiette D. Abnormal sleep and sleepiness in Parkinson's disease. *Curr Opin Neurol.* 2008;21(4):472–477.

60. Paus S, Brecht HM, Köster J, Seeger G, Klockgether T, Wüllner U. Sleep attacks, daytime sleepiness, and dopamine agonists in Parkinson's disease. *Mov Disord.* 2003;18(6):659–667.

61. Beaulieu J-M, Gainetdinov RR. The physiology, signaling, and pharmacology of dopamine receptors. *Pharmacol Rev.* 2011;63(1):182–217.

62. Lin JS, Sakai K, Jouvet M. Evidence for histaminergic arousal mechanisms in the hypothalamus of cat. *Neuropharmacology.* 1988;27(2):111–122. *http://www.ncbi.nlm.nih.gov/pubmed/2965315.*

63. Monti JM, Pellejero T, Jantos H. Effects of H1- and H2-histamine receptor agonists and antagonists on sleep and wakefulness in the rat. *J Neural Transm.* 1986;66(1):1–11. *http://www.ncbi.nlm.nih.gov/pubmed/3734773.*

64. Sakai K, El Mansari M, Lin J, Zhang J, Vanni-Mercier G. The Posterior Hypothalamus in the regulation of wakefulness and paradoxical sleep. In: Mancia M, Marini G, eds. *The Diencephaon and Sleep.* New York: Raven Press Ltd; 1990:171–198.

65. Mochizuki T, Yamatodani A, Okakura K, Horii A, Inagaki N, Wada H. Circadian rhythm of histamine release from the hypothalamus of freely moving rats. *Physiol Behav.* 1992;51(2):391–394. *http://www.ncbi.nlm.nih.gov/pubmed/1313592.*

66. Parmentier R, Ohtsu H, Djebbara-Hannas Z, Valatx J-L, Watanabe T, Lin J-S. Anatomical, physiological, and pharmacological characteristics of histidine decarboxylase knock-out mice: evidence for the role of brain histamine in behavioral and sleep-wake control. *J Neurosci.* 2002;22(17):7695–7711. *http://www.ncbi.nlm.nih.gov/pubmed/12196593.*

67. Passani MB, Blandina P, Torrealba F. The histamine H3 receptor and eating behavior. *J Pharmacol Exp Ther.* 2011;336(1):24–29.

68. Van Ruitenbeek P, Vermeeren A, Riedel WJ. Cognitive domains affected by histamine H(1)-antagonism in humans: a literature review. *Brain Res Rev.* 2010;64(2):263–282.

69. Roehrs TA, Tietz EI, Zorick FJ, Roth T. Daytime sleepiness and antihistamines. *Sleep.* 1984;7(2):137–141. *http://www.ncbi.nlm.nih.gov/pubmed/6146180.*

70. Huang Z-L, Mochizuki T, Qu W-M, Hong Z-Y, Watanabe T, Urade Y, Hayaishi O. Altered sleep-wake characteristics and lack of arousal response to H3 receptor antagonist in histamine H1 receptor knockout mice. *Proc Natl Acad Sci USA.* 2006;103(12):4687–4692.

71. Williams RH, Chee MJS, Kroeger D, Ferrari LL, Maratos-Flier E, Scammell TE, Arrigoni E. Optogenetic-mediated release of histamine reveals distal and autoregulatory mechanisms for controlling arousal. *J Neurosci.* 2014;34(17):6023–6029.

72. Leurs R, Bakker RA, Timmerman H, de Esch IJP. The histamine H3 receptor: from gene cloning to H3 receptor drugs. *Nat Rev Drug Discov.* 2005;4(2):107–120.

73. Leu-Semenescu S, Nittur N, Golmard J-L, Arnulf I. Effects of pitolisant, a histamine H3 inverse agonist, in drug-resistant idiopathic and symptomatic hypersomnia: a chart review. *Sleep Med.* 2014;15(6):681–687.

74. Portas CM, Bjorvatn B, Fagerland S, Grønli J, Mundal V, Sørensen E, Ursin R. On-line detection of extracellular levels of serotonin in dorsal raphe nucleus and frontal cortex over the sleep/wake cycle in the freely moving rat. *Neuroscience.* 1998;83(3):807–814. *http://www.ncbi.nlm.nih.gov/pubmed/9483564.*

75. Trulson ME, Jacobs BL. Raphe unit activity in freely moving cats: correlation with level of behavioral arousal. *Brain Res.* 1979;163(1):135–150. *http://www.ncbi.nlm.nih.gov/pubmed/218676.*

76. Wilson S, Argyropoulos S. Antidepressants and sleep: a qualitative review of the literature. *Drugs.* 2005;65(7):927–947. *http://www.ncbi.nlm.nih.gov/pubmed/15892588.*

77. Hoque R, Chesson AL. Pharmacologically induced/exacerbated restless legs syndrome, periodic limb movements of sleep, and REM behavior disorder/REM sleep without atonia: literature review, qualitative scoring, and comparative analysis. *J Clin Sleep Med.* 2010;6(1):79–83.

78. Billiard M, Narcolepsy:. current treatment options and future approaches. *Neuropsychiatr Dis Treat.* 2008;4(3):557–566. *http://www.pubmedcentral.nih.gov/articlerender.fcgi?artid=2526380&tool=pmcentrez&rendertype=abstract.*

79. Bjorvatn B, Ursin R. Effects of the selective 5-HT1B agonist, CGS 12066B, on sleep/waking stages and EEG power spectrum in rats. *J Sleep Res.* 1994;3(2):97–105. *http://www.ncbi.nlm.nih.gov/pubmed/10607113.*

80. Boutrel B, Franc B, Hen R, Hamon M, Adrien J. Key role of 5-HT1B receptors in the regulation of paradoxical sleep as evidenced in 5-HT1B knock-out mice. *J Neurosci.* 1999;19(8):3204–3212. *http://www.ncbi.nlm.nih.gov/pubmed/10191333.*

81. Dugovic C, Wauquier A, Leysen JE, Marrannes R, Janssen PA. Functional role of 5-HT2 receptors in the regulation of sleep and wakefulness in the rat. *Psychopharmacology (Berl).* 1989;97(4):436–442. *http://www.ncbi.nlm.nih.gov/pubmed/2524856.*

82. Dzoljic MR, Ukponmwan OE, Saxena PR. 5-HT1-like receptor agonists enhance wakefulness. *Neuropharmacology.* 1992;31(7):623–633. *http://www.ncbi.nlm.nih.gov/pubmed/1407402.*

83. Ponzoni A, Monti JM, Jantos H. The effects of selective activation of the 5-HT3 receptor with m-chlorophenylbiguanide on sleep and wakefulness in the rat. *Eur J Pharmacol.* 1993;249(3):259–264. *http://www.ncbi.nlm.nih.gov/pubmed/8287912.*

84. Lemoine P, Guilleminault C, Alvarez E. Improvement in subjective sleep in major depressive disorder with a novel antidepressant, agomelatine: randomized, double-blind comparison with venlafaxine. *J Clin Psychiatry.* 2007;68(11):1723–1732. *http://www.ncbi.nlm.nih.gov/pubmed/18052566.*

85. Monti JM. Serotonin 5-HT(2A) receptor antagonists in the treatment of insomnia: present status and future prospects. *Drugs Today (Barc).* 2010;46(3):183–193.

86. Teegarden BR, Al Shamma H, Xiong Y. 5-HT(2A) inverse-agonists for the treatment of insomnia. *Curr Top Med Chem.* 2008;8(11):969–976. *http://www.ncbi.nlm.nih.gov/pubmed/18673166.*

87. Crocker A, España RA, Papadopoulou M, Saper CB, Faraco J, Sakurai T, Honda M, Mignot E, Scammell TE. Concomitant loss of dynorphin, NARP, and orexin in narcolepsy. *Neurology.* 2005;65(8):1184–1188.

88. Peyron C, Faraco J, Rogers W, Ripley B, Overeem S, Charnay Y, Nevsimalova S, Aldrich M, Reynolds D, Albin R, Li R, Hungs M, Pedrazzoli M, Padigaru M, Kucherlapati M, Fan J, Maki R, Lammers GJ, Bouras C, Kucherlapati R, Nishino S, Mignot E. A mutation in a case of early onset narcolepsy and a generalized absence of hypocretin peptides in human narcoleptic brains. *Nat Med*. 2000;6(9):991–997.

89. Thannickal TC, Moore RY, Nienhuis R, Ramanathan L, Gulyani S, Aldrich M, Cornford M, Siegel JM. Reduced number of hypocretin neurons in human narcolepsy. *Neuron*. 2000;27(3):469–474. *http://www.ncbi.nlm.nih.gov/ pubmed/11055430*.

90. Peyron C, Tighe DK, van den Pol AN, de Lecea L, Heller HC, Sutcliffe JG, Kilduff TS. Neurons containing hypocretin (orexin) project to multiple neuronal systems. *J Neurosci*. 1998;18(23):9996–10015. *http://www.ncbi.nlm.nih. gov/pubmed/9822755*.

91. Sakurai T, Amemiya A, Ishii M, Matsuzaki I, Chemelli RM, Tanaka H, Williams SC, Richarson JA, Kozlowski GP, Wilson S, Arch JR, Buckingham RE, Haynes AC, Carr SA, Annan RS, McNulty DE, Liu WS, Terrett JA, Elshourbagy NA, Bergsma DJ, Yanagisawa M. Orexins and orexin receptors: a family of hypothalamic neuropeptides and G protein-coupled receptors that regulate feeding behavior. *Cell*. 1998; 92(5):1 page following 696. http://www.ncbi.nlm.nih.gov/pubmed/9527442.

92. Bourgin P, Huitrón-Résendiz S, Spier AD, Fabre V, Morte B, Criado JR, Sutcliffe JG, Henriksen SJ, de Lecea L. Hypocretin-1 modulates rapid eye movement sleep through activation of locus coeruleus neurons. *J Neurosci*. 2000;20(20):7760–7765. *http://www.ncbi.nlm.nih.gov/pubmed/11027239*.

93. España RA, Baldo BA, Kelley AE, Berridge CW. Wake-promoting and sleep-suppressing actions of hypocretin (orexin): basal forebrain sites of action. *Neuroscience*. 2001;106(4):699–715. *http://www.ncbi.nlm.nih.gov/ pubmed/11682157*.

94. Hagan JJ, Leslie RA, Patel S, Evans ML, Wattam TA, Holmes S, Benham CD, Taylor SG, Routledge C, Hemmati P, Munton RP, Ashmeade TE, Shah AS, Hatcher JP, Hatcher PD, Jones DN, Smith MI, Piper DC, Hunter AJ, Porter RA, Upton N, Orexin A. activates locus coeruleus cell firing and increases arousal in the rat. *Proc Natl Acad Sci USA*. 1999;96(19):10911–10916.

95. Lee MG, Hassani OK, Jones BE. Discharge of identified orexin/hypocretin neurons across the sleep-waking cycle. *J Neurosci*. 2005;25(28):6716–6720.

96. Mileykovskiy BY, Kiyashchenko LI, Siegel JM. Behavioral correlates of activity in identified hypocretin/orexin neurons. *Neuron*. 2005;46(5):787–798.

97. Adamantidis AR, Zhang F, Aravanis AM, Deisseroth K, de Lecea L. Neural substrates of awakening probed with optogenetic control of hypocretin neurons. *Nature*. 2007;450(7168):420–424.

98. Carter ME, Adamantidis A, Ohtsu H, Deisseroth K, de Lecea L. Sleep homeostasis modulates hypocretin-mediated sleep-to-wake transitions. *J Neurosci*. 2009;29(35):10939–10949.

99. Baumann CR, Bassetti CL, Valko PO, Haybaeck J, Keller M, Clark E, Stocker R, Tolnay M, Scammell TE. Loss of hypocretin (orexin) neurons with traumatic brain injury. *Ann Neurol*. 2009;66(4):555–559.

100. Benarroch EE, Schmeichel AM, Sandroni P, Low PA, Parisi JE. Involvement of hypocretin neurons in multiple system atrophy. *Acta Neuropathol*. 2007;113(1):75–80.

101. Fronczek R, Baumann CR, Lammers GJ, Bassetti CL, Overeem S. Hypocretin/orexin disturbances in neurological disorders. *Sleep Med Rev*. 2009;13(1):9–22.

102. Fronczek R, Overeem S, Lee SYY, Hegeman IM, van Pelt J, van Duinen SG, Lammers GJ, Swaab DF. Hypocretin (orexin) loss in Parkinson's disease. *Brain*. 2007;130(Pt 6):1577–1585.

103. Thannickal TC, Lai Y-Y, Siegel JM. Hypocretin (orexin) cell loss in Parkinson's disease. *Brain*. 2007;130(Pt 6):1586–1595.

104. Mochizuki T, Crocker A, McCormack S, Yanagisawa M, Sakurai T, Scammell TE. Behavioral state instability in orexin knock-out mice. *J Neurosci*. 2004;24(28):6291–6300.

105. Aston-Jones G, Smith RJ, Moorman DE, Richardson KA. Role of lateral hypothalamic orexin neurons in reward processing and addiction. *Neuropharmacology*. 2009;56(1 suppl 1):112–121.

106. Borgland SL, Chang S-J, Bowers MS, Thompson JL, Vittoz N, Floresco SB, Chou J, Chen BT, Bonci A. Orexin A/ hypocretin-1 selectively promotes motivation for positive reinforcers. *J Neurosci*. 2009;29(36):11215–11225.

107. Burdakov D, Gerasimenko O, Verkhratsky A. Physiological changes in glucose differentially modulate the excitability of hypothalamic melanin-concentrating hormone and orexin neurons in situ. *J Neurosci*. 2005;25(9):2429–2433.

108. España RA, Melchior JR, Roberts DCS, Jones SR. Hypocretin 1/orexin A in the ventral tegmental area enhances dopamine responses to cocaine and promotes cocaine self-administration. *Psychopharmacology (Berl)*. 2011;214(2):415–426.

109. España RA, Oleson EB, Locke JL, Brookshire BR, Roberts DCS, Jones SR. The hypocretin-orexin system regulates cocaine self-administration via actions on the mesolimbic dopamine system. *Eur J Neurosci*. 2010;31(2):336–348.

110. Moriguchi T, Sakurai T, Nambu T, Yanagisawa M, Goto K. Neurons containing orexin in the lateral hypothalamic area of the adult rat brain are activated by insulin-induced acute hypoglycemia. *Neurosci Lett*. 1999;264(1–3):101–104. *http://www.ncbi.nlm.nih.gov/pubmed/10320024*.

111. Rhyne DN, Anderson SL. Suvorexant in insomnia: efficacy, safety and place in therapy. *Ther Adv drug Saf*. 2015;6(5):189–195.

112. Gaus SE, Strecker RE, Tate BA, Parker RA, Saper CB. Ventrolateral preoptic nucleus contains sleep-active, galaninergic neurons in multiple mammalian species. *Neuroscience*. 2002;115(1):285–294. *http://www.ncbi.nlm.nih.gov/pubmed/12401341*.

113. Sherin JE, Elmquist JK, Torrealba F, Saper CB. Innervation of histaminergic tuberomammillary neurons by GABAergic and galaninergic neurons in the ventrolateral preoptic nucleus of the rat. *J Neurosci*. 1998;18(12):4705–4721. *http://www.ncbi.nlm.nih.gov/pubmed/9614245*.

114. Lu J, Greco MA, Shiromani P, Saper CB. Effect of lesions of the ventrolateral preoptic nucleus on NREM and REM sleep. *J Neurosci*. 2000;20(10):3830–3842. *http://www.ncbi.nlm.nih.gov/pubmed/10804223*.

115. Suntsova N, Szymusiak R, Md. Alam N, Guzman-Marin R, McGinty D. Sleep-waking discharge patterns of median preoptic nucleus neurons in rats. *J Physiol*. 2002;543(2):665–677.

116. Takahashi K, Lin J-S, Sakai K. Characterization and mapping of sleep-waking specific neurons in the basal forebrain and preoptic hypothalamus in mice. *Neuroscience*. 2009;161(1):269–292.

117. Szymusiak R, Alam N, Steininger TL, McGinty D. Sleep-waking discharge patterns of ventrolateral preoptic/anterior hypothalamic neurons in rats. *Brain Res*. 1998;803(1–2):178–188. *http://www.ncbi.nlm.nih.gov/pubmed/9729371*.

118. Hassani OK, Henny P, Lee MG, Jones BE. GABAergic neurons intermingled with orexin and MCH neurons in the lateral hypothalamus discharge maximally during sleep. *Eur J Neurosci*. 2010;32(3):448–457.

119. Hassani OK, Lee MG, Henny P, Jones BE. Discharge profiles of identified GABAergic in comparison to cholinergic and putative glutamatergic basal forebrain neurons across the sleep-wake cycle. *J Neurosci*. 2009;29(38): 11828–11840.

120. Manns ID, Alonso A, Jones BE. Discharge profiles of juxtacellularly labeled and immunohistochemically identified GABAergic basal forebrain neurons recorded in association with the electroencephalogram in anesthetized rats. *J Neurosci*. 2000;20(24):9252–9263. *http://www.ncbi.nlm.nih.gov/pubmed/11125003*.

121. Sakai K, Sastre JP, Salvert D, Touret M, Tohyama M, Jouvet M. Tegmentoreticular projections with special reference to the muscular atonia during paradoxical sleep in the cat: an HRP study. *Brain Res*. 1979;176(2):233–254. *http://www.ncbi.nlm.nih.gov/pubmed/227527*.

122. Lu J, Sherman D, Devor M, Saper CB. A putative flip-flop switch for control of REM sleep. *Nature*. 2006;441(7093):589–594.

123. Boissard R, Gervasoni D, Schmidt MH, Barbagli B, Fort P, Luppi P-H. The rat ponto-medullary network responsible for paradoxical sleep onset and maintenance: a combined microinjection and functional neuroanatomical study. *Eur J Neurosci*. 2002;16(10):1959–1973. *http://www.ncbi.nlm.nih.gov/pubmed/12453060*.

124. Hendricks JC, Morrison AR, Mann GL. Different behaviors during paradoxical sleep without atonia depend on pontine lesion site. *Brain Res*. 1982;239(1):81–105. *http://www.ncbi.nlm.nih.gov/pubmed/7093693*.

125. Sastre JP, Jouvet M. [Oneiric behavior in cats]. *Physiol Behav*. 1979;22(5):979–989. *http://www.ncbi.nlm.nih.gov/pubmed/228328*.

126. Shouse MN, Siegel JM. Pontine regulation of REM sleep components in cats: integrity of the pedunculopontine tegmentum (PPT) is important for phasic events but unnecessary for atonia during REM sleep. *Brain Res*. 1992;571(1):50–63. *http://www.ncbi.nlm.nih.gov/pubmed/1611494*.

127. Weber F, Chung S, Beier KT, Xu M, Luo L, Dan Y. Control of REM sleep by ventral medulla GABAergic neurons. *Nature*. 2015;526(7573):435–438.

128. Prabakaran S. GABA, the time keeper. *Sci Signal*. 2015;8(388):ec213–ec1213.

129. Gottesmann C. GABA mechanisms and sleep. *Neuroscience*. 2002;111(2):231–239. *http://www.ncbi.nlm.nih.gov/pubmed/11983310*.

130. Sanger DJ. The pharmacology and mechanisms of action of new generation, non-benzodiazepine hypnotic agents. *CNS Drugs*. 2004;18(suppl 1):9–15.

131. Vienne J, Bettler B, Franken P, Tafti M. Differential effects of GABAB receptor subtypes, {gamma}-hydroxybutyric Acid, and Baclofen on EEG activity and sleep regulation. *J Neurosci*. 2010;30(42):14194–14204.

132. Stege G, Vos PJE, van den Elshout FJJ, Richard Dekhuijzen PN, van de Ven MJT, Heijdra YF. Sleep, hypnotics and chronic obstructive pulmonary disease. *Respir Med.* 2008;102(6):801–814.

133. el Mansari M, Sakai K, Jouvet M. Unitary characteristics of presumptive cholinergic tegmental neurons during the sleep-waking cycle in freely moving cats. *Exp brain Res.* 1989;76(3):519–529. *http://www.ncbi.nlm.nih.gov/pubmed/2551709.*

134. Kayama Y, Ohta M, Jodo E. Firing of "possibly" cholinergic neurons in the rat laterodorsal tegmental nucleus during sleep and wakefulness. *Brain Res.* 1992;569(2):210–220. *http://www.ncbi.nlm.nih.gov/pubmed/1540827.*

135. Steriade M, Datta S, Paré D, Oakson G, Curró Dossi RC. Neuronal activities in brain-stem cholinergic nuclei related to tonic activation processes in thalamocortical systems. *J Neurosci.* 1990;10(8):2541–2559. *http://www.ncbi.nlm.nih.gov/pubmed/2388079.*

136. Hu B, Steriade M, Deschênes M. The cellular mechanism of thalamic ponto-geniculo-occipital waves. *Neuroscience.* 1989;31(1):25–35. *http://www.ncbi.nlm.nih.gov/pubmed/2771060.*

137. Lim AS, Lozano AM, Moro E, Hamani C, Hutchison WD, Dostrovsky JO, Lang AE, Wennberg RA, Murray BJ. Characterization of REM-sleep associated ponto-geniculo-occipital waves in the human pons. *Sleep.* 2007;30(7):823–827. *http://www.pubmedcentral.nih.gov/articlerender.fcgi?artid=1978372&tool=pmcentrez&rendertype=abstract.*

138. Fernández-Mendoza J, Lozano B, Seijo F, Santamarta-Liébana E, Ramos-Platón MJ, Vela-Bueno A, Fernández-González F. Evidence of subthalamic PGO-like waves during REM sleep in humans: a deep brain polysomnographic study. *Sleep.* 2009;32(9):1117–1126. *http://www.pubmedcentral.nih.gov/articlerender.fcgi?artid=2737569&tool=pmcentrez&rendertype=abstract.*

139. Jouvet M. Paradoxical sleep—a study of its nature and mechanisms. *Prog Brain Res.* 1965;18:20–62. *http://www.ncbi.nlm.nih.gov/pubmed/14329040.*

140. Ramírez-Salado I, Rivera-García AP, Moctezuma JV, Anguiano AJ, Pellicer F. GABAA receptor agonist at the caudo-lateral peribrachial area suppresses ponto-geniculo-occipital waves and its related states. *Pharmacol Biochem Behav.* 2014;124:333–340.

141. Morales FR, Engelhardt JK, Soja PJ, Pereda AE, Chase MH. Motoneuron properties during motor inhibition produced by microinjection of carbachol into the pontine reticular formation of the decerebrate cat. *J Neurophysiol.* 1987;57(4):1118–1129. *http://www.ncbi.nlm.nih.gov/pubmed/3585456.*

142. Webster HH, Jones BE. Neurotoxic lesions of the dorsolateral pontomesencephalic tegmentum-cholinergic cell area in the cat. II. Effects upon sleep-waking states. *Brain Res.* 1988;458(2):285–302. *http://www.ncbi.nlm.nih.gov/pubmed/2905197.*

143. Alam MN, Gong H, Alam T, Jaganath R, McGinty D, Szymusiak R. Sleep-waking discharge patterns of neurons recorded in the rat perifornical lateral hypothalamic area. *J Physiol.* 2002;538(Pt 2):619–631. *http://www.pubmedcentral.nih.gov/articlerender.fcgi?artid=2290077&tool=pmcentrez&rendertype=abstract.*

144. Bittencourt JC, Presse F, Arias C, Peto C, Vaughan J, Nahon JL, Vale W, Sawchenko PE. The melanin-concentrating hormone system of the rat brain: an immuno- and hybridization histochemical characterization. *J Comp Neurol.* 1992;319(2):218–245.

145. Kilduff TS, De Lecea L. Mapping of the mRNAs for the hypocretin/orexin and melanin-concentrating hormone receptors: networks of overlapping peptide systems. *J Comp Neurol.* 2001;435(1):1–5.

146. Koyama Y, Takahashi K, Kodama T, Kayama Y. State-dependent activity of neurons in the perifornical hypothalamic area during sleep and waking. *Neuroscience.* 2003;119(4):1209–1219. *http://www.ncbi.nlm.nih.gov/pubmed/12831874.*

147. Verret L, Goutagny R, Fort P, Cagnon L, Salvert D, Léger L, Boissard R, Salin P, Peyron C, Luppi P-H. A role of melanin-concentrating hormone producing neurons in the central regulation of paradoxical sleep. *BMC Neurosci.* 2003;4:19.

148. Ahnaou A, Drinkenburg WHIM, Bouwknecht JA, Alcazar J, Steckler T, Dautzenberg FM. Blocking melanin-concentrating hormone MCH1 receptor affects rat sleep-wake architecture. *Eur J Pharmacol.* 2008;579(1–3):177–188.

149. Willie JT, Sinton CM, Maratos-Flier E, Yanagisawa M. Abnormal response of melanin-concentrating hormone deficient mice to fasting: hyperactivity and rapid eye movement sleep suppression. *Neuroscience.* 2008;156(4):819–829.

150. Legendre R, Pieron H. Recherches sur le besoin de sommeil consecutif a une veille prolongee. *Z Allgem Physiol.* 1913;14:235–262.

151. Ishimori K. True cause of sleep: A hypnogenic substance as evidenced in the brain of sleep-deprived animals. *Tokyo Igakkai Zasshi.* 1909;23:429–457.

152. Pappenheimer JR, Koski G, Fencl V, Karnovsky ML, Krueger J. Extraction of sleep-promoting factor S from cerebrospinal fluid and from brains of sleep-deprived animals. *J Neurophysiol.* 1975;38(6):1299–1311. *http://www. ncbi.nlm.nih.gov/pubmed/1221075.*

153. Krueger JM, Szentirmai E, Kapas L. Biochemistry of sleep function: A paradigm for brain organization of sleep. In: Amlaner C, Fuller P, eds. *Basics of Sleep, Guide.* Westchester: Sleep Research Society; 2015:69–74.

154. Radulovacki M, Virus RM, Djuricic-Nedelson M, Green RD. Adenosine analogs and sleep in rats. *J Pharmacol Exp Ther.* 1984;228(2):268–274. *http://www.ncbi.nlm.nih.gov/pubmed/6694111.*

155. Basheer R, Strecker RE, Thakkar MM, McCarley RW. Adenosine and sleep-wake regulation. *Prog Neurobiol.* 2004;73(6):379–396.

156. Benington JH, Kodali SK, Heller HC. Stimulation of A1 adenosine receptors mimics the electroencephalographic effects of sleep deprivation. *Brain Res.* 1995;692(1–2):79–85. *http://www.ncbi.nlm.nih.gov/pubmed/8548323.*

157. Porkka-Heiskanen T, Strecker RE, Thakkar M, Bjorkum AA, Greene RW, McCarley RW. Adenosine: a mediator of the sleep-inducing effects of prolonged wakefulness. *Science.* 1997;276(5316):1265–1268.

158. Penetar D, McCann U, Thorne D, Kamimori G, Galinski C, Sing H, Thomas M, Belenky G. Caffeine reversal of sleep deprivation effects on alertness and mood. *Psychopharmacology (Berl).* 1993;112(2–3):359–365. *http://www. ncbi.nlm.nih.gov/pubmed/7871042.*

159. Huang Z-L, Urade Y, Hayaishi O. Prostaglandins and adenosine in the regulation of sleep and wakefulness. *Curr Opin Pharmacol.* 2007;7(1):33–38.

160. Oishi Y, Huang Z-L, Fredholm BB, Urade Y, Hayaishi O. Adenosine in the tuberomammillary nucleus inhibits the histaminergic system via A1 receptors and promotes non-rapid eye movement sleep. *Proc Natl Acad Sci USA.* 2008;105(50):19992–19997.

161. Scammell TE, Gerashchenko DY, Mochizuki T, McCarthy MT, Estabrooke IV, Sears CA, Saper CB, Urade Y, Hayaishi O. An adenosine A2a agonist increases sleep and induces Fos in ventrolateral preoptic neurons. *Neuroscience.* 2001;107(4):653–663. *http://www.ncbi.nlm.nih.gov/pubmed/11720788.*

162. Halassa MM, Florian C, Fellin T, Munoz JR, Lee S-Y, Abel T, Haydon PG, Frank MG. Astrocytic modulation of sleep homeostasis and cognitive consequences of sleep loss. *Neuron.* 2009;61(2):213–219.

163. Rector DM, Topchiy IA, Carter KM, Rojas MJ. Local functional state differences between rat cortical columns. *Brain Res.* 2005;1047(1):45–55.

164. Hanlon EC, Faraguna U, Vyazovskiy VV, Tononi G, Cirelli C. Effects of skilled training on sleep slow wave activity and cortical gene expression in the rat. *Sleep.* 2009;32(6):719–729. *http://www.pubmedcentral.nih.gov/articlerender.fcgi?artid=2690558&tool=pmcentrez&rendertype=abstract.*

165. Huber R, Ghilardi MF, Massimini M, Tononi G. Local sleep and learning. *Nature.* 2004;430(6995):78–81.

166. Burnstock G. Purinergic nerves. *Pharmacol Rev.* 1972;24(3):509–581. *http://www.ncbi.nlm.nih.gov/pubmed/4404211.*

167. Pollak KA, Swenson JD, Vanhaitsma TA, Hughen RW, Jo D, White AT, Light KC, Schweinhardt P, Amann M, Light AR. Exogenously applied muscle metabolites synergistically evoke sensations of muscle fatigue and pain in human subjects. *Exp Physiol.* 2014;99(2):368–380.

168. Mullington J, Opp M. Immunology. In: Amlaner C, Fuller PM, eds. *Basics of Sleep Guide.* Westchester: Sleep Research Society; 2015:169–177.

169. Bredow S, Guha-Thakurta N, Taishi P, Obál F, Krueger JM. Diurnal variations of tumor necrosis factor alpha mRNA and alpha-tubulin mRNA in rat brain. *Neuroimmunomodulation..* 1997;4(2):84–90. *http://www.ncbi.nlm.nih. gov/pubmed/9483199.*

170. Floyd RA, Krueger JM. Diurnal variation of TNF alpha in the rat brain. *Neuroreport.* 1997;8(4):915–918. *http://www.ncbi.nlm.nih.gov/pubmed/9141064.*

171. Taishi P, Bredow S, Guha-Thakurta N, Obál F, Krueger JM. Diurnal variations of interleukin-1 beta mRNA and beta-actin mRNA in rat brain. *J Neuroimmunol.* 1997;75(1-2):69–74. *http://www.ncbi.nlm.nih.gov/pubmed/9143239.*

172. Krueger JM, Majde JA. Microbial products and cytokines in sleep and fever regulation. *Crit Rev Immunol.* 1994;14(3–4):355–379. *http://www.ncbi.nlm.nih.gov/pubmed/7755878.*

173. Pandey HP, Ram A, Matsumura H, Satoh S, Hayaishi O. Circadian variations of prostaglandins D2, E2, and F2 alpha in the cerebrospinal fluid of anesthetized rats. *Biochem Biophys Res Commun.* 1995;213(2):625–629.

174. Ram A, Pandey HP, Matsumura H, Kasahara-Orita K, Nakajima T, Takahata R, Satoh S, Terao A, Hayaishi O. CSF levels of prostaglandins, especially the level of prostaglandin D2, are correlated with increasing propensity towards sleep in rats. *Brain Res.* 1997;751(1):81–89. *http://www.ncbi.nlm.nih.gov/pubmed/9098570.*

175. Mizoguchi A, Eguchi N, Kimura K, Kiyohara Y, Qu WM, Huang ZL, Mochizuki T, Lazarus M, Kobayashi T, Kaneko T, Narumiya S, Urade Y, Hayaishi O. Dominant localization of prostaglandin D receptors on arachnoid

trabecular cells in mouse basal forebrain and their involvement in the regulation of non-rapid eye movement sleep. *Proc Natl Acad Sci USA.* 2001;98(20):11674–11679.

176. Scammell T, Gerashchenko D, Urade Y, Onoe H, Saper C, Hayaishi O. Activation of ventrolateral preoptic neurons by the somnogen prostaglandin D2. *Proc Natl Acad Sci USA.* 1998;95(13):7754–7759.

177. Ueno R, Honda K, Inoué S, Hayaishi O. Prostaglandin D2, a cerebral sleep-inducing substance in rats. *Proc Natl Acad Sci USA.* 1983;80(6):1735–1737.

178. Inoué S, Honda K, Komoda Y, Uchizono K, Ueno R, Hayaishi O. Differential sleep-promoting effects of five sleep substances nocturnally infused in unrestrained rats. *Proc Natl Acad Sci USA.* 1984;81(19):6240–6244.

179. Onoe H, Ueno R, Fujita I, Nishino H, Oomura Y, Hayaishi O. Prostaglandin D2, a cerebral sleep-inducing substance in monkeys. *Proc Natl Acad Sci USA.* 1988;85(11):4082–4086.

180. Kantha SS. Histamine-interleukin-prostaglandin pathway: a hypothesis for a biochemical cycle regulating sleep and wakefulness. *Med Hypotheses.* 1994;42(5):335–339. *http://www.ncbi.nlm.nih.gov/pubmed/7935077.*

181. Pentreath VW, Rees K, Owolabi OA, Philip KA, Doua F. The somnogenic T lymphocyte suppressor prostaglandin D2 is selectively elevated in cerebrospinal fluid of advanced sleeping sickness patients. *Trans R Soc Trop Med Hyg.* 1990;84(6):795–799. *http://www.ncbi.nlm.nih.gov/pubmed/2096510.*

182. McCormick DA. Neurotransmitter actions in the thalamus and cerebral cortex and their role in neuromodulation of thalamocortical activity. *Prog Neurobiol.* 1992;39(4):337–388. *http://www.ncbi.nlm.nih.gov/pubmed/1354387.*

183. Picciotto MR, Higley MJ, Mineur YS. Acetylcholine as a neuromodulator: cholinergic signaling shapes nervous system function and behavior. *Neuron.* 2012;76(1):116–129.

184. Gerashchenko D, Wisor JP, Burns D, Reh RK, Shiromani PJ, Sakurai T, de la Iglesia HO, Kilduff TS. Identification of a population of sleep-active cerebral cortex neurons. *Proc Natl Acad Sci USA.* 2008;105(29):10227–10232.

185. Steriade M, Domich L, Oakson G, Deschênes M. The deafferented reticular thalamic nucleus generates spindle rhythmicity. *J Neurophysiol.* 1987;57(1):260–273. *http://www.ncbi.nlm.nih.gov/pubmed/3559675.*

186. Adams JH, Graham DI, Jennett B. The neuropathology of the vegetative state after an acute brain insult. *Brain.* 2000;123(Pt 7):1327–1338. *http://www.ncbi.nlm.nih.gov/pubmed/10869046.*

187. Buzsaki G, Bickford RG, Ponomareff G, Thal LJ, Mandel R, Gage FH. Nucleus basalis and thalamic control of neocortical activity in the freely moving rat. *J Neurosci.* 1988;8(11):4007–4026. *http://www.ncbi.nlm.nih.gov/pubmed/3183710.*

188. Fuller PM, Fuller P, Sherman D, Pedersen NP, Saper CB, Lu J. Reassessment of the structural basis of the ascending arousal system. *J Comp Neurol.* 2011;519(5):933–956.

189. Kinney HC, Korein J, Panigrahy A, Dikkes P, Goode R. Neuropathological findings in the brain of Karen Ann Quinlan. The role of the thalamus in the persistent vegetative state. *N Engl J Med.* 1994;330(21):1469–1475.

190. Vanderwolf CH, Robinson TE. Reticulo-cortical activity and behavior: a critique of the arousal theory and a new synthesis. *Behav Brain Sci.* 2010;4(03):459.

191. Chemelli RM, Willie JT, Sinton CM, Elmquist JK, Scammell T, Lee C, Richardson JA, Williams SC, Xiong Y, Kisanuki Y, Fitch TE, Nakazato M, Hammer RE, Saper CB, Yanagisawa M. Narcolepsy in orexin knockout mice: molecular genetics of sleep regulation. *Cell.* 1999;98(4):437–451. *http://www.ncbi.nlm.nih.gov/pubmed/10481909.*

192. Adamantidis A, Salvert D, Goutagny R, Lakaye B, Gervasoni D, Grisar T, Luppi P-H, Fort P. Sleep architecture of the melanin-concentrating hormone receptor 1-knockout mice. *Eur J Neurosci.* 2008;27(7):1793–1800.

193. Ranson SW. Somnolence caused by hypothalamic lesions in the monkey. *Arch Neurol Psychiatry.* 1939;41(1):1.

194. Gerashchenko D, Kohls MD, Greco M, Waleh NS, Salin-Pascual R, Kilduff TS, Lappi DA, Shiromani PJ. Hypocretin-2-saporin lesions of the lateral hypothalamus produce narcoleptic-like sleep behavior in the rat. *J Neurosci.* 2001;21(18):7273–7283. *http://www.ncbi.nlm.nih.gov/pubmed/11549737.*

195. Lee MG, Hassani OK, Alonso A, Jones BE. Cholinergic basal forebrain neurons burst with theta during waking and paradoxical sleep. *J Neurosci.* 2005;25(17):4365–4369.

196. Berridge CW, Foote SL. Enhancement of behavioral and electroencephalographic indices of waking following stimulation of noradrenergic beta-receptors within the medial septal region of the basal forebrain. *J Neurosci.* 1996;16(21):6999–7009. *http://www.ncbi.nlm.nih.gov/pubmed/8824336.*

197. Sherin JE, Shiromani PJ, McCarley RW, Saper CB. Activation of ventrolateral preoptic neurons during sleep. *Science.* 1996;271(5246):216–219. *http://www.ncbi.nlm.nih.gov/pubmed/8539624.*

198. McCarley RW, Hobson JA. Neuronal excitability modulation over the sleep cycle: a structural and mathematical model. *Science.* 1975;189(4196):58–60. *http://www.ncbi.nlm.nih.gov/pubmed/1135627.*

199. Sakai K. Central mechanisms of paradoxical sleep. *Brain Dev.* 1986;8(4):402–407. *http://www.ncbi.nlm.nih.gov/pubmed/3799909.*

200. Xi M-C, Morales FR, Chase MH. Interactions between GABAergic and cholinergic processes in the nucleus pontis oralis: neuronal mechanisms controlling active (rapid eye movement) sleep and wakefulness. *J Neurosci.* 2004;24(47):10670–10678.

201. Brown RE, Sergeeva O, Eriksson KS, Haas HL. Orexin A excites serotonergic neurons in the dorsal raphe nucleus of the rat. *Neuropharmacology.* 2001;40(3):457–459. *http://www.ncbi.nlm.nih.gov/pubmed/11166339.*

202. Eggermann E, Serafin M, Bayer L, Machard D, Saint-Mleux B, Jones BE, Mühlethaler M. Orexins/hypocretins excite basal forebrain cholinergic neurones. *Neuroscience.* 2001;108(2):177–181. *http://www.ncbi.nlm.nih.gov/pubmed/11734353.*

203. Marcus JN, Aschkenasi CJ, Lee CE, Chemelli RM, Saper CB, Yanagisawa M, Elmquist JK. Differential expression of orexin receptors 1 and 2 in the rat brain. *J Comp Neurol.* 2001;435(1):6–25. *http://www.ncbi.nlm.nih.gov/pubmed/11370008.*

204. Braak H, Del Tredici K, Rüb U, de Vos RAI, Jansen Steur ENH, Braak E. Staging of brain pathology related to sporadic Parkinson's disease. *Neurobiol Aging.* 2003;24(2):197–211. *http://www.ncbi.nlm.nih.gov/pubmed/12498954.*

205. Iranzo A, Molinuevo JL, Santamaría J, Serradell M, Martí MJ, Valldeoriola F, Tolosa E. Rapid-eye-movement sleep behaviour disorder as an early marker for a neurodegenerative disorder: a descriptive study. *Lancet Neurol.* 2006;5(7):572–577.

206. Borbely A, Tobler I. Homeostatic and circadian principles in sleep regulation in the rat. In: McGinty DJ, ed. *Brain Mechanisms of Sleep.* New York: Raven Press; 1985:35–44.

207. Chou TC, Scammell TE, Gooley JJ, Gaus SE, Saper CB, Lu J. Critical role of dorsomedial hypothalamic nucleus in a wide range of behavioral circadian rhythms. *J Neurosci.* 2003;23(33):10691–10702. *http://www.ncbi.nlm.nih.gov/pubmed/14627654.*

208. Dijk DJ, Czeisler CA. Contribution of the circadian pacemaker and the sleep homeostat to sleep propensity, sleep structure, electroencephalographic slow waves, and sleep spindle activity in humans. *J Neurosci.* 1995;15(5 Pt 1):3526–3538. *http://www.ncbi.nlm.nih.gov/pubmed/7751928.*

209. Lu J, Zhang YH, Chou TC, Gaus SE, Elmquist JK, Shiromani P, Saper CB. Contrasting effects of ibotenate lesions of the paraventricular nucleus and subparaventricular zone on sleep-wake cycle and temperature regulation. *J Neurosci.* 2001;21(13):4864–4874. *http://www.pubmedcentral.nih.gov/articlerender.fcgi?artid=3508730&tool=pmcentrez&rendertype=abstract.*

210. Bass J, Takahashi JS. Circadian integration of metabolism and energetics. *Science.* 2010;330(6009):1349–1354.

211. Buhr ED, Yoo S-H, Takahashi JS. Temperature as a universal resetting cue for mammalian circadian oscillators. *Science.* 2010;330(6002):379–385.

Anatomy and Physiology of the Circadian System

R.P. Najjar*,**,†, J.M. Zeitzer*,**

*Department of Psychiatry and Behavioral Sciences, Stanford University, Stanford, CA, USA
**Mental Illness Research, Education and Clinical Center, Veterans Affairs Palo Alto Health Care System, Palo Alto, CA, USA
†Department of Visual Neurosciences, Singapore Eye Research Institute (SERI), The Academia, Singapore

BIOLOGICAL RHYTHMS AT THE ORIGIN OF THE CIRCADIAN TIMING SYSTEM

The earth's rotation around its polar axis gives rise to the day and night cycle, the revolutions of the eartharound the sun give rise to the unfailing seasonal procession, and the more complicated movements of the moon in relation to the earth and the sun give rise to the lunar month and to the tidal cycles. Given the occurrence of all these cycles, it is not surprising

Sleep and Neurologic Disease. http://dx.doi.org/10.1016/B978-0-12-804074-4.00002-9

to find that most organisms show rhythmic modulations in their bodily processes and behavior as predictive and adaptive mechanisms. Chronobiology is devoted to the study of these biological rhythms [from ancient Greek χρόνoς (chrónos, meaning "time") βιoζ (bios, meaning "life"), and λoγoζ (logos, meaning "study")]. Such rhythms are categorized based on the length of their time period (*tau* or *τ*). Ultradian rhythms have a *tau* <24 h, examples of which include the 90-min rapid eye movement (REM) cycle in sleep, the 4 h nasal cycle, and the 3 h cycle of growth hormone secretion. Circadian rhythms [(from Latin *circa* (meaning "approximately") and *dies* (meaning "day")] have a *tau* of nearly 24 h, examples of which include melatonin secretion, core body temperature (CBT) and cortisol fluctuations. Infradian rhythms have a *tau* longer than 24 h and include circannual rhythms (*tau* ~1 year, e.g., annual migration), circatidal rhythms (*tau* ~12.4 days, e.g., fiddler crab feeding behavior), and circalunar rhythms (*tau* ~29.5 days, e.g., human menstrual cycle). Of the biological rhythms, circadian rhythms have been the most intensively studied in a large number of organisms, ranging from single-celled organisms to humans.

ENDOGENOUS CIRCADIAN RHYTHMS

Periodic leaf movements in plants were first described by Androsthenes Thasius 325 BC. Many centuries later, the French geophysicist Jean-Jacques d'Ortous de Mairan (1729) observed that the mimosa plant, even when kept in total darkness, opened and closed its leaves according to the day-night cycle. Although not stated by de Mairan himself, his work implied that the force driving the plant's rhythms was endogenously generated. de Mairan's work eventually led to the creation of the "flower clock" by Linnaeus in 1751. The first observation of a daily rhythm in humans was made by William Ogle in 1866, who noticed that his own body temperature followed a daily rhythm.[1]

Circadian rhythms exist in organisms ranging from cyanobacteria[2] to humans.[3] These rhythms allow the anticipation of environmental changes and are consequently important adaptive mechanisms for species survival and interaction.[4,5] In mammals and humans in particular, various physiological (e.g., neuroendocrine, autonomic, cardiovascular, cognitive) and behavioral processes (e.g., sleep–wake cycle) follow an endogenous circadian rhythm that is under the control of a central pacemaker oscillating at its own endogenous period.

THE CIRCADIAN TIMING SYSTEM

In humans, the circadian timing system (CTS) can be schematically simplified as a three-stage processor. At the core of the processor is the central pacemaker. This pacemaker receives photic input from the eye and sends entraining signals to its outputs in various parts of the organism. Outputs also feed back to the central clock,[6,7] rendering the CTS an *ouroboros* (from the Greek, tail-devouring snake) that is modulated by its own signaling (Fig. 2.1).

The Central Oscillator

In mammals, the central pacemaker is located in the basal hypothalamus, dorsal to the optic chiasm on both sides of the third ventricle.[8] This structure, called the suprachiasmatic

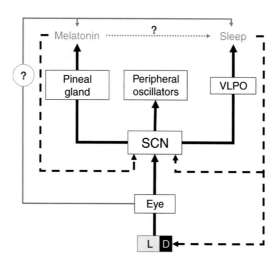

FIGURE 2.1 **A modified eskinogram of the suprachiasmatic nucleus (SCN) inputs and outputs.** The central clock located in the *SCN* is entrained to the environmental light (*L*)/dark (*D*) cycle by the neurotransduced photic input from the eye. The *SCN* controls the rhythmicity of multiple cerebral and noncerebral outputs including the ventrolateral preoptic nucleus (*VLPO*) implicated in sleep, the pineal gland responsible for melatonin secretion, and peripheral oscillators located in other organs, such as, the kidneys and lungs. In return, these outputs of the *SCN* feed back and influence the *SCN* via receptors on its surface (e.g., melatonin) or indirectly by modulating the exposure to the light/dark cycle (e.g., sleep). Environmental light might also bypass the *SCN* to affect its outputs independently, including a direct influence on sleep and melatonin production.

nucleus (SCN), is composed of two nuclei, each the size of a needle pin head (1 mm^3) containing around 10,000 neurons each. Nearly all of the cells constituting the SCN can act as individual circadian oscillators[9] with a high electrical firing rate during the day and low activity during the night.[10]

The SCN can be divided into two compartments, a shell and a core, based on differential expression of neuropeptides along with differences in the afferent and efferent connections of each partition.[11] The dorsomedial compartment, or shell, comprises small arginine vasopressin (AVP) and neurotensin containing neurons that are poor in dendritic arbors. The ventrolateral region of the nucleus, or core, is formed of relatively large neurons that have extensive dendritic arbors and express vasoactive intestinal polypeptide (VIP), gastrin-releasing peptide, neurotensin, neuropeptide Y, substance P and calbindin.[12–15] SCN projections to the rest of the brain originate from both AVP- and VIP-expressing neurons.[16,17] Almost all SCN neurons express γ-aminobutyric acid (GABA) as a primary neurotransmitter.[15,18] In humans, it is worth noting that there is a sex difference in the shape of the AVP subnucleus and in VIP-expressing cells numbers,[8,19–21] which are greater in males than females by the age of ten. By age 40, however, this sex difference seems to be reversed.[21]

Genetics of the Clockwork in Mammals

Advancements in molecular genetics have played a compelling role in elucidating the molecular core of SCN clockwork in mammals and other species. Briefly, the molecular clock is based on interlocked transcription–translation feedback loops in which the expression of clock genes is periodically repressed by their protein products. The canonical clock genes period *(Per)* and cryptochrome *(Cry)* make up a self-sustaining circadian oscillator[22] in which the critical mechanism is a delayed negative feedback. These genes are switched on by the proteins CLOCK and BMAL1, and periodically switched off by a complex of their own encoded proteins, PER and CRY. An unfailing daily loop of gene turn-on unavoidably followed by gene turn-off is thus set in place. Mutations that affect this loop's stability,[23] as well as polymorphisms in genotypes,[24–26] are tightly linked to inherited human sleep disorders and morningness-eveningness preferences.

Peripheral Oscillators

Although the SCN is the primary circadian oscillator in mammals, molecular oscillators with 24-h periodicity are also found in other regions of the brain, the retina and many peripheral organs.[27–32] For example, fibroblast explants stimulated with serum or specific chemical compounds can exhibit a rhythmic expression of circadian genes that persist for several days.[33,34] These peripheral oscillators conserve a molecular conformation that is closely similar to that of the SCN[35] and cycle accordingly to their own (~ 24 h) period.[36] In the mammalian circadian timing system, most (e.g., liver clock, retina clock) peripheral oscillators are under the entraining coordination of the SCN. While these peripheral clocks have the machinery to express a 24-h oscillation, their synchrony to the outside world and to other tissues is under the control of the SCN. The SCN synchronizes the timing of these peripheral oscillators according to the light input received from the environment and, in turn, peripheral oscillators regulate local physiological and behavioral rhythms.[37,38] Local clocks are also likely influenced by tissue-specific mechanisms. For example, the timing of peripheral oscillations in specific peripheral tissues can be influenced by feeding time.[29,31,39] Interestingly, under severe food restriction the phase of the SCN and its entrainment by light are also modified,[40] although feeding is not thought to be a major influence on the timing of the human circadian clock.

The Photic Input of the Circadian Timing System in Mammals

Light is the primary stimulus by which the mammalian circadian system remains synchronized to the light/dark cycle. In 1972, however, lesion studies demonstrated that damage to the various components of the rat visual system [lateral geniculate nucleus (LGN), primary optic tracts, accessory optic system], failed to disrupt entrainment of the central clock by light.[41–43] These unforeseen results, at that time, led to the hypothesis that a direct retinohypothalamic pathway conveyed photic information to the SCN. Autoradiographic tracing methods revealed a direct monosynaptic retinohypothalamic tract (RHT) terminating in the SCN.[44,45] This tract, also identified in humans,[46] plays the dominant role in coupling the circadian system to the external light/dark cycle. The RHT projects from a distinct subset of retinal ganglion cells in the retina[47] to the core of the SCN, where its terminals contain glutamate, which codes chemically for "light" information, as well as pituitary adenylate cyclase-activating polypeptide (PACAP), which codes chemically for "darkness" information.[48] It should be noted that the extent to which the retina projects to the SCN is highly variable across species.[49–51] The SCN also receives indirect photic information from the intergeniculate leaflet nucleus of the thalamus (IGL), which integrates information about light and motor activity, via the geniculo-hypothalamic tract (GHT).[52] The GHT contributes to photic entrainment but is not absolutely necessary for it to occur.[53]

All light information to the human SCN appears to come from retinal photoreception, as bilaterally enucleated or blindfolded individuals fail to entrain and exhibit a non-24 h (free-running) circadian rhythm.[54,55] In 1998 however, Campbell and Murphy, challenged the belief that mammals are incapable of extra-ocular circadian phototransduction, claiming that light pulses presented to the popliteal region (behind the knee) are capable of photoentraining the circadian system.[56] These results were never replicated or confirmed by other scientists.[57–59]

Circadian and Nonimage Forming Photoreception

Conventional visual photoreceptors (rods and cones) located in the outer mammalian retina have, since 1854, been considered to be the only light sensitive cells in the eye. Photoreceptors owe their light sensitivity to the photopigment they enclose in their outer segments. In humans, rhodopsin is the photopigment expressed within the rod outer segment and has a peak sensitivity (λ_{max}) around 505 nm (blue-green). Rods are very sensitive to light and mediate colorless vision in very dim light (scotopic) conditions.[60] Color vision in humans is achieved by a multimodal integration of signals from three types of cone photoreceptors. S-cones are a minority (5–10% of all cones[61]) and contain short-wavelength-sensitive (λ_{max} ~430 nm, blue) pigments; M-cones and L-cones contain respectively middle-wavelength (λ_{max} ~530 nm, green) and long-wavelength sensitive (λ_{max} ~560 nm, yellow) pigments.[62] The presence of these three types of cones in the retina makes human color vision "trichromatic." In spite of these beliefs, unattenuated circadian entrainment and melatonin suppression by light in blind individuals[54] and in rodents lacking rods and cones[63–65] drew attention toward an unidentified photopigment in the mammalian eye.

The photopigment melanopsin was first identified in melanophores, brain and eyes of the African clawed frog (*Xenopus laevis*) by Provencio et al. in 1998.[66] Subsequently, melanopsin was found in the mammalian inner retina in a subset of retinal ganglion cells (RGC), the cells that project to the brain to convey information about environmental light.[67,68] In rodents, 1–2% of RGC contain melanopsin[68], whereas in humans only 0.2–0.8% of RGC contain this photopigment[69]. Remarkably, in 2002, Berson and coworkers demonstrated that RGC expressing melanopsin could directly respond to light independent of any input from rods or cones.[70] Given their innate ability to respond to light, these RGC have been called intrinsically photosensitive RGC (ipRGC). Peak sensitivity of melanopsin and subsequently ipRGCs is around 480 nm (blue/indigo) in rodents, primates and humans.[69–72] In addition to their intrinsic sensitivity to light, that generates a sluggish yet long lasting response at high irradiances, ipRGCs synapse with bipolar and amacrine cells and receive synaptic input from rods and cones (extrinsic light input). The rods and cones generate a fast and short response to light at low and high irradiances, respectively.[70,73–79] Therefore, an integration of nervous signals incoming from rods, cones, and melanopsin takes place at the level of the ipRGC before being conveyed to nonimage forming (NIF) brain structures, such as the SCN for circadian photoentrainment,[48,75,80] the olivary pretectal nucleus (OPN), pupillary light reflex (PLR), and various sleep centers.[81,82] ipRGCs have been shown to relay light and dark information from both rod-cone and melanopsin-based pathways to modulate sleep and wakefulness in rodents[77] and light aversion in preterm or neonatal mice.[83] Projections that arise from the ipRGCs, possibly modulate the activity of dura-sensitive thalamocortical neurons[84] and shed light on the conservation of light-induced migraine in blind individuals with light perception.[85] Melanopsin has also been recently reported to be involved in the eye's development[86] and axonal regeneration in the central nervous system after optic nerve crush.[87] Aside from nonimage forming centers in the brain, ipRGCs have been found to project to image forming areas of the brain, such as the dLGN and visual cortex in rodents[88,89] and LGN in primates.[69,90] The latter studies showed that ipRGC not only have a role in sustaining firing in the dLGN neurons but also provide information about spatial brightness to the thalamo-cortical visual projections. Conscious visual awareness elicited by ipRGCs has also been elegantly demonstrated in humans with outer retinal disease.[91,92] Due to its projection to these visual centers and its implication in numerous visually guided behaviors, melanopsin containing ipRGC are now emerging as a possible therapeutic option in outer-retinal degenerative pathologies.

To date, five subtypes of ipRGCs (M1, M2, M3, M4, and M5) have been identified in mice[93,94] and rats.[95] These subtypes differ by dendritic morphology, intraretinal dendritic ramification and signaling, and axonal projections. In terms of retinal dendritic ramification: M1 cells stratify in the OFF sublamina of the inner plexiform layer (IPL); M2, M4, and M5 cells stratify in the ON sublamina of the IPL and M3 cells are variable, stratifying in the ON and OFF sublamina of the IPL.[90,96] Morphologically, M2 and M3 cells are closely similar with larger and more complex dendritic arbors and somas than M1 cells;[97] M4 cells have the largest somas and dendritic arbors of all subtypes, whereas M5 have small and bushy dendritic arbors. Functionally, compared to M2-M5 cells, M1 cells are very sensitive and have the largest intrinsic light-evoked responses, widest action potentials and spike at lower frequencies.[97,98] Using multiphoton microscopy to visualize whole-cell recording from fluorescently labeled ipRGCs, Zhao and coworkers also showed an increased intrinsic photo-sensitivity in M1 cells compared to other subtypes and a center surround-organized receptive field allowing spatial contrast detection in M2–M5 cells [99].

Due to their diverse projections, ipRGC subtypes are likely to have different roles in NIF responses to light. While projections of the M1 cells include the OPN (shell) and the SCN, responsible for PLR and circadian photoentrainment, projections of M2-M5 cells innervate the dLGN, OPN (core), and superior colliculus, and are involved in rudimentary, low-acuity visual function or brightness discrimination in mice and humans.[90,100] Moreover, M1 cells are divided into two subpopulations with ipRGCs containing the transcription factor Brn3b[101] that project to the OPN and other known M1 targets except the SCN. Ablation of these Brn3b-positive M1 cells alters PLR but not photoentrainment. Non-M1 subtypes also project throughout the brain, directly influencing activity, sleep/wake states, nociception and anxiety.[81,102,103] While not shown in humans, mice have different melanopsin isoforms that have been recently reported to mediate different behavioral responses to light such as PLR (OPN4S) and negative masking (OPN4L).[104]

Photoreceptors' Contribution to Nonimage Forming Responses to Light

The conventional visual photoreception system and the recently discovered melanopsin based system are complementary.[105–107] Rods, cones and melanopsin all participate to some extent in circadian photoentrainment,[79,108,109] melatonin suppression[110] and PLR.[111,112] Conventional photoreceptors' contribution to NIF responses to light is mediated through the ipRGCs, since rodents lacking ipRGCs retain pattern vision, but show severe deficits in both PLR and circadian photoentrainment.[113] Rods are the most sensitive to light and respond quickly in dim light conditions, whereas cones and melanopsin require high irradiances. Cones mediate an initial fast response to various light levels, whereas melanopsin cells participate in a long lasting sustained physiological responses to bright light.[74,79,88] Animal research suggests that rods are mostly implicated in circadian photoentrainment[77] whereas cones provide a high input to illicit PLR.[109] Both groups of classical photoreceptors are also essential in illuminance detection by the SCN.[114] S-cones have a poorly known function in both vision and circadian photoreception. These short wavelength sensitive cones might contribute to the off-responses[69] or the alerting effect of light.[115a]

OUTPUTS OF THE CENTRAL CLOCK

Anatomical Projections of the SCN

The SCN exerts its controls over its outputs via humoral and neural pathways. The SCN's direct efferent projections are sparse and mostly confined within the medial hypothalamus, reaching the medial preoptic area (MPO), the paraventricular nucleus of the hypothalamus (PVN), the subparaventricular nucleus (sPVN or SPZ) and the dorsomedial nucleus of the hypothalamus (DMH). Outside the hypothalamus, the SCN projects to the paraventricular nucleus of the thalamus, the IGL and indirectly to the ventral tegmentum through the median preoptic nucleus of the anterior hypothalamus. This latter connection could be responsible for circadian regulation of behavioral processes that include arousal and motivation.[115b]

At least three different types of neuronal targets of the SCN fibers can be discriminated within the medial hypothalamus: (1) endocrine neurons [such as those containing corticotropin-releasing hormone (CRH), thyrotropin-releasing hormone (TRH) and gonadotropin-releasing hormone (GnRH)], (2) projections to the autonomic PVN to influence autonomic neurons projecting to preganglionic parasympathetic and sympathetic neurons in the spinal cord[116–119], and (3) neurons reached by the direct SCN connections in the MPO, SPZ, and DMH, also called intermediate neurons, that likely integrate circadian information with other hypothalamic inputs before the information is passed on to endocrine and possibly autonomic neurons.

Some of the Functions Under the Control of the SCN

Locomotor Activity

The most apparent output of the circadian clock in animals is locomotor activity, which has also been used to define the temporal niche of an animal (diurnal, nocturnal or crepuscular). As such, locomotor activity is the most widely used tool to assess clock functioning in nonhuman animals. The regulation of circadian locomotor rhythms involves relays in the ventral subparaventricular zone (vSPZ) and DMH.[120,121] DMH neurons are also involved in the transmission of the circadian timing signal from the SCN to structures regulating feeding.[18] There also may be humoral mechanisms that are involved in the control of circadian locomotor rhythms, possibly through direct release of paracrine agents into the third ventricle, because neurally-isolated SCN explants placed into the base of the third ventricle have the capacity to restore locomotor rhythms in SCN-lesioned rats.[122]

Heart Rate

Resting heart rate in rodents and humans follows a circadian pattern, being high at phases corresponding to the active part of the daily cycle.[123,124] This circadian modulation of heart rate is due to a multisynaptic autonomic connection between SCN neurons and the heart with a relay at the autonomic part of the PVN.[124,125] Blood pressure also follows an endogenous circadian rhythm that peaks around 9:00 p.m. in humans.[126] Light can also directly impact heart rate such that light in the middle of the night and particularly in the early morning, when heart rate is low, induces an increase in heart rate, while light during the day has little or no effect.[127–129] This acute effect of light on heart rate requires an intact SCN and is modulated by melatonin.[124,130]

Temperature

Homeothermic creatures (examples include birds, rodents, and humans) maintain a stable internal core body temperature (CBT) regardless of external thermal fluctuations. In such animals, endogenous body temperature follows regular 24-h variations and is under both homeostatic and circadian regulation. When assessed independently of sleep in humans, CBT clearly shows an endogenous circadian profile with a nadir approximately 2 h before habitual wake time and a maximum at the end of the normal day.[123,131–133] The SCN rhythmically regulates CBT via the dSPZ neurons and the MPO.[18] In humans, exposure to light at night increases CBT,[134,135] an effect that likely requires an intact SCN.[136] In addition to the SCN-dependent light input, retinal projections to the SPZ[137] could provide a direct pathway for an acute effect of light on CBT. Such an SCN-independent pathway could account for the observation that 6 months after SCN lesioning, the acute response to light is regained, suggesting that after lesioning of the primary relay station for light information, secondary relays can take over this function.[136]

Sleep

Sleep is regulated by two main processes: homeostasis (often referred to as "process S") and circadian (often referred to as "process C"). Process S represents an appetitive process such that the longer one is awake, the more tired they get, and the more they sleep, the less tired they become. The physiologic correlate for this homeostatic mechanism remains unclear, however, one theory is that it is related to increased adenosine release in specific brain regions during the wake period.[138–140] Process C is generated from the SCN such that a peak circadian drive for wake occurs near the end of the normal waking day and a peak circadian drive for sleep occurs near the end of the normal sleep period.[141] The SCN and the SPZ have, however, very modest direct projection to sleep centers such as the ventrolateral preoptic nucleus (VLPO) and the arousing lateral hypothalamic orexin neurons (LHA).[121,142] The SPZ's main projections, however, are to the DMH which projects extensively to the VLPO and the LHA.[143] Therefore, the SCN mainly via a SPZ–DMH pathway could control the circadian process of sleep.[144] Lesion to the DMH profoundly diminishes circadian drive of sleep and wakefulness, as well as locomotor activity, corticosteroid secretion, and feeding.[121]

The Corticosteroid System

The endogenous circadian rhythm is integral to the release of cortisol from the adrenal glands, as evidenced by lesion studies that demonstrated the lack of rhythmic secretion of these hormones after lesions of the SCN.[41] The cortisol peak precedes the activity period in both diurnal and nocturnal animals.[145] Such a secretion pattern is thought to prepare the organism for awakening and the active period that follows. The circadian control of the cortisol rhythm appears to involve SCN connections to both neuroendocrine and autonomic systems. The SCN controls the activity of the hypothalamo-pituitary-adrenal (HPA) axis by affecting the release of cortisol via projections to the corticotrophin-releasing hormone (CRH) producing neurons of the medial parvocellular part of the PVH,[146] and by adjusting the sensitivity of the adrenal cortex to adrenocorticotropic hormone (ACTH) via SCN projections to the autonomic neurons of the PVH.[116] Lesioning of the vSPZ or the DMH leads to a near total loss in circadian rhythmicity of cortisol secretion. These results suggest that the cortisol rhythm cannot be fully maintained by the direct SCN pathways to the PVH, and mainly depends on indirect pathways via the vSPZ and the DMH.[120,121,147]

Cortisol levels are mainly affected by different stress stimuli. In addition, light has an acute impact on corticosteroid levels. When applied at the beginning of the active period, light increases cortisol levels in diurnal humans, and decreases levels in nocturnal rats.[148] As the cortisol peak is thought to prepare an organism for the active period and as light is the signal for the active period in diurnal animals, and the signal for the inactive period in nocturnal animals, the acute response of light on cortisol levels in both nocturnal and diurnal animals is adaptively and evolutionary significant. The fast and acute light-induced change in plasma cortisol requires an intact SCN[116] and appears to depend on the circadian phase, thus suggesting an active involvement of the SCN in the transmission of the light signal.[116,148] It is worth noting, however, that the SPZ also receives direct retinal projections, suggesting a direct effect of light on the pathway downstream the SCN.[137,149]

Cognitive Performance and Alertness

Cognitive processes (e.g., attention, executive functions, memory) and alertness follow a circadian pattern.[150–153] The circadian fluctuation of cognitive performances and alertness follows that of CBT, with minimum performance and alertness occurring near minimum of CBT (i.e., near the end of the habitual sleep period).[150,151,154] Sleep homeostasis also has a strong effect on performance and alertness, with a decline in these behaviors with increased time awake. The circadian and homeostatic systems interact in their regulation of alertness and performance such that these are maintained at a stable level throughout a 16-h day.[150] Light can acutely enhance alertness and cognitive performance,[134,155–159] even during daytime.[160–162] This direct effect of light may be due to its effects on alertness-related subcortical structures, such as the locus coeruleus,[163] hypothalamus,[164] and hippocampus/amygdala.[163,165] Clinically, light has also been used to attenuate cognitive deterioration, as well as ameliorate depressive symptoms, in elderly residents of group care facilities.[166]

Melatonin Secretion

Characterized and isolated for the first time from bovine pineal tissue by Aaron Lerner in 1958,[167] melatonin (N-acetyl-5-methoxytryptamine) is the main hormone synthesized by the pineal gland, located just below the splenium of the corpus callosum on the body side of the blood-brain barrier. Melatonin's secretion by the pineal gland is under direct control of the SCN[168] via projections to the autonomic division of the PVN that reach the pineal gland through a multisynaptic pathway involving relays in the intermediolateral column of the thoracic spinal cord and the superior cervical ganglion.[117] This cerebrospinal pathway is the main reason why individuals with neurologically complete damage to their cervical spinal cord have been demonstrated to produce little to no melatonin.[169] Transneuronal tract tracing from the pineal gland has demonstrated that the majority of the SCN neurons involved in this pathway are located in the dorsomedial portion of the SCN.[117] Melatonin can be found in cerebrospinal fluid, plasma, and saliva, and its main metabolite, 6-sulfatoxymelatonin (aMT6), can be detected in urine. There has been much written on potential functions of endogenous melatonin, including its involvement in regulating sleep cycles, thermoregulation, nocturnal alertness, cancer risk, and aging processes, among others. Much of this evidence has been derived from epidemiologic studies, in vitro studies, and animal studies, and direct evidence for these roles of melatonin in humans is lacking. Melatonin secretion may be used by the SCN to distribute rhythmic information across an organism and control other physiological functions (e.g., sleep-wake cycle, immune functions); secreted melatonin can also feed back at the level of the SCN and alter SCN function via specific melatonin receptors[170,171](Fig. 2.1). The duration of the nocturnal melatonin secretory episode increases with the duration of nocturnal darkness, thereby providing a measure of day length that can be used to regulate seasonal cycles in reproduction and other functions in photoperiodic species.[172]

Circadian Characteristics of Melatonin

Although locomotor activity is considered to be a reliable marker in animals, it is subject to many biases in humans. Melatonin production, however, has been shown to be relatively insensitive to nonphotic environmental and physiological factors (e.g., sleep, postural change, temperature, food intake) and is less variable than other circadian markers such as cortisol and core body temperature.[173] When measured under proper conditions (notably, when the individual is in dim light, as even moderate illuminances can suppress pineal melatonin production), melatonin is considered to be the most reliable marker of the circadian clock in humans. Melatonin can be acutely inhibited by a nocturnal light exposure of sufficient irradiance and duration.[174] Nocturnal light exposure can also change the timing of the melatonin profile on subsequent days. These effects of light on melatonin (suppression and phase shifting) appear to follow a similar yet not identical dose-response relationship to light intensity.[175] As it is far easier to study melatonin suppression, this measure has often been used as a proxy of the impact of light on circadian rhythms in humans. Recent studies, however, caution against the use of such proxy measures of circadian responses to light because the physiologic mechanisms underlying these responses may not be the same.[176]

ipRGCs, utilizing both intrinsic (melanopsin) and extrinsic (rods, cones) inputs are thought to convey the necessary light information that is involved in the suppression of melatonin.[177,178] The degree of inhibition of melatonin synthesis is dependent upon the wavelength,[177,179,180] intensity,[175] pattern,[176] and circadian phase of light administration.[181] Suppression of melatonin can occur even at room light intensity, but this may be due to the artificial laboratory environment in which this is measured, as the lengthy dim light that preceded such measurements may increase the sensitivity of melatonin suppression to light.[182,183]

ENTRAINMENT OF THE CIRCADIAN TIMING SYSTEM

A fundamental property of all circadian oscillators is that they run at a period of approximately, but not exactly, 24 h and require an external signal to semi-continuously adjust this non-24-h rhythm to the external 24 h light/dark cycle. In humans, the circadian timing system has a period that is close to (24.2 h) but not exactly 24 h,[3] and tends to be shorter in women (24.09 ± 0.2 h) than in men (24.19 ± 0.2 h).[184] The external time cues that are capable of entraining organisms are referred to as *zeitgebers* (from the German for "time givers"). For an environmental factor to be considered a *zeitgeber* it should fulfill four conditions. First, the absence of the *zeitgeber* should lead to a free running situation in which the central pacemaker follows its endogenous period. Second, during exposure to the *zeitgeber*, the period of the rhythm should become equal to that of the *zeitgeber*. Third, the phase angle of entrainment (i.e., the time between the time cue and the phase of the circadian oscillator) should be stable. And finally, the internal oscillator will retain its own period upon removal of the *zeitgeber*.

Photo-entrainment

For many transparent and semitransparent organisms (e.g., bacteria and zebrafish), light can directly reset the molecular clockwork of each cell.[185,186] In higher vertebrates such as mammals, however, this entrainment process remains hierarchical as peripheral oscillators do not directly respond to light but require a neural or humoral synchronizing signal from the SCN.[30,187–189] Human entrainment studies in early years led to the belief that the primary entraining factor for humans was not light, but rather social interaction.[190] This belief is now considered incorrect and the dominant time cue to the circadian clock in humans and other mammals is now known to be the light/dark cycle. The phase shifting effects of light appear to be brought about by an initial shift of a relatively small group of neurons in the SCN that becomes highly synchronized following a shift in the external light cycle.[191] The specific effects of light on the timing of the SCN are dependent upon timing,[192,193] intensity,[175,194] duration,[195,196] wavelength[177,179,180,197] and pattern[176,198–202] of the light exposure. Inappropriate or inadequate entrainment, as in jet lag for example, is associated with temperature, cardiovascular, immunological, sleep, vigilance, memory, and neurocognitive alterations.[141,154,200,203–207] The most compelling piece of evidence supporting the dangers of circadian misalignment comes from a study done by Davidson and coworkers in which old mice subjected to temporal disruptions equivalent to a flight from Washington to Paris, once a week for 8 weeks, died prematurely.[208] Similarly, though less severe, flight attendants traveling regularly across time zones undergo cognitive alterations associated with reductions in temporal lobe structures[209] and shift workers show an increased prevalence for cancer,[210] diabetes,[211] ulcers,[212] hypertension and cardiovascular disease,[213] and psychological disorders.[214]

Phase Sensitivity of the Circadian Timing System

The amount and direction (advance or delay) of a phase shift is dependent upon the timing of the light exposure relative to the timing of the circadian system. The relationship between the timing of the light and its effects on circadian phase can be illustrated using a phase response curve (PRC), which was first described in 1958.[215] To derive a PRC, the response of an output of the circadian clock (using markers such as locomotor activity onset or clock genes in animals and melatonin or CBT timing in humans) to a specific stimulus is evaluated thoroughly and systematically over the entire circadian cycle.[216,217] The shape of a PRC provides precise information concerning the overall relationship between phase shift magnitude and circadian phase of the stimulus. The shape of the PRC is similar in many organisms such that light elicits phase delays in the early subjective night and phase advances in late subjective

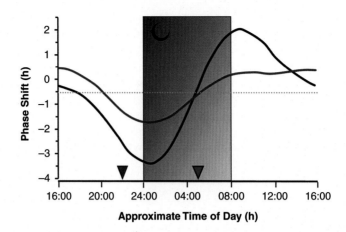

FIGURE 2.2   **The human type-1 PRC to bright light.** The impact of bright light on the circadian system is dependent upon the circadian phase/timing of light administration. Circadian phase shift of the melatonin secretion profiles are plotted here as a function of the timing of light administration. The *red descending triangle* represents the approximate onset of melatonin secretion (22:00), while the *blue descending triangle* represents the approximate CBT minimum (05:00) in an individual with a sleep episode typically occurring between 00:00 and 08:00 *(gray shaded area)*. Whether the stimulus is a 6.7 h *(black curve)* or 1 h *(red curve)* light pulse, if administered prior to bedtime or during the first hours of sleep, the exposure elicits a circadian phase delay (negative values). If administered in the hours prior to or after wake time, the light exposure elicits a phase advance (positive values). The figure was replotted from Khalsa et al.[192] and St-Hilaire et al.[193] conducted in controlled constant routine conditions. *The horizontal dashed line* at 0.54 h represents the anticipated average delay of the circadian pacemaker between the pre- and poststimulus phase assessments.

night. The region that lies between the phase delay and phase advance portions of the PRC, called the "critical" region, is of major importance to determine whether a PRC is type 0 (strong resetting, with phase shifts as large as 12 h in the critical region) or type 1 (slow resetting, with phase shifts as small as 0 h in the critical region).

The relationship between normal light exposure and phase shifts in humans is described by a type 1 PRC.[192,193,218] Although it is possible to elicit a type 0 PRC and 12 h phase shifts in humans,[195] this requires three successive days of exquisitely timed light in a laboratory environment. From early to late subjective day there is a decrease in phase advances and a gradual increase in phase delays elicited by light. In the human PRC, the transition during the subjective night from delays to advances is rapid, while the transition from phase advances to phase delays during the subjective day is more gradual (Fig. 2.2). Unlike most nocturnal rodents, light continuously affects the human circadian system during subjective day.[219] The absence of a "dead zone" in diurnal mammals, like humans, may allow the circadian system to rapidly accomplish stable entrainment in response to exposure to typical environmental light stimuli occurring during the subjective day.

Irradiance Sensitivity of the Circadian Timing System

In humans, the CTS has a threshold irradiance below which no response is engendered and a saturation irradiance above which elicited responses remain similar. Between these two asymptotes, the response to light follows a logarithmic rise. This pattern appears as a sigmoid shape between an output marker of the CTS (melatonin suppression, phase shift) and different light intensities.[175] Human studies by Zeitzer and coworkers showed that half of the maximal phase-delaying response and acute melatonin suppression in response to a single episode of evening bright light (~9000 lux) can be obtained with just over 1% of this light (room light of ~100 lux).[175,194,200] Similarly, half of the maximum alerting response to bright light of ~9000 lux was obtained with a room light of 100 lux.[157] These results indicate that even small changes in ordinary light exposure within the range of typical, ambient, room light (90–180 lux) during the late evening hours have the capacity to produce a significant impact on physiological functions such as plasma melatonin concentrations, circadian phase entrainment, subjective alertness and its electrophysiological correlates. Recent findings also showed that light exposures of 40 lux, whether from normal indoor light or even the light emitted by computer screens around the same illuminance range, can have an impact on circadian physiology, melatonin secretion, alertness, and cognitive performance levels.[220–222] It is important to note, however, that these laboratory experiments were conducted on a background of very dim light that can sensitize the circadian system (the dimmer the light during the day, the greater the effects of light at night) and should be extrapolated with caution when considering how light effects people in their natural environments.[223]

Duration and Pattern Sensitivity of the Circadian Timing System

In humans, exposure to very long pulses of light over the course of multiple hours is typically used in laboratory experiments to elicit significant changes in circadian timing. While producing less change in phase, short light exposure (0.2 h, 1 h) is more efficient, on a minute-to-minute basis, than long light exposure (2.5–4.0 h) in phase shifting the clock, suppressing melatonin, and inducing changes in alertness.[196] Sequences of short light exposure with interspersed darkness can even be more efficient than continuous light exposure.

In humans, a single sequence of intermittent bright light pulses (six 15 min bright light pulses of approximately 9500 lux separated by 60 min of very dim light of <1 lux) can phase delay the human CTS to an extent comparable to that measured after continuous bright light exposure (6.5 h exposure of 9500 lux).[199] Recent studies suggest that phase shifting of the human CTS can even be observed in response to a sequence of ultra-short (2 ms) flashes of light.[201] When presented in a sequence such that flashes are separated by 5–10 s, these flashes of light elicit two- to threefold larger phase shifts of the CTS,[176] and do so without significantly affecting alertness or melatonin suppression. These data are consistent with in vitro[224] and behavioral data exposures[225,226] in rodents, suggesting that the CTS can temporally integrate very brief light exposures.

Furthermore, flashes of light can readily pass through the eyelids and phase shift the CTS during sleep.[202,227] As the CTS is most sensitive to light during the timing of normal sleep, the use of a sequence of light flashes during sleep holds significant promise in the treatment of circadian-based sleep disorders. Modulations of the duration and pattern of light exposure have also been used to aid workers in extreme illumination environments. Gronfier and coworkers showed that a modulated light exposure protocol (10 h of 25 lux light followed by 100 lux light for the remainder of the day and two 45-min bright light pulses of 9500 lux near the end of wakefulness) is more efficient in entraining the CTS than a continuous 25 lux or 100 lux light exposure.[200] Such lighting designs using flashes and intermittent lighting can be used to entrain astronauts and submariners in conditions where natural light cycles are atypical or even missing.

Spectral Sensitivity of the Circadian Timing System

Unlike the visual system that has a peak sensitivity to green (555 nm) light in photopic conditions, the circadian system's peak sensitivity is shifted to the blue region of the visible light spectrum (460–480 nm). This peak spectral sensitivity is in physiological agreement with the peak sensitivity of the melanopsin containing ipRGCs. In humans, blue light (~480 nm) is more effective than green light (~550 nm) in increasing CBT and heart rate,[129] alertness and vigilance,[129,158] cognitive performances,[228,229] phase shift[230–232] and melatonin suppression.[177,179,180,197] This blue-shifted sensitivity of the CTS has led to the successful implementation of blue-enriched white light in indoor lighting to enhance entrainment of the circadian system and alleviate sleep and attentional lapses in urban workplaces[233] or extreme lighting environments such in polar regions where natural sunlight is missing during winter.[234–236] As previously noted, the effects of light on NIF functions of the brain are mediated by the ipRGC, which have both an intrinsic (melanopsin) and extrinsic (rods, cones) component to photoreception. Rods are responsive to very low intensity light, with a peak sensitivity near 505 nm. Cones are responsive to moderate to high intensity of light, with a peak sensitivity near 550 nm. Melanopsin is thought to be mainly responsive to higher intensity of light with a peak sensitivity near 480 nm. As such, the relationship between the spectral content of light and circadian phase shifting, among other NIF light functions, is likely moderated by the intensity of light. Further research is essential to determine the exact dynamics of photoreceptor contribution in nonimage forming responses to light.

Nonphotic *Zeitgebers*

In the absence of light, other cues may act as *zeitgebers*.[237,238] The most commonly studied nonphotic *zeitgebers* are temperature,[239] food intake,[240] social cues,[241] exercise[242,243] and melatonin.[244] Nonphotic *zeitgebers'* effectiveness upon the human circadian system remain minor compared to that of the light/dark cycle. Nonphotic *zeitgebers* can, however, be effectively used to reduce entrainment issues in visually impaired or blind individuals. One compelling piece of evidence for nonphotic entrainment in a totally blind individual was reported by Klerman et al. in 1998.[245] In this experiment, the participant had absent nonvisual responses to bright light and an endogenous period of 24.1 h, as estimated during a forced desynchrony protocol. Without any photic entrainment, the visually impaired participant was therefore expected to phase delay 0.1 h (6 min) every day, yet he remained entrained to the 24-h day. In this case, entrainment was presumably mediated by one or more nonphotic cues that the protocol had included, such as scheduled sleep, mealtimes and exercise. Moreover, nonphotic cues were able to entrain this blind individual to a non-24-h (23.8-h) sleep-wake schedule while living in constant near darkness. Nonphotic synchronizers might then be beneficial in visually impaired individuals with a *tau* that is close to 24-h, while a relatively long or short *tau* may be out of the range of entrainment for weak nonphotic *zeitgebers*. In fact, bilaterally enucleated blind individuals studied over a long duration display free-running circadian rhythms despite living on a 24-h day with nonphotic time cues such as those of social structure, family, employment, guide dogs, alarm clocks, food, caffeine, and regular activies.[54,55,245–247]

Exogenous Melatonin as a Zeitgeber

There are two general uses for exogenous melatonin, as a hypnotic and a chronobiotic (phase shifting) agent. A metaanalysis on the effects of exogenous melatonin administration indicated that when melatonin is administered during the daytime, it acts as a very mild hypnotic.[248] It has been hypothesized that this effect of melatonin is likely mediated through suppression of SCN firing activity via MT1 receptors.[170] Exogenous melatonin can also be used as a chronobiotic, acting to advance or delay the timing of the circadian system through a direct interaction with SCN MT2 receptors.[249,250] Its action can be described by a PRC[251] such that melatonin administration in the early part of the subjective night advances the clock, and melatonin in the early morning produces phase delays (see review by Skene et al.[252]). Recent data, however, suggest a melatonin PRC with a less distinguishable phase delay region.[253] Daily administration of melatonin has been used to entrain free-running patients who presumably lack visual input to their SCN.[254–256] In addition, sleep disorders in blind individuals have also been reported to improve with melatonin administration.[247]

CONCLUSIONS

The circadian timing system is central to the organization and harmonization of bodily functions. Overseen by the central circadian pacemaker of the SCN, this system requires fine tuning by environmental photic time cues detected at the ocular level to optimally entrain its physiological and behavioral outputs. Whether due to social constraints, long haul travel, disease states, or suboptimal lighting environments, the dysregulation of the circadian system can lead to severe neurological, cardiovascular, and metabolic disturbances. Understanding the neurobiology of the circadian timing system, with the electrophysiological and molecular machineries of its dynamic interactions with its afferent and efferent structures in humans, is critical for the optimization of light therapy strategies and the development of personalized chronotherapeutic medicine.

References

1. Ogle W. *On the Diurnal Variations in the Temperature of the Human Body in Health*. London: St. George's Hospital Reports; 1866:221–245.
2. Williams SB. A circadian timing mechanism in the cyanobacteria. *Adv Microb Physiol*. 2007;52:229–296.
3. Czeisler CA, Duffy JF, Shanahan TL, Brown EN, Mitchell JF, Rimmer DW, Ronda JM, Silva EJ, Allan JS, Emens JS, Dijk DJ, Kronauer RE. Stability, precision, and near-24-hour period of the human circadian pacemaker. *Science*. 1999;284(5423):2177–2181.
4. Aschoff J, Daan S, Groos GA. *Vertebrate Circadian Systems: Structure and Physiology*. Berlin Heidelberg: Springer; 1982.
5. Morin LP. The circadian visual system. *Brain Res Brain Res Rev*. 1994;19(1):102–127.
6. Antle MC, Mistlberger RE. Circadian clock resetting by sleep deprivation without exercise in the Syrian hamster. *J Neurosci*. 2000;20(24):9326–9332.
7. Deboer T, Vansteensel MJ, Détári L, Meijer JH. Sleep states alter activity of suprachiasmatic nucleus neurons. *Nat Neurosci*. 2003;6(10):1086–1090.
8. Hofman MA, Fliers E, Goudsmit E, Swaab DF. Morphometric analysis of the suprachiasmatic and paraventricular nuclei in the human brain: sex differences and age-dependent changes. *J Anat*. 1988;160:127–143.

9. Welsh DK, Logothetis DE, Meister M, Reppert SM. Individual neurons dissociated from rat suprachiasmatic nucleus express independently phased circadian firing rhythms. *Neuron*. 1995;14(4):697–706.

10. Green DJ, Gillette R. Circadian rhythm of firing rate recorded from single cells in the rat suprachiasmatic brain slice. *Brain Res*. 1982;245(1):198–200.

11. Antle MC, Silver R. Orchestrating time: arrangements of the brain circadian clock. *Trends Neurosci*. 2005;28(3):145–151.

12. Mai JK, Kedziora O, Teckhaus L, Sofroniew MV. Evidence for subdivisions in the human suprachiasmatic nucleus. *J Comp Neurol*. 1991;305(3):508–525.

13. Moore RY. The fourth C.U. Ariëns Kappers lecture. The organization of the human circadian timing system. *Prog Brain Res*. 1992;93:99–115.

14. Romijn HJ, van Uum JF, Emmering J, Goncharuk V, Buijs RM. Colocalization of VIP with AVP in neurons of the human paraventricular, supraoptic and suprachiasmatic nucleus. *Brain Res*. 1999;832(1–2):47–53.

15. Moore RY, Speh JC, Leak RK. Suprachiasmatic nucleus organization. *Cell Tissue Res*. 2002;309(1):89–98.

16. Dai J, Swaab DF, Buijs RM. Distribution of vasopressin and vasoactive intestinal polypeptide (VIP) fibers in the human hypothalamus with special emphasis on suprachiasmatic nucleus efferent projections. *J Comp Neurol*. 1997;383(4):397–414.

17. Buijs RM, Kalsbeek A. Hypothalamic integration of central and peripheral clocks. *Nat Rev Neurosci*. 2001;2(7):521–526.

18. Saper CB, Scammell TE, Lu J. Hypothalamic regulation of sleep and circadian rhythms. *Nature*. 2005;437(7063):1257–1263.

19. Hofman MA, Zhou JN, Swaab DF. Suprachiasmatic nucleus of the human brain: an immunocytochemical and morphometric analysis. *Anat Rec*. 1996;244(4):552–562.

20. Swaab DF, Fliers E, Partiman TS. The suprachiasmatic nucleus of the human brain in relation to sex, age and senile dementia. *Brain Res*. 1985;342(1):37–44.

21. Zhou JN, Hofman MA, Swaab DF. VIP neurons in the human SCN in relation to sex, age, and Alzheimer's disease. *Neurobiol Aging*. 1995;16(4):571–576.

22. Reppert SM, Weaver DR. Molecular analysis of mammalian circadian rhythms. *Annu Rev Physiol*. 2001;63:647–676.

23. Toh KL, Jones CR, He Y, Eide EJ, Hinz WA, Virshup DM, Ptácek LJ, Fu YH. An hPer2 phosphorylation site mutation in familial advanced sleep phase syndrome. *Science*. 2001;291(5506):1040–1043.

24. Archer SN, Robilliard DL, Skene DJ, Smits M, Williams A, Arendt J, von Schantz M. A length polymorphism in the circadian clock gene Per3 is linked to delayed sleep phase syndrome and extreme diurnal preference. *Sleep*. 2003;26(4):413–415.

25. Viola AU, Chellappa SL, Archer SN, Pugin F, Götz T, Dijk DJ, Cajochen C. Interindividual differences in circadian rhythmicity and sleep homeostasis in older people: effect of a PER3 polymorphism. *Neurobiol Aging*. 2012;33(5):1010.e17–1010.e27.

26. Hida A, Kitamura S, Katayose Y, Kato M, Ono H, Kadotani H, Uchiyama M, Ebisawa T, Inoue Y, Kamei Y, Okawa M, Takahashi K, Mishima K. Screening of clock gene polymorphisms demonstrates association of a PER3 polymorphism with morningness-eveningness preference and circadian rhythm sleep disorder. *Sci Rep*. 2014;4:6309.

27. Emery IF, Noveral JM, Jamison CF, Siwicki KK. Rhythms of Drosophila period gene expression in culture. *Proc Natl Acad Sci USA*. 1997;94(8):4092–4096.

28. Krishnan B, Dryer SE, Hardin PE. Circadian rhythms in olfactory responses of Drosophila melanogaster. *Nature*. 1999;400(6742):375–378.

29. Damiola F, Le Minh N, Preitner N, Kornmann B, Fleury-Olela F, Schibler U. Restricted feeding uncouples circadian oscillators in peripheral tissues from the central pacemaker in the suprachiasmatic nucleus. *Genes Dev*. 2000;14(23):2950–2961.

30. Yamazaki S, Numano R, Abe M, Hida A, Takahashi R, Ueda M, Block GD, Sakaki Y, Menaker M, Tei H. Resetting central and peripheral circadian oscillators in transgenic rats. *Science*. 2000;288(5466):682–685.

31. Stokkan KA, Yamazaki S, Tei H, Sakaki Y, Menaker M. Entrainment of the circadian clock in the liver by feeding. *Science*. 2001;291(5503):490–493.

32. Tosini G, Pozdeyev N, Sakamoto K, Iuvone PM. The circadian clock system in the mammalian retina. *Bioessays*. 2008;30(7):624–633.

33. Balsalobre A, Brown SA, Marcacci L, Tronche F, Kellendonk C, Reichardt HM, Schütz G, Schibler U. Resetting of circadian time in peripheral tissues by glucocorticoid signaling. *Science*. 2000;289(5488):2344–2347.

34. Yagita K, Okamura H. Forskolin induces circadian gene expression of rPer1, rPer2 and dbp in mammalian rat-1 fibroblasts. *FEBS Lett.* 2000;465(1):79–82.

35. Yagita K, Tamanini F, van Der Horst GT, Okamura H. Molecular mechanisms of the biological clock in cultured fibroblasts. *Science.* 2001;292(5515):278–281.

36. Zylka MJ, Shearman LP, Weaver DR, Reppert SM. Three period homologs in mammals: differential light responses in the suprachiasmatic circadian clock and oscillating transcripts outside of brain. *Neuron.* 1998;20(6):1103–1110.

37. Reppert SM, Weaver DR. Coordination of circadian timing in mammals. *Nature.* 2002;418(6901):935–941.

38. Schibler U. Circadian rhythms. Liver regeneration clocks on. *Science.* 2003;302(5643):234–235.

39. Feillet CA, Albrecht U, Challet E. "Feeding time" for the brain: a matter of clocks. *J Physiol Paris.* 2006;100(5–6):252–260.

40. Challet E. Interactions between light, mealtime and calorie restriction to control daily timing in mammals. *J Comp Physiol B.* 2010;180(5):631–644.

41. Moore RY, Eichler VB. Loss of a circadian adrenal corticosterone rhythm following suprachiasmatic lesions in the rat. *Brain Res.* 1972;42(1):201–206.

42. Stephan FK, Zucker I. Rat drinking rhythms: central visual pathways and endocrine factors mediating responsiveness to environmental illumination. *Physiol Behav.* 1972;8(2):315–326.

43. Dark JG, Asdourian D. Entrainment of the rat's activity rhythm by cyclic light following lateral geniculate nucleus lesions. *Physiol Behav.* 1975;15(3):295–301.

44. Hendrickson A, Moe L, Noble B. Staining for autoradiography of the central nervous system. *Stain Technol.* 1972;47(6):283–290.

45. Moore RY, Lenn NJ. A retinohypothalamic projection in the rat. *J Comp Neurol.* 1972;146(1):1–14.

46. Dai J, Swaab DF, Van der Vliet J, Buijs RM. Postmortem tracing reveals the organization of hypothalamic projections of the suprachiasmatic nucleus in the human brain. *J Comp Neurol.* 1998;400(1):87–102.

47. Moore RY, Speh JC, Card JP. The retinohypothalamic tract originates from a distinct subset of retinal ganglion cells. *J Comp Neurol.* 1995;352(3):351–366.

48. Hannibal J. Neurotransmitters of the retino-hypothalamic tract. *Cell Tissue Res.* 2002;309(1):73–88.

49. Smale L, Blanchard J, Moore RY, Morin LP. Immunocytochemical characterization of the suprachiasmatic nucleus and the intergeniculate leaflet in the diurnal ground squirrel, *Spermophilus lateralis*. *Brain Res.* 1991;563(1–2):77–86.

50. Smale L, Boverhof J. The suprachiasmatic nucleus and intergeniculate leaflet of *Arvicanthis niloticus*, a diurnal murid rodent from East Africa. *J Comp Neurol.* 1999;403(2):190–208.

51. Major DE, Rodman HR, Libedinsky C, Karten HJ. Pattern of retinal projections in the California ground squirrel (*Spermophilus beecheyi*): anterograde tracing study using cholera toxin. *J Comp Neurol.* 2003;463(3):317–340.

52. Zhang DX, Rusak B. Photic sensitivity of geniculate neurons that project to the suprachiasmatic nuclei or the contralateral geniculate. *Brain Res.* 1989;504(1):161–164.

53. Harrington ME, Rusak B. Lesions of the thalamic intergeniculate leaflet alter hamster circadian rhythms. *J Biol Rhythms.* 1986;1(4):309–325.

54. Czeisler CA, Shanahan TL, Klerman EB, Martens H, Brotman DJ, Emens JS, Klein T, Rizzo 3rd JF. Suppression of melatonin secretion in some blind patients by exposure to bright light. *N Engl J Med.* 1995;332(1):6–11.

55. Lockley SW, Skene DJ, Arendt J, Tabandeh H, Bird AC, Defrance R. Relationship between melatonin rhythms and visual loss in the blind. *J Clin Endocrinol Metab.* 1997;82(11):3763–3770.

56. Campbell SS, Murphy PJ. Extraocular circadian phototransduction in humans. *Science.* 1998;279(5349):396–399.

57. Lockley SW, Skene DJ, Thapan K, English J, Ribeiro D, Haimov I, Hampton S, Middleton B, von Schantz M, Arendt J. Extraocular light exposure does not suppress plasma melatonin in humans. *J Clin Endocrinol Metab.* 1998;83(9):3369–3372.

58. Eastman CI, Martin SK, Hebert M. Failure of extraocular light to facilitate circadian rhythm reentrainment in humans. *Chronobiol Int.* 2000;17(6):807–826.

59. Wright Jr KP, Czeisler CA. Absence of circadian phase resetting in response to bright light behind the knees. *Science.* 2002;297(5581):571.

60. Makous W, Scotopic vision. Challupa L, Werner JS, eds. *The Visual Neurosciences*, vol. 1. Cambridge, MA: MIT Press; 2004:838–850.

61. Calkins D, Linking retinal circuits to color opponency. Chalupa LM, Werner JS, eds. *The Visual Neurosciences*, vol. 2. Cambridge, MA: MIT Press; 2003.

62. Merbs SL, Nathans J. Absorption spectra of human cone pigments. *Nature.* 1992;356(6368):433–435.

63. Foster RG, Provencio I, Hudson D, Fiske S, De Grip W, Menaker M. Circadian photoreception in the retinally degenerate mouse (rd/rd). *J Comp Physiol A*. 1991;169(1):39–50.

64. Freedman MS, Lucas RJ, Soni B, von Schantz M, Muñoz M, David-Gray Z, Foster R. Regulation of mammalian circadian behavior by non-rod, non-cone, ocular photoreceptors. *Science*. 1999;284(5413):502–504.

65. Keeler CE, Sutcliffe E, Chaffee EL. Normal and "rodless" retinae of the house mouse with respect to the electromotive force generated through stimulation by light. *Proc Natl Acad Sci USA*. 1928;14(6):477–484.

66. Provencio I, Jiang G, De Grip WJ, Hayes WP, Rollag MD, Melanopsin:. an opsin in melanophores, brain, and eye. *Proc Natl Acad Sci USA*. 1998;95(1):340–345.

67. Provencio I, Rodriguez IR, Jiang G, Hayes WP, Moreira EF, Rollag MD. A novel human opsin in the inner retina. *J Neurosci*. 2000;20(2):600–605.

68. Hattar S, Liao HW, Takao M, Berson DM, Yau KW. Melanopsin-containing retinal ganglion cells: architecture, projections, and intrinsic photosensitivity. *Science*. 2002;295(5557):1065–1070.

69. Dacey DM, Liao H-W, Peterson BB, Robinson FR, Smith VC, Pokorny J, Yau K-W, Gamlin PD. Melanopsin-expressing ganglion cells in primate retina signal colour and irradiance and project to the LGN. *Nature*. 2005;433(7027):749–754.

70. Berson DM, Dunn FA, Takao M. Phototransduction by retinal ganglion cells that set the circadian clock. *Science*. 2002;295(5557):1070–1073.

71. Qiu X, Kumbalasiri T, Carlson SM, Wong KY, Krishna V, Provencio I, Berson DM. Induction of photosensitivity by heterologous expression of melanopsin. *Nature*. 2005;433(7027):745–749.

72. Bailes HJ, Lucas RJ. Human melanopsin forms a pigment maximally sensitive to blue light ($\lambda_{max} \approx 479$ nm) supporting activation of G(q/11) and G(i/o) signalling cascades. *Proc Biol Sci*. 2013;280(1759):20122987.

73. Panda S, Sato TK, Castrucci AM, Rollag MD, DeGrip WJ, Hogenesch JB, Provencio I, Kay SA. Melanopsin (Opn4) requirement for normal light-induced circadian phase shifting. *Science*. 2002;298(5601):2213–2216.

74. Lucas RJ, Hattar S, Takao M, Berson DM, Foster RG, Yau K-W. Diminished pupillary light reflex at high irradiances in melanopsin-knockout mice. *Science*. 2003;299(5604):245–247.

75. Drouyer E, Rieux C, Hut RA, Cooper HM. Responses of suprachiasmatic nucleus neurons to light and dark adaptation: relative contributions of melanopsin and rod-cone inputs. *J Neurosci*. 2007;27(36):9623–9631.

76. Güler AD, Altimus CM, Ecker JL, Hattar S. Multiple photoreceptors contribute to nonimage-forming visual functions predominantly through melanopsin-containing retinal ganglion cells. *Cold Spring Harb Symp Quant Biol*. 2007;72:509–515.

77. Altimus CM, Güler AD, Villa KL, McNeill DS, Legates TA, Hattar S. Rods-cones and melanopsin detect light and dark to modulate sleep independent of image formation. *Proc Natl Acad Sci USA*. 2008;105(50):19998–20003.

78. Altimus CM, Güler AD, Alam NM, Arman AC, Prusky GT, Sampath AP, Hattar S. Rod photoreceptors drive circadian photoentrainment across a wide range of light intensities. *Nat Neurosci*. 2010;13(9):1107–1112.

79. Gooley JJ, Rajaratnam SMW, Brainard GC, Kronauer RE, Czeisler CA, Lockley SW. Spectral responses of the human circadian system depend on the irradiance and duration of exposure to light. *Sci Transl Med*. 2010;2(31):31ra33.

80. Gooley JJ, Lu J, Chou TC, Scammell TE, Saper CB. Melanopsin in cells of origin of the retinohypothalamic tract. *Nat Neurosci*. 2001;4(12):1165.

81. Hattar S, Kumar M, Park A, Tong P, Tung J, Yau K-W, Berson DM. Central projections of melanopsin-expressing retinal ganglion cells in the mouse. *J Comp Neurol*. 2006;497(3):326–349.

82. Hankins MW, Peirson SN, Foster RG. Melanopsin: an exciting photopigment. *Trends Neurosci*. 2008;31(1):27–36.

83. Johnson J, Wu V, Donovan M, Majumdar S, Rentería RC, Porco T, Van Gelder RN, Copenhagen DR. Melanopsin-dependent light avoidance in neonatal mice. *Proc Natl Acad Sci USA*. 2010;107(40):17374–17378.

84. Noseda R, Kainz V, Jakubowski M, Gooley JJ, Saper CB, Digre K, Burstein R. A neural mechanism for exacerbation of headache by light. *Nat Neurosci*. 2010;13(2):239–245.

85. Kawasaki A, Kardon RH. Intrinsically photosensitive retinal ganglion cells. *J Neuro-Ophthalmol*. 2007;27(3):195–204.

86. Rao S, Chun C, Fan J, Kofron JM, Yang MB, Hegde RS, Ferrara N, Copenhagen DR, Lang RA. A direct and melanopsin-dependent fetal light response regulates mouse eye development. *Nature*. 2013;494(7436):243–246.

87. Li S, Yang C, Zhang L, Gao X, Wang X, Liu W, Wang Y, Jiang S, Wong YH, Zhang Y, Liu K. Promoting axon regeneration in the adult CNS by modulation of the melanopsin/GPCR signaling. *Proc Natl Acad Sci USA*. 2016;113(7):1937–1942.

88. Brown TM, Gias C, Hatori M, Keding SR, Semo M 'ayan, Coffey PJ, Gigg J, Piggins HD, Panda S, Lucas RJ. Melanopsin contributions to irradiance coding in the thalamo-cortical visual system. *PLoS Biol*. 2010;8(12):e1000558.

89. Brown TM, Wynne J, Piggins HD, Lucas RJ. Multiple hypothalamic cell populations encoding distinct visual information. *J Physiol*. 2011;589(Pt 5):1173–1194.

90. Ecker JL, Dumitrescu ON, Wong KY, Alam NM, Chen S-K, LeGates T, Renna JM, Prusky GT, Berson DM, Hattar S. Melanopsin-expressing retinal ganglion-cell photoreceptors: cellular diversity and role in pattern vision. *Neuron*. 2010;67(1):49–60.

91. Vandewalle G, Collignon O, Hull JT, Daneault V, Albouy G, Lepore F, Phillips C, Doyon J, Czeisler CA, Dumont M, Lockley SW, Carrier J. Blue light stimulates cognitive brain activity in visually blind individuals. *J Cogn Neurosci*. 2013;25(12):2072–2085.

92. Zaidi FH, Hull JT, Peirson SN, Wulff K, Aeschbach D, Gooley JJ, Brainard GC, Gregory-Evans K, Rizzo JF, Czeisler CA, Foster RG, Moseley MJ, Lockley SW. Short-wavelength light sensitivity of circadian, pupillary, and visual awareness in humans lacking an outer retina. *Curr Biol*. 2007;17(24):2122–2128.

93. Schmidt TM, Chen S-K, Hattar S. Intrinsically photosensitive retinal ganglion cells: many subtypes, diverse functions. *Trends Neurosci*. 2011;34(11):572–580.

94. Sand A, Schmidt TM, Kofuji P. Diverse types of ganglion cell photoreceptors in the mammalian retina. *Prog Retin Eye Res*. 2012;31(4):287–302.

95. Reifler AN, Chervenak AP, Dolikian ME, Benenati BA, Meyers BS, Demertzis ZD, Lynch AM, Li BY, Wachter RD, Abufarha FS, Dulka EA, Pack W, Zhao X, Wong KY. The rat retina has five types of ganglion-cell photoreceptors. *Exp Eye Res*. 2015;130:17–28.

96. Schmidt TM, Kofuji P. Differential cone pathway influence on intrinsically photosensitive retinal ganglion cell subtypes. *J Neurosci*. 2010;30(48):16262–16271.

97. Schmidt TM, Kofuji P. Functional and morphological differences among intrinsically photosensitive retinal ganglion cells. *J Neurosci*. 2009;29(2):476–482.

98. Hu C, Hill DD, Wong KY. Intrinsic physiological properties of the five types of mouse ganglion-cell photoreceptors. *J Neurophysiol*. 2013;109(7):1876–1889.

99. Zhao X, Stafford BK, Godin AL, King WM, Wong KY. Photoresponse diversity among the five types of intrinsically photosensitive retinal ganglion cells. *J Physiol*. 2014;592(7):1619–1636.

100. Brown TM, Tsujimura S-I, Allen AE, Wynne J, Bedford R, Vickery G, Vugler A, Lucas RJ. Melanopsin-based brightness discrimination in mice and humans. *Curr Biol*. 2012;22(12):1134–1141.

101. Chen S-K, Badea TC, Hattar S. Photoentrainment and pupillary light reflex are mediated by distinct populations of ipRGCs. *Nature*. 2011;476(7358):92–95.

102. Maren S, Fanselow MS. The amygdala and fear conditioning: has the nut been cracked?. *Neuron*. 1996;16(2):237–240.

103. Schmidt TM, Do MTH, Dacey D, Lucas R, Hattar S, Matynia A. Melanopsin-positive intrinsically photosensitive retinal ganglion cells: from form to function. *J Neurosci*. 2011;31(45):16094–16101.

104. Jagannath A, Hughes S, Abdelgany A, Pothecary CA, Di Pretoro S, Pires SS, Vachtsevanos A, Pilorz V, Brown LA, Hossbach M, MacLaren RE, Halford S, Gatti S, Hankins MW, Wood MJA, Foster RG, Peirson SN. Isoforms of melanopsin mediate different behavioral responses to light. *Curr Biol*. 2015;25(18):2430–2434.

105. Hattar S, Lucas RJ, Mrosovsky N, Thompson S, Douglas RH, Hankins MW, Lem J, Biel M, Hofmann F, Foster RG, Yau K-W. Melanopsin and rod-cone photoreceptive systems account for all major accessory visual functions in mice. *Nature*. 2003;424(6944):76–81.

106. Panda S, Provencio I, Tu DC, Pires SS, Rollag MD, Castrucci AM, Pletcher MT, Sato TK, Wiltshire T, Andahazy M, Kay SA, Van Gelder RN, Hogenesch JB. Melanopsin is required for non-image-forming photic responses in blind mice. *Science*. 2003;301(5632):525–527.

107. Dkhissi-Benyahya O, Gronfier C, De Vanssay W, Flamant F, Cooper HM. Modeling the role of mid-wavelength cones in circadian responses to light. *Neuron*. 2007;53(5):677–687.

108. Dollet A, Albrecht U, Cooper HM, Dkhissi-Benyahya O. Cones are required for normal temporal responses to light of phase shifts and clock gene expression. *Chronobiol Int*. 2010;27(4):768–781.

109. Lall GS, Revell VL, Momiji H, Al Enezi J, Altimus CM, Güler AD, Aguilar C, Cameron MA, Allender S, Hankins MW, Lucas RJ. Distinct contributions of rod, cone, and melanopsin photoreceptors to encoding irradiance. *Neuron*. 2010;66(3):417–428.

110. Revell VL, Skene DJ. Light-induced melatonin suppression in humans with polychromatic and monochromatic light. *Chronobiol Int*. 2007;24(6):1125–1137.

111. McDougal DH, Gamlin PD. The influence of intrinsically-photosensitive retinal ganglion cells on the spectral sensitivity and response dynamics of the human pupillary light reflex. *Vision Res*. 2010;50(1):72–87.

112. Gooley JJ, Mien IH, Hilaire MAS, Yeo S-C, Chua EC-P, Reen Evan, Hanley CJ, Hull JT, Czeisler CA, Lockley SW. Melanopsin and rod–cone photoreceptors play different roles in mediating pupillary light responses during exposure to continuous light in humans. *J Neurosci*. 2012;32(41):14242–14253.

113. Güler AD, Ecker JL, Lall GS, Haq S, Altimus CM, Liao H-W, Barnard AR, Cahill H, Badea TC, Zhao H, Hankins MW, Berson DM, Lucas RJ, Yau K-W, Hattar S. Melanopsin cells are the principal conduits for rod-cone input to non-image-forming vision. *Nature*. 2008;453(7191):102–105.

114. van Diepen HC, Ramkisoensing A, Peirson SN, Foster RG, Meijer JH. Irradiance encoding in the suprachiasmatic nuclei by rod and cone photoreceptors. *FASEB J*. 2013;27(10):4204–4212.

115a. Revell VL, Arendt J, Fogg LF, Skene DJ. Alerting effects of light are sensitive to very short wavelengths. *Neurosci Lett*. 2006;399(1–2):96–100.

115b. Luo AH, Aston-Jones G. Circuit projection from suprachiasmatic nucleus to ventral tegmental area: a novel circadian output pathway. *Eur J Neurosci*. 2009;29:748–760.

116. Buijs RM, Wortel J, Van Heerikhuize JJ, Feenstra MG, Ter Horst GJ, Romijn HJ, Kalsbeek A. Anatomical and functional demonstration of a multisynaptic suprachiasmatic nucleus adrenal (cortex) pathway. *Eur J Neurosci*. 1999;11(5):1535–1544.

117. Teclemariam-Mesbah R, Ter Horst GJ, Postema F, Wortel J, Buijs RM. Anatomical demonstration of the suprachiasmatic nucleus-pineal pathway. *J Comp Neurol*. 1999;406(2):171–182.

118. Kalsbeek A, Fliers E, Franke AN, Wortel J, Buijs RM. Functional connections between the suprachiasmatic nucleus and the thyroid gland as revealed by lesioning and viral tracing techniques in the rat. *Endocrinology*. 2000;141(10):3832–3841.

119. Bartness TJ, Song CK, Demas GE. SCN efferents to peripheral tissues: implications for biological rhythms. *J Biol Rhythms*. 2001;16(3):196–204.

120. Lu J, Zhang YH, Chou TC, Gaus SE, Elmquist JK, Shiromani P, Saper CB. Contrasting effects of ibotenate lesions of the paraventricular nucleus and subparaventricular zone on sleep-wake cycle and temperature regulation. *J Neurosci*. 2001;21(13):4864–4874.

121. Chou TC, Scammell TE, Gooley JJ, Gaus SE, Saper CB, Lu J. Critical role of dorsomedial hypothalamic nucleus in a wide range of behavioral circadian rhythms. *J Neurosci*. 2003;23(33):10691–10702.

122. Silver R, LeSauter J, Tresco PA, Lehman MN. A diffusible coupling signal from the transplanted suprachiasmatic nucleus controlling circadian locomotor rhythms. *Nature*. 1996;382(6594):810–813.

123. Kräuchi K, Wirz-Justice A. Circadian rhythm of heat production, heart rate, and skin and core temperature under unmasking conditions in men. *Am J Physiol*. 1994;267(3 Pt 2):R819–R829.

124. Scheer FA, Ter Horst GJ, van Der Vliet J, Buijs RM. Physiological and anatomic evidence for regulation of the heart by suprachiasmatic nucleus in rats. *Am J Physiol*. 2001;280(3):H1391–H1399.

125. Scheer FA, Kalsbeek A, Buijs RM. Cardiovascular control by the suprachiasmatic nucleus: neural and neuroendocrine mechanisms in human and rat. *Biol Chem*. 2003;384(5):697–709.

126. Shea SA, Hilton MF, Hu K, Scheer FAJL. Existence of an endogenous circadian blood pressure rhythm in humans that peaks in the evening. *Circ Res*. 2011;108(8):980–984.

127. Scheer FA, van Doornen LJ, Buijs RM. Light and diurnal cycle affect human heart rate: possible role for the circadian pacemaker. *J Biol Rhythms*. 1999;14(3):202–212.

128. Scheer FA, Van Doornen LJP, Buijs RM. Light and diurnal cycle affect autonomic cardiac balance in human; possible role for the biological clock. *Auton Neurosci*. 2004;110(1):44–48.

129. Cajochen C, Münch M, Kobialka S, Kräuchi K, Steiner R, Oelhafen P, Orgül S, Wirz-Justice A. High sensitivity of human melatonin, alertness, thermoregulation, and heart rate to short wavelength light. *J Clin Endocrinol Metab*. 2005;90(3):1311–1316.

130. Mutoh T, Shibata S, Korf H-W, Okamura H. Melatonin modulates the light-induced sympathoexcitation and vagal suppression with participation of the suprachiasmatic nucleus in mice. *J Physiol*. 2003;547(Pt 1):317–332.

131. Aschoff J. Circadian control of body temperature. *J Therm Biol*. 1983;8(1–2):143–147.

132. Czeisler CA, Allan JS, Strogatz SH, Ronda JM, Sánchez R, Ríos CD, Freitag WO, Richardson GS, Kronauer RE. Bright light resets the human circadian pacemaker independent of the timing of the sleep-wake cycle. *Science*. 1986;233(4764):667–671.

133. Dawson D, Lushington K, Lack L, Campbell S, Matthews C. The variability in circadian phase and amplitude estimates derived from sequential constant routines. *Chronobiol Int*. 1992;9(5):362–370.

134. Badia P, Myers B, Boecker M, Culpepper J, Harsh JR. Bright light effects on body temperature, alertness, EEG and behavior. *Physiol Behav.* 1991;50(3):583–588.

135. Cajochen C, Dijk DJ, Borbély AA. Dynamics of EEG slow-wave activity and core body temperature in human sleep after exposure to bright light. *Sleep.* 1992;15(4):337–343.

136. Scheer FAJL, Pirovano C, Van Someren EJW, Buijs RM. Environmental light and suprachiasmatic nucleus interact in the regulation of body temperature. *Neuroscience.* 2005;132(2):465–477.

137. Hannibal J, Fahrenkrug J. Target areas innervated by PACAP-immunoreactive retinal ganglion cells. *Cell Tissue Res.* 2004;316(1):99–113.

138. Benington JH, Heller HC. Restoration of brain energy metabolism as the function of sleep. *Prog Neurobiol.* 1995;45(4):347–360.

139. Porkka-Heiskanen T, Strecker RE, Thakkar M, Bjorkum AA, Greene RW, McCarley RW. Adenosine: a mediator of the sleep-inducing effects of prolonged wakefulness. *Science.* 1997;276(5316):1265–1268.

140. Strecker RE, Morairty S, Thakkar MM, Porkka-Heiskanen T, Basheer R, Dauphin LJ, Rainnie DG, Portas CM, Greene RW, McCarley RW. Adenosinergic modulation of basal forebrain and preoptic/anterior hypothalamic neuronal activity in the control of behavioral state. *Behav Brain Res.* 2000;115(2):183–204.

141. Dijk DJ, Czeisler CA. Contribution of the circadian pacemaker and the sleep homeostat to sleep propensity, sleep structure, electroencephalographic slow waves, and sleep spindle activity in humans. *J Neurosci.* 1995;15(5 Pt 1):3526–3538.

142. Yoshida K, McCormack S, España RA, Crocker A, Scammell TE. Afferents to the orexin neurons of the rat brain. *J Comp Neurol.* 2006;494(5):845–861.

143. Chou TC, Bjorkum AA, Gaus SE, Lu J, Scammell TE, Saper CB. Afferents to the ventrolateral preoptic nucleus. *J Neurosci.* 2002;22(3):977–990.

144. Deurveilher S, Semba K. Indirect projections from the suprachiasmatic nucleus to major arousal-promoting cell groups in rat: implications for the circadian control of behavioral state. *Neuroscience.* 2005;130(1):165–183.

145. Kalsbeek A, Buijs RM. Output pathways of the mammalian suprachiasmatic nucleus: coding circadian time by transmitter selection and specific targeting. *Cell Tissue Res.* 2002;309(1):109–118.

146. Vrang N, Larsen PJ, Mikkelsen JD. Direct projection from the suprachiasmatic nucleus to hypophysiotropic corticotropin-releasing factor immunoreactive cells in the paraventricular nucleus of the hypothalamus demonstrated by means of Phaseolus vulgaris-leucoagglutinin tract tracing. *Brain Res.* 1995;684(1):61–69.

147. Kalsbeek A, Drijfhout WJ, Westerink BH, van Heerikhuize JJ, van der Woude TP, van der Vliet J, Buijs RM. GABA receptors in the region of the dorsomedial hypothalamus of rats are implicated in the control of melatonin and corticosterone release. *Neuroendocrinol.* 1996;63(1):69–78.

148. Scheer FA, Buijs RM. Light affects morning salivary cortisol in humans. *J Clin Endocrinol Metab.* 1999;84(9):3395–3398.

149. Gooley JJ, Lu J, Fischer D, Saper CB. A broad role for melanopsin in nonvisual photoreception. *J Neurosci.* 2003;23(18):7093–7106.

150. Dijk DJ, Duffy JF, Czeisler CA. Circadian and sleep/wake dependent aspects of subjective alertness and cognitive performance. *J Sleep Res.* 1992;1(2):112–117.

151. Wyatt JK, Cecco AR-D, Czeisler CA, Dijk D-J. Circadian temperature and melatonin rhythms, sleep, and neurobehavioral function in humans living on a 20-h day. *Am J Physiol.* 1999;277(4):R1152–R1163.

152. Carrier J, Monk TH. Circadian rhythms of performance: new trends. *Chronobiol Int.* 2000;17(6):719–732.

153. Blatter K, Cajochen C. Circadian rhythms in cognitive performance: methodological constraints, protocols, theoretical underpinnings. *Physiol Behav.* 2007;90(2–3):196–208.

154. Wright Jr KP, Hull JT, Czeisler CA. Relationship between alertness, performance, and body temperature in humans. *Am J Physiol.* 2002;283(6):R1370–R1377.

155. Campbell SS, Dawson D. Enhancement of nighttime alertness and performance with bright ambient light. *Physiol Behav.* 1990;48(2):317–320.

156. Daurat A, Aguirre A, Foret J, Gonnet P, Keromes A, Benoit O. Bright light affects alertness and performance rhythms during a 24-h constant routine. *Physiol Behav.* 1993;53(5):929–936.

157. Cajochen C, Zeitzer JM, Czeisler CA, Dijk DJ. Dose-response relationship for light intensity and ocular and electroencephalographic correlates of human alertness. *Behav Brain Res.* 2000;115(1):75–83.

158. Lockley SW, Evans EE, Scheer FAJL, Brainard GC, Czeisler CA, Aeschbach D. Short-wavelength sensitivity for the direct effects of light on alertness, vigilance, and the waking electroencephalogram in humans. *Sleep.* 2006;29(2):161–168.

159. Chellappa SL, Steiner R, Blattner P, Oelhafen P, Götz T, Cajochen C. Non-visual effects of light on melatonin, alertness and cognitive performance: can blue-enriched light keep us alert?. *PLoS One.* 2011;6(1):e16429.

160. Phipps-Nelson J, Redman JR, Dijk D-J, Rajaratnam SMW. Daytime exposure to bright light, as compared to dim light, decreases sleepiness and improves psychomotor vigilance performance. *Sleep.* 2003;26(6):695–700.

161. Rüger M, Gordijn MCM, Beersma DGM, de Vries B, Daan S. Time-of-day-dependent effects of bright light exposure on human psychophysiology: comparison of daytime and nighttime exposure. *Am J Physiol.* 2006;290(5):R1413–R1420.

162. Vandewalle G, Maquet P, Dijk D-J. Light as a modulator of cognitive brain function. *Trends Cogn Sci.* 2009;13(10):429–438.

163. Vandewalle G, Schmidt C, Albouy G, Sterpenich V, Darsaud A, Rauchs G, Berken P-Y, Balteau E, Degueldre C, Luxen A, Maquet P, Dijk D-J. Brain responses to violet, blue, and green monochromatic light exposures in humans: prominent role of blue light and the brainstem. *PLoS One.* 2007;2(11):e1247.

164. Perrin F, Peigneux P, Fuchs S, Verhaeghe S, Laureys S, Middleton B, Degueldre C, Del Fiore G, Vandewalle G, Balteau E, Poirrier R, Moreau V, Luxen A, Maquet P, Dijk D-J. Nonvisual responses to light exposure in the human brain during the circadian night. *Curr Biol.* 2004;14(20):1842–1846.

165. Vandewalle G, Balteau E, Phillips C, Degueldre C, Moreau V, Sterpenich V, Albouy G, Darsaud A, Desseilles M, Dang-Vu TT, Peigneux P, Luxen A, Dijk D-J, Maquet P. Daytime light exposure dynamically enhances brain responses. *Curr Biol.* 2006;16(16):1616–1621.

166. Riemersma-van der Lek RF, Swaab DF, Twisk J, Hol EM, Hoogendijk WJG, Van Someren EJW. Effect of bright light and melatonin on cognitive and noncognitive function in elderly residents of group care facilities: a randomized controlled trial. *JAMA.* 2008;299(22):2642–2655.

167. Lerner AB, Case JD, Takahashi Y, Lee TH, Mori W. Isolation of melatonin, the pineal gland factor that lightens melanocytes. *J Am Chem Soc.* 1958;80(10):2587.

168. Moore RY, Klein DC. Visual pathways and the central neural control of a circadian rhythm in pineal serotonin N-acetyltransferase activity. *Brain Res.* 1974;71(1):17–33.

169. Zeitzer JM, Ayas NT, Shea SA, Brown R, Czeisler CA. Absence of detectable melatonin and preservation of cortisol and thyrotropin rhythms in tetraplegia. *J Clin Endocrinol Metab.* 2000;85(6):2189–2196.

170. Liu C, Weaver DR, Jin X, Shearman LP, Pieschl RL, Gribkoff VK, Reppert SM. Molecular dissection of two distinct actions of melatonin on the suprachiasmatic circadian clock. *Neuron.* 1997;19(1):91–102.

171. Reppert SM, Weaver DR, Godson C. Melatonin receptors step into the light: cloning and classification of subtypes. *Trends Pharmacol Sci.* 1996;17(3):100–102.

172. Pévet P. Melatonin in animal models. *Dialogues Clin Neurosci.* 2003;5(4):343–352.

173. Klerman EB, Gershengorn HB, Duffy JF, Kronauer RE. Comparisons of the variability of three markers of the human circadian pacemaker. *J Biol Rhythms.* 2002;17(2):181–193.

174. Lewy AJ, Wehr TA, Goodwin FK, Newsome DA, Markey SP. Light suppresses melatonin secretion in humans. *Science.* 1980;210(4475):1267–1269.

175. Zeitzer JM, Dijk DJ, Kronauer R, Brown E, Czeisler CA. Sensitivity of the human circadian pacemaker to nocturnal light: melatonin phase resetting and suppression. *J Physiol.* 2000;526(Pt 3):695–702.

176. Najjar RP, Zeitzer JM. Temporal integration of light flashes by the human circadian system. *J Clin Invest.* 2016;126(3):938–947.

177. Brainard GC, Hanifin JP, Greeson JM, Byrne B, Glickman G, Gerner E, Rollag MD. Action spectrum for melatonin regulation in humans: evidence for a novel circadian photoreceptor. *J Neurosci.* 2001;21(16):6405–6412.

178. Foster RG, Hankins MW. Non-rod, non-cone photoreception in the vertebrates. *Prog Retin Eye Res.* 2002;21(6):507–527.

179. Thapan K, Arendt J, Skene DJ. An action spectrum for melatonin suppression: evidence for a novel non-rod, non-cone photoreceptor system in humans. *J Physiol.* 2001;535(Pt 1):261–267.

180. Najjar RP, Chiquet C, Teikari P, Cornut P-L, Claustrat B, Denis P, Cooper HM, Gronfier C. Aging of non-visual spectral sensitivity to light in humans: compensatory mechanisms?. *PLoS One.* 2014;9(1):e85837.

181. McIntyre IM, Norman TR, Burrows GD, Armstrong SM. Human melatonin response to light at different times of the night. *Psychoneuroendocrinology.* 1989;14(3):187–193.

182. Hébert M, Martin SK, Lee C, Eastman CI. The effects of prior light history on the suppression of melatonin by light in humans. *J Pineal Res.* 2002;33(4):198–203.

183. Smith KA, Schoen MW, Czeisler CA. Adaptation of human pineal melatonin suppression by recent photic history. *J Clin Endocrinol Metab.* 2004;89(7):3610–3614.

184. Duffy JF, Cain SW, Chang A-M, Phillips AJK, Münch MY, Gronfier C, Wyatt JK, Dijk DJ, Wright KP, Czeisler CA. Sex difference in the near-24-hour intrinsic period of the human circadian timing system. *Proc Natl Acad Sci USA*. 2011;108(suppl 3):15602–15608.

185. Plautz JD, Kaneko M, Hall JC, Kay SA. Independent photoreceptive circadian clocks throughout Drosophila. *Science*. 1997;278(5343):1632–1635.

186. Whitmore D, Foulkes NS, Sassone-Corsi P. Light acts directly on organs and cells in culture to set the vertebrate circadian clock. *Nature*. 2000;404(6773):87–91.

187. Balsalobre A, Damiola F, Schibler U. A serum shock induces circadian gene expression in mammalian tissue culture cells. *Cell*. 1998;93(6):929–937.

188. McNamara P, Seo SB, Rudic RD, Sehgal A, Chakravarti D, FitzGerald GA. Regulation of CLOCK and MOP4 by nuclear hormone receptors in the vasculature: a humoral mechanism to reset a peripheral clock. *Cell*. 2001;105(7):877–889.

189. Pando MP, Morse D, Cermakian N, Sassone-Corsi P. Phenotypic rescue of a peripheral clock genetic defect via SCN hierarchical dominance. *Cell*. 2002;110(1):107–117.

190. Wever RA. *The Circadian System of Man: Results of Experiments Under Temporal Isolation*. 1st ed. New York, NY: Springer-Verlag; 1979.

191. Rohling JHT, vanderLeest HT, Michel S, Vansteensel MJ, Meijer JH. Phase resetting of the mammalian circadian clock relies on a rapid shift of a small population of pacemaker neurons. *PLoS One*. 2011;6(9):e25437.

192. Khalsa SBS, Jewett ME, Cajochen C, Czeisler CA. A phase response curve to single bright light pulses in human subjects. *J Physiol*. 2003;549(Pt 3):945–952.

193. St Hilaire MA, Gooley JJ, Khalsa SBS, Kronauer RE, Czeisler CA, Lockley SW. Human phase response curve (PRC) to a 1-hour pulse of bright white light. *J Physiol*. 2012;590(Pt 13):3035–3045.

194. Zeitzer JM, Khalsa SBS, Boivin DB, Duffy JF, Shanahan TL, Kronauer RE, Czeisler CA. Temporal dynamics of late-night photic stimulation of the human circadian timing system. *Am J Physiol*. 2005;289(3):R839–R844.

195. Czeisler CA, Kronauer RE, Allan JS, Duffy JF, Jewett ME, Brown EN, Ronda JM. Bright light induction of strong (type 0) resetting of the human circadian pacemaker. *Science*. 1989;244(4910):1328–1333.

196. Chang A-M, Santhi N, St Hilaire MA, Gronfier C, Bradstreet DS, Duffy JF, Lockley SW, Kronauer RE, Czeisler CA. Human responses to bright light of different durations. *J Physiol*. 2012;590(13):3103–3112.

197. Lockley SW, Brainard GC, Czeisler CA. High sensitivity of the human circadian melatonin rhythm to resetting by short wavelength light. *J Clin Endocrinol Metab*. 2003;88(9):4502–4505.

198. Rimmer DW, Boivin DB, Shanahan TL, Kronauer RE, Duffy JF, Czeisler CA. Dynamic resetting of the human circadian pacemaker by intermittent bright light. *Am J Physiol*. 2000;279(5):R1574–R1579.

199. Gronfier C, Wright Jr KP, Kronauer RE, Jewett ME, Czeisler CA. Efficacy of a single sequence of intermittent bright light pulses for delaying circadian phase in humans. *Am J Physiol*. 2004;287(1):E174–E181.

200. Gronfier C, Wright KP, Kronauer RE, Czeisler CA. Entrainment of the human circadian pacemaker to longer-than-24-h days. *Proc Natl Acad Sci USA*. 2007;104(21):9081–9086.

201. Zeitzer JM, Ruby NF, Fisicaro RA, Heller HC. Response of the human circadian system to millisecond flashes of light. *PLoS One*. 2011;6(7):e22078.

202. Zeitzer JM, Fisicaro RA, Ruby NF, Heller HC. Millisecond flashes of light phase delay the human circadian clock during sleep. *J Biol Rhythms*. 2014;29(5):370–376.

203. Weibel L, Brandenberger G. Disturbances in hormonal profiles of night workers during their usual sleep and work times. *J Biol Rhythms*. 1998;13(3):202–208.

204. Van Cauter E, Leproult R, Plat L. Age-related changes in slow wave sleep and REM sleep and relationship with growth hormone and cortisol levels in healthy men. *JAMA*. 2000;284(7):861–868.

205. Spiegel K, Sheridan JF, Van Cauter E. Effect of sleep deprivation on response to immunization. *JAMA*. 2002;288(12):1471–1472.

206. Wright Jr KP, Hull JT, Hughes RJ, Ronda JM, Czeisler CA. Sleep and wakefulness out of phase with internal biological time impairs learning in humans. *J Cogn Neurosci*. 2006;18(4):508–521.

207. Young ME, Bray MS. Potential role for peripheral circadian clock dyssynchrony in the pathogenesis of cardio-vascular dysfunction. *Sleep Med*. 2007;8(6):656–667.

208. Davidson AJ, Sellix MT, Daniel J, Yamazaki S, Menaker M, Block GD. Chronic jet-lag increases mortality in aged mice. *Curr Biol*. 2006;16(21):R914–R916.

209. Cho K. Chronic "jet lag" produces temporal lobe atrophy and spatial cognitive deficits. *Nat Neurosci*. 2001;4(6):567–568.

210. Conlon M, Lightfoot N, Kreiger N. Rotating shift work and risk of prostate cancer. *Epidemiol Camb Mass.* 2007;18(1):182–183.

211. Morikawa Y, Nakagawa H, Miura K, Soyama Y, Ishizaki M, Kido T, Naruse Y, Suwazono Y, Nogawa K. Shift work and the risk of diabetes mellitus among Japanese male factory workers. *Scand J Work Environ Health.* 2005;31(3):179–183.

212. Koda S, Yasuda N, Sugihara Y, Ohara H, Udo H, Otani T, Hisashige A, Ogawa T, Aoyama H. Analyses of work-relatedness of health problems among truck drivers by questionnaire survey. *J Occup Health (Sangyō Eiseigaku Zasshi).* 2000;42(1):6–16.

213. Kivimäki M, Virtanen M, Elovainio M, Väänänen A, Keltikangas-Järvinen L, Vahtera J. Prevalent cardiovascular disease, risk factors and selection out of shift work. *Scand J Work Environ Health.* 2006;32(3):204–208.

214. Bildt C, Michélsen H. Gender differences in the effects from working conditions on mental health: a 4-year follow-up. *Int Arch Occup Environ Health.* 2002;75(4):252–258.

215. Hastings JW, Sweeney BM. A persistent diurnal rhythm of luminescence in *Gonyaulax polyedra. Biol Bull.* 1958;115(3):440–458.

216. Bruce VG, Pittendrigh CS. Resetting the euglena clock with a single light stimulus. *Am Nat.* 1958;92(866): 295–306.

217. Spoelstra K, Albrecht U, van der Horst GTJ, Brauer V, Daan S. Phase responses to light pulses in mice lacking functional per or cry genes. *J Biol Rhythms.* 2004;19(6):518–529.

218. Van Cauter E, Sturis J, Byrne MM, Blackman JD, Leproult R, Ofek G, L'Hermite-Balériaux M, Refetoff S, Turek FW, Van Reeth O. Demonstration of rapid light-induced advances and delays of the human circadian clock using hormonal phase markers. *Am J Physiol.* 1994;266(6 Pt 1):E953–E963.

219. Jewett ME, Rimmer DW, Duffy JF, Klerman EB, Kronauer RE, Czeisler CA. Human circadian pacemaker is sensitive to light throughout subjective day without evidence of transients. *Am J Physiol.* 1997;273(5 Pt 2):R1800–R1809.

220. Cajochen C, Frey S, Anders D, Späti J, Bues M, Pross A, Mager R, Wirz-Justice A, Stefani O. Evening exposure to a light-emitting diodes (LED)-backlit computer screen affects circadian physiology and cognitive performance. *J Appl Physiol.* 2011;110(5):1432–1438.

221. Chang A-M, Aeschbach D, Duffy JF, Czeisler CA. Evening use of light-emitting eReaders negatively affects sleep, circadian timing, and next-morning alertness. *Proc Natl Acad Sci.* 2015;112(4):1232–1237.

222. Chellappa SL, Viola AU, Schmidt C, Bachmann V, Gabel V, Maire M, Reichert CF, Valomon A, Götz T, Landolt H-P, Cajochen C. Human melatonin and alerting response to blue-enriched light depend on a polymorphism in the clock gene PER3. *J Clin Endocrinol Metab.* 2012;97(3):E433–E437.

223. Zeitzer JM. Real life trumps laboratory in matters of public health. *Proc Natl Acad Sci USA.* 2015;112(13): E1513.

224. Walch OJ, Zhang LS, Reifler AN, Dolikian ME, Forger DB, Wong KY. Characterizing and modeling the intrinsic light response of rat ganglion-cell photoreceptors. *J Neurophysiol.* 2015;114(5):2955–2966.

225. Van Den Pol AN, Cao V, Heller HC. Circadian system of mice integrates brief light stimuli. *Am J Physiol.* 1998;275(2 Pt 2):R654–R657.

226. Arvanitogiannis A, Amir S. Resetting the rat circadian clock by ultra-short light flashes. *Neurosci Lett.* 1999;261(3):159–162.

227. Figueiro MG, Plitnick B, Rea MS. Pulsing blue light through closed eyelids: effects on acute melatonin suppression and phase shifting of dim light melatonin onset. *Nat Sci Sleep.* 2014;6:149–156.

228. Vandewalle G, Schwartz S, Grandjean D, Wuillaume C, Balteau E, Degueldre C, Schabus M, Phillips C, Luxen A, Dijk DJ, Maquet P. Spectral quality of light modulates emotional brain responses in humans. *Proc Natl Acad Sci USA.* 2010;107(45):19549–19554.

229. Vandewalle G, Gais S, Schabus M, Balteau E, Carrier J, Darsaud A, Sterpenich V, Albouy G, Dijk DJ, Maquet P. Wavelength-dependent modulation of brain responses to a working memory task by daytime light exposure. *Cereb Cortex.* 2007;17(12):2788–2795.

230. Wright HR, Lack LC, Kennaway DJ. Differential effects of light wavelength in phase advancing the melatonin rhythm. *J Pineal Res.* 2004;36(2):140–144.

231. Revell VL, Arendt J, Terman M, Skene DJ. Short-wavelength sensitivity of the human circadian system to phase-advancing light. *J Biol Rhythms.* 2005;20(3):270–272.

232. Sletten TL, Revell VL, Middleton B, Lederle KA, Skene DJ. Age-related changes in acute and phase-advancing responses to monochromatic light. *J Biol Rhythms.* 2009;24(1):73–84.

233. Viola AU, James LM, Schlangen LJM, Dijk D-J. Blue-enriched white light in the workplace improves self-reported alertness, performance and sleep quality. *Scand J Work Environ Health*. 2008;34(4):297–306.

234. Francis G, Bishop L, Luke C, Middleton B, Williams P, Arendt J. Sleep during the Antarctic winter: preliminary observations on changing the spectral composition of artificial light. *J Sleep Res*. 2008;17(3):354–360.

235. Mottram V, Middleton B, Williams P, Arendt J. The impact of bright artificial white and "blue-enriched" light on sleep and circadian phase during the polar winter. *J Sleep Res*. 2011;20(1 Pt 2):154–161.

236. Najjar RP, Wolf L, Taillard J, Schlangen LJM, Salam A, Cajochen C, Gronfier C. Chronic artificial blue-enriched white light is an effective countermeasure to delayed circadian phase and neurobehavioral decrements. *PLoS One*. 2014;9(7):e102827.

237. Honma K, Hashimoto S, Nakao M, Honma S. Period and phase adjustments of human circadian rhythms in the real world. *J Biol Rhythms*. 2003;18(3):261–270.

238. Mistlberger RE, Skene DJ. Nonphotic entrainment in humans?. *J Biol Rhythms*. 2005;20(4):339–352.

239. Miyasako Y, Umezaki Y, Tomioka K. Separate sets of cerebral clock neurons are responsible for light and temperature entrainment of Drosophila circadian locomotor rhythms. *J Biol Rhythms*. 2007;22(2):115–126.

240. Fuller PM, Lu J, Saper CB. Differential rescue of light- and food-entrainable circadian rhythms. *Science*. 2008;320(5879):1074–1077.

241. Aschoff J, Fatranská M, Giedke H, Doerr P, Stamm D, Wisser H. Human circadian rhythms in continuous darkness: entrainment by social cues. *Science*. 1971;171(3967):213–215.

242. Van Reeth O, Sturis J, Byrne MM, Blackman JD, L'Hermite-Balériaux M, Leproult R, Oliner C, Refetoff S, Turek FW, Van Cauter E. Nocturnal exercise phase delays circadian rhythms of melatonin and thyrotropin secretion in normal men. *Am J Physiol*. 1994;266(6 Pt 1):E964–E974.

243. Buxton OM, Lee CW, L'Hermite-Baleriaux M, Turek FW, Van Cauter E. Exercise elicits phase shifts and acute alterations of melatonin that vary with circadian phase. *Am J Physiol*. 2003;284(3):R714–R724.

244. Burgess HJ, Revell VL, Eastman CI. A three pulse phase response curve to three milligrams of melatonin in humans. *J Physiol*. 2008;586(2):639–647.

245. Klerman EB, Rimmer DW, Dijk DJ, Kronauer RE, Rizzo 3rd JF, Czeisler CA. Nonphotic entrainment of the human circadian pacemaker. *Am J Physiol*. 1998;274(4 Pt 2):R991–R996.

246. Miles LE, Raynal DM, Wilson MA. Blind man living in normal society has circadian rhythms of 24.9 hours. *Science*. 1977;198(4315):421–423.

247. Skene DJ, Lockley SW, Arendt J. Melatonin in circadian sleep disorders in the blind. *Biol Signals Recept*. 1999;8(1–2):90–95.

248. Buscemi N, Vandermeer B, Hooton N, Pandya R, Tjosvold L, Hartling L, Baker G, Klassen TP, Vohra S. The efficacy and safety of exogenous melatonin for primary sleep disorders. A meta-analysis. *J Gen Intern Med*. 2005;20(12):1151–1158.

249. Dubocovich ML, Yun K, Al-Ghoul WM, Benloucif S, Masana MI. Selective MT2 melatonin receptor antagonists block melatonin-mediated phase advances of circadian rhythms. *FASEB J*. 1998;12(12):1211–1220.

250. Hunt AE, Al-Ghoul WM, Gillette MU, Dubocovich ML. Activation of MT(2) melatonin receptors in rat suprachiasmatic nucleus phase advances the circadian clock. *Am J Physiol*. 2001;280(1):C110–C118.

251. Lewy AJ, Ahmed S, Jackson JM, Sack RL. Melatonin shifts human circadian rhythms according to a phase-response curve. *Chronobiol Int*. 1992;9(5):380–392.

252. Skene DJ. Optimization of light and melatonin to phase-shift human circadian rhythms. *J Neuroendocrinol*. 2003;15(4):438–441.

253. Burgess HJ, Revell VL, Molina TA, Eastman CI. Human phase response curves to three days of daily melatonin: 0.5 mg versus 3.0 mg. *J Clin Endocrinol Metab*. 2010;95(7):3325–3331.

254. Lockley SW, Skene DJ, James K, Thapan K, Wright J, Arendt J. Melatonin administration can entrain the free-running circadian system of blind subjects. *J Endocrinol*. 2000;164(1):R1–R6.

255. Sack RL, Brandes RW, Kendall AR, Lewy AJ. Entrainment of free-running circadian rhythms by melatonin in blind people. *N Engl J Med*. 2000;343(15):1070–1077.

256. Hack LM, Lockley SW, Arendt J, Skene DJ. The effects of low-dose 0.5-mg melatonin on the free-running circadian rhythms of blind subjects. *J Biol Rhythms*. 2003;18(5):420–429.

The Functions of Sleep and the Effects of Sleep Deprivation

E.H. During*, **, M. Kawai**

*Division of Sleep Medicine, Stanford University, Redwood City, CA, USA
**The Stanford Center for Sleep Sciences and Medicine, Redwood City, CA, USA

INTRODUCTION

As humans, we spend approximately one third of our lives sleeping. Sleep is not only critical for physiological homeostasis and recovery also several essential metabolic, psychological, cognitive and immune functions. Sleep deprivation is a relative concept because it compares actual with ideal sleep duration, which is not universally agreed upon, and may vary from person to person. Expert consensus from the American Academy of

Sleep and Neurologic Disease. http://dx.doi.org/10.1016/B978-0-12-804074-4.00003-0

Sleep Medicine has recently estimated that a minimum of 7 h of sleep is needed on average to support optimal health in adults.[1] The National Sleep Foundation has also published sleep duration recommendations for each age group.[2] According to these recommendations, adults require 7–9 h of sleep, whereas teenagers require 8–10 h, and children ages 6–13 require 9–11 h. Although 7 h may be sufficient for most adults, it may not be sufficient for other age groups, especially younger populations. Despite these recommendations, a recent report from the Centers for Disease Control and Prevention estimates that over one third of the US population sleeps less than 7 h.[3] Unfortunately, sleep is generally perceived by our society as a nonproductive state. This creates a double-bind situation in which the individual often opts to increase wake time over sleep, which paradoxically leads to reduced performance. Although many studies have demonstrated the adverse consequences of insufficient sleep, many individuals remain strikingly unaware of them.

To understand sleep deprivation, it is necessary to first measure actual sleep duration. This can be accomplished in several ways. One simple method is to have the patient keep a sleep diary (Fig. 3.1), though this method involves some level of subjective interpretation and

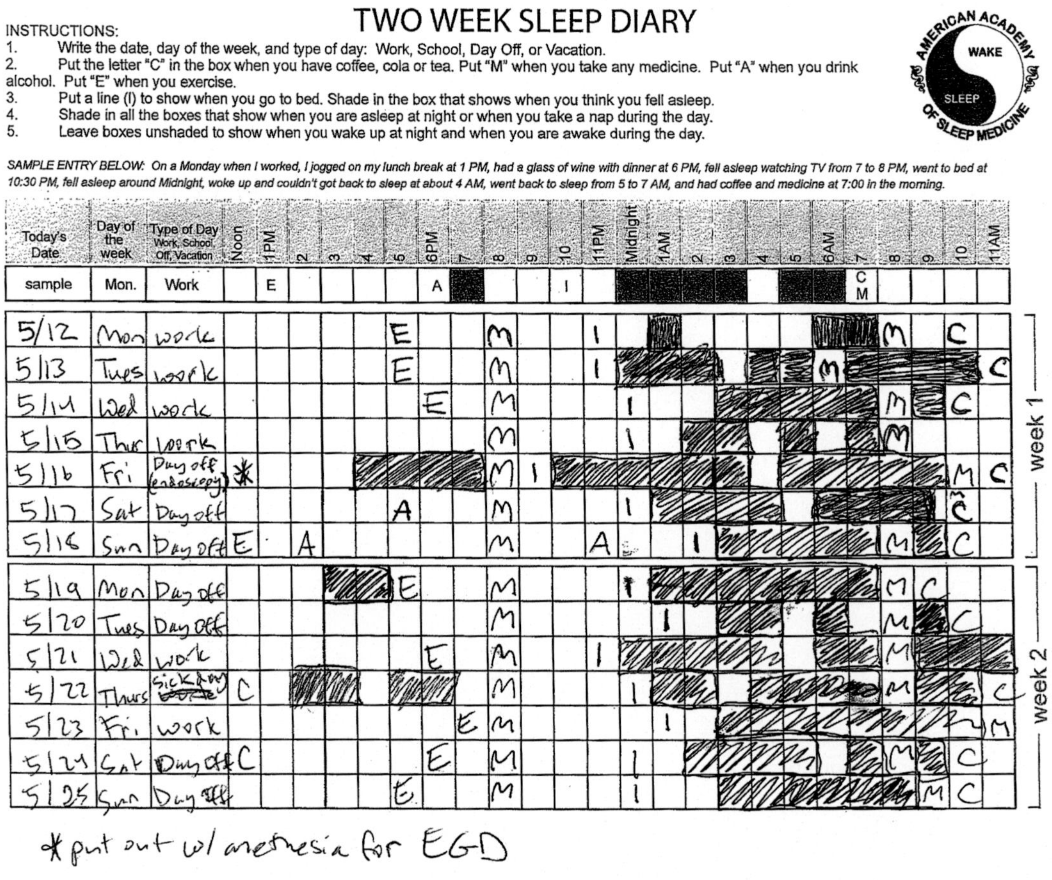

FIGURE 3.1   **An example of a 2-week sleep diary (American Academy of Sleep Medicine).**

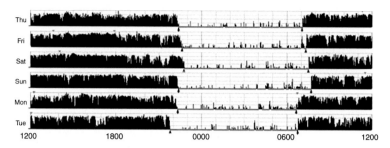

FIGURE 3.2 **A typical actigraph recording for six consecutive days and nights.** *Arrowheads* indicate the beginning and end of time in bed. The time with reduced activity is considered as a rough estimate of sleep time.[4]

recall, which may be impaired if the patient is sleep deprived. Actigraphy provides a more objective measurement by using an accelerometer to measure body movement (Fig. 3.2), and thus estimates sleep time and sleep efficiency. In clinical practice, however, measuring precise sleep duration is not always necessary, and self-reported sleep duration is often sufficient. Exceptions do exist, including patients with sleep-state misperception, shift workers, and professional drivers who may have a motive for misreporting.

SLEEP AND STRESS

Wakefulness is a state of high metabolic demand, due to the frequent need to adapt and respond to sensory inputs. In contrast, sleep is a state of energy conservation and recovery that contributes to the homeostasis of many physiological and psychological functions. This net economy in energy expenditure is more pronounced in nonrapid eye movement (NREM) sleep, as demonstrated by decreased cortical (prefrontal cortex, anterior cingulate cortex, precuneus) and subcortical (thalami, basal ganglia and basal forebrain) glucose metabolism in regions that are usually hypermetabolic.[5] This reduced metabolic state reaches its lowest point in slow wave sleep (SWS), showing a 40% global decrement.[6]

Sleep deprivation, in turn, results in hyperactivation of the hypothalamic-pituitary-adrenal (HPA) axis, as evidenced by the fact that adrenocorticotropic hormone (ACTH), corticotropin-releasing hormone (CRH) and glucocorticoid levels all increase after sleep deprivation. Sleep restriction to 4 h for 6 consecutive nights in young men results in increased levels of cortisol in the afternoon and early evening, as well as impaired glucose tolerance, a response similar to the response seen in those with noninsulin dependent diabetes.[7]

Stress and the metabolic consequences of sleep deprivation may increase glucocorticoid secretion, and it is often difficult to separate the two. This distinction is complicated by the fact that the sleep restriction of research protocols and in vivo ecological situations that can lead to sleep loss may both constitute a stress. Furthermore, insufficient sleep and stress have reciprocal consequences, which may explain why chronic short sleepers have higher levels of cortisol when compared to chronic long sleepers.[8] There is some data to suggest that this bidirectional relationship may be regulated by hypocretin/orexin, and hypocretinergic projections are found in a broad network responsive to CRH, including the locus coeruleus, paraventricular nucleus of the hypothalamus, bed nucleus of the stria terminalis, and central amygdala.[9]

REGULATION OF APPETITE

With the exception of rapid eye movement (REM) sleep, which represents a relative hypermetabolic state, sleep is a period of fasting and energy conservation.[10] Leptin is a hormone that reflects the physiological state of satiety released by adipocytes in response to food intake,[11] while ghrelin is an appetite stimulant hormone. Sleep restriction results in decreased leptin[12] and increased ghrelin serum levels,[13,14] both of which result in increased food intake, with a predilection for energy-rich foods with high carbohydrate content.[14]

Chronic disruption of regular sleep patterns in shift workers is associated with higher body mass indices (BMI),[15] perhaps due to the combined impairment of the HPA axis and appetite stimulation. A metaanalysis across all age groups demonstrated an inverse relationship between hours of sleep and BMI, with each hour of sleep reduction corresponding to a 0.35 kg/m^2 increase in BMI.[16]

Increased levels of TNF-α have been associated with insulin resistance,[17] linking sleep loss, insulin resistance and weight gain. Vgontzas et al.[18] showed increased TNF-α levels after 1 week of mild sleep deprivation in male participants. Adiponectin has anti TNF-α properties and, similar to leptin, enhances insulin sensitivity.[19] In addition, shorter sleep duration is associated with decreased levels of adiponectine.[20]

GH, PRL, TSH, GnRH, AND TESTOSTERONE

The growth hormone (GH) axis is closely linked to SWS. GH plasma levels peak within minutes during each SWS period and remain significantly elevated during SWS throughout the night. Furthermore, pharmacological increases in SWS with ritanserin[21] or gamma-hydroxybutyric acid (GHB)[22] are associated with an increase in GH plasma levels. Prolactin (PRL) secretion is subject to inhibitory dopaminergic tone, which is higher during wakefulness. Shortly after sleep onset, PRL levels increase and reach a twofold elevation during sleep, as compared to trough levels during wakefulness.[23] Sleep exerts an inhibitory effect on thyroid-stimulating hormone (TSH) secretion,[24] however, TSH secretion is more closely linked to the circadian rhythm than to sleep. TSH peaks in the evening and early night and reaches its trough in the afternoon. Nocturnal elevation of testosterone levels is temporally linked to the latency of the first REM period[25] and is also observed during daytime naps.[26] Conversely, experimental fragmentation of sleep, particularly REM sleep, leads to decreased testosterone secretion.[27] In women, sleep has been associated with inhibition of the pulsatile gonadotropin-releasing hormone (GnRH) release in early follicular and early luteal phases.[28]

MOOD, EMOTIONAL STABILITY AND REWARD CIRCUITS

Sleep serves an important role in mood and emotional homeostasis. Although total acute sleep deprivation has shown to have a transient but robust antidepressive effect, likely via elevation of brain-derived neurotrophic factor BDNF[29] and serotonin, taurine and tryptophan levels,[30] prolonged sleep deprivation in turn has deleterious effects on depression and mood stability. Subjective perception of mood[31] and severity of unipolar depression[32] are significantly affected by even minimal misalignment between circadian and sleep–wake cycle.

Sleep is also involved in the processing and modulation of emotional stimuli. One experiment required two groups of participants to rate the same set of photographs showing facial expressions of positive (happiness) and negative (fear, sadness, anger) emotions during two sessions, 5 h apart, with or without an intervening nap. In participants who napped, results showed lower ratings of negative fear and anger, and higher ratings of happiness between sessions. In addition, only those individuals who achieved REM sleep during their nap displayed this modulation of affective reactivity.[33] A study recorded skin conductance response (sweating) and corrugator supercilii muscle tone via electromyography (frowning reaction) to measure unconscious reactivity to negative emotional pictures. Results showed decreased responses (habituation) when retested 2.5 h later in the group allowed to nap, compared to the group requested to stay awake.[34]

Function MRI (fMRI) studies have demonstrated that the neurophysiological correlates of sleep-dependent emotional brain homeostasis may involve enhanced activity in the ventromedial PFC, coupled with decreased amygdala activation.[35] Reciprocally, sleep deprivation is associated with heightened amygdala activation in response to negative stimuli[36] and exaggerated appreciation of pleasure-evoking stimuli, correlating with increased activity in reward, and emotion-related circuits.[37] Both findings are associated with decreased coupling on fMRI in medial and orbitofrontal cortices, areas known to be involved in behavioral inhibitory control and appreciation of rewarding experiences.[38] Sleep deprivation generally elevates expectations of higher reward on gambling task, as nucleus accumbens activation increased with risky choices, and it attenuates neural responses to losses in the insular and orbitofrontal cortices.[39] Taken together, these findings may explain how sleep-deprived individuals show higher risk-taking in reward-related situations with decreased insight into their negative consequences.[40]

EMOTIONAL MATURATION

Smiles during sleep in neonates consistently emerge before smiles during wakefulness, usually observed around 2 months of age. These spontaneous "Duchenne smiles"[41] and other emotional expressions are associated with the active sleep of neonates, the equivalent of adult REM sleep, and can be seen as early as 13th week of gestation.[42] These facial expressions of emotions during sleep have been hypothesized to foster complex motor programs, thereby facilitating the development of the emotional and social smile.[43]

DREAMING

Although dream activity and REM sleep have been closely associated, dreams occur in all stages of sleep, as evidenced by persistent reports of dreams after reduction or suppression of REM sleep.[44] The quality of dreams, however, differs between REM and NREM sleep. Dreams arising from NREM sleep are usually less emotionally charged, less vivid, and less complex when compared to dreams associated with REM sleep. With the exception of frightening images often reported in NREM parasomnias, NREM mentations usually consist of images, events or situations of low emotional valence and do not have the narrative complexity of dreams arising from REM sleep.[45]

In each civilization, symbols and meanings have been associated with dream contents and have given rise to a wide body of explanatory models, from animistic, spiritual, and religious belief systems to psychoanalytical and neurocognitive theories. In the early 20th century, Sigmund Freud and Carl Jung put forth two psychoanalytical frameworks that have since become major theories of dreaming in our modern society. While Freud hypothesized that dreams are a window into our individual unconscious psyche,[46] Jung emphasized their relation with universal archetypes that transcend individuals, culture, and beliefs.[47] An alternative explanatory theory suggests that experiences in dreams could provide an offline environment for simulating situations similar to the real perceptual world. According to the threat simulation theory, dreams often contain challenging and novel situations (e.g., missing an important appointment, arriving late or forgetting to study for an exam, losing money or valuables, etc.) whose purpose might be to rehearse and improve our responses in similar waking conditions.[48] This could explain the abundance of fear-related experiences in dreams, in some respect comparable to exposure therapy.[48]

Integrating neuroimaging, neurophysiological, and clinical evidence, the reward activation model proposed by Perogamvros and Schwartz[49] places the mesolimbic dopaminergic pathway at the center of a circuit involved in: (1) the generation of REM via activation of the sublaterodorsal nucleus of the pons; (2) dream contents; and (3) the offline activation and consolidation of memories of high emotional or motivational values. Sex is a prominent theme in human dreams. The prominent activity of the dopaminergic reward circuits in REM sleep described previously provides a neurophysiological explanation of the universality of sex-related dreams. On a psychodynamic level, Freud viewed our dreams as a free expression of unconscious wishes. On a neurophysiological level, sexual mentation in REM sleep may be facilitated by the specific dynamics of testosterone secretion during sleep. As described previously, hormonal levels are coupled with the first nocturnal REM period, sustained during sleep,[25] and also promoted during daytime naps.[26] Beyond the interconnection between sexual dreams and testosterone levels, sleep has a potent effect on the activity of the neuroendocrine reproductive system and is ultimately implicated in multiple aspects of sexual satisfaction.[50]

REGULATION OF PAIN

Sleep may also be involved in the regulation of pain via involvement of the anterior cingulate cortex and limbic pathways, which are highly connected with the emotional neural network. Four studies have shown that sleep duration below 6 h is associated with increased subjective[51–53] and objective[54] measures of pain.

COGNITIVE PERFORMANCE AND THE EFFECTS OF SLEEP DEPRIVATION

A large body of neuropsychological literature has applied the paradigm of sleep deprivation to evaluate the neurocognitive functions of sleep. These studies have some limitations, namely that sleep deprivation itself can result in a stress response with multiple metabolic, neuroendocrine, behavioral and cognitive consequences. It is also difficult to delineate single-function deficits due to the fact that sleep deprivation affects multiple cognitive domains and behavioral aspects such as mood and affect, which determine the level of engagement for a given task. Other potential confounding factors are intersubject variability (heterogeneity of susceptibility to sleep loss between individuals) and intrasubject variability (improvement of performance on a task can reflect learning effect).[55]

The psychomotor vigilance task (PVT) measures speed of responses to visual stimuli via a thumb-operated device and is not subject to learning effect. In the PVT, sleepiness manifests as microsleep or instability of the wake state. These translate to lapses of response (error of omission) or erroneous responses (error of commission), especially when the person requires significant effort in staying awake.[56] In other words, this represents a balance between the homeostatic drive to fall asleep and a resistance from wake-promoting mechanisms.[57] Regardless of the task, sleep deprivation generally results in a progressive degradation of the performance during extended tasks (fatigue effect) exacerbated by sleep loss.[58,59] While tasks involving higher cognitive functions have been relatively less sensitive to sleep loss, possibly due to a higher level of engagement and compensatory effort,[40] a large body of literature describes the consequences of sleep deprivation on sustained attention and other executive functions, procedural and episodic memory formation, and consolidation, as well as insight and creativity.

In an experiment studying reaction time over 7 days of variable degrees of sleep restriction, sleep deprivation was shown to correlate with longer reaction time on PVT testing. Longer reaction times were noted after only one night of sleep restriction (2 h of sleep reduction) compared to volunteers sleeping on average 8.5 h.[60] Performance continued to diminish with each subsequent day and was directly correlated with the degree of sleep restriction. In the groups allowed to sleep for a maximum of 7 and 5 h, vigilance deficit stabilized after 5 nights, whereas in participants sleeping 3 h or less, vigilance continued to deteriorate in a linear fashion showing a twofold decrease compared to the control group at the end of the sleep restriction phase. After three nights with 8 h spent in bed, none of the sleep restriction groups returned to their baseline reaction times, suggesting that recovery from even mild sleep deprivation may last several days. This experiment also suggests that there may be an adaptive mechanism to mild and moderate chronic sleep deprivation (7 and 5 h groups) that may be insufficient to maintain performance levels in situations of severe chronic sleep deprivation (3 h group). Other studies have confirmed the finding that impairment in PVT testing increases in proportion to the degree of sleep deprivation (Fig. 3.3).[61]

Sleep deprivation is known to decrease insight into one's own performance,[40,62] and some studies suggest that in conditions of total sleep deprivation, wake promoting agents could be associated with decreased self-monitoring,[63] possibly due to lack of compensatory effort when compared to placebo.[64] However, some evidence suggests that subjective sleepiness does not correlate well with the magnitude of cognitive impairment. In an experiment conducted by Van Dongen et al.,[61] PVT, working memory, cognitive throughput measured by the digit symbol substitution task[65] and serial addition/subtraction tasks (SAST),[66] demonstrated a comparable degree of impairment in conditions of acute total and mild chronic sleep deprivation (sleep restriction to 6 h over 14 days), supporting the concept of a "cumulative sleep debt" with regard to executive functions.[61] After 14 days, although cognitive performance reached their lowest levels, participants in the 4 h and 6 h groups reported feeling only slightly sleepy on the Stanford sleepiness scale, suggesting that individuals subjected to

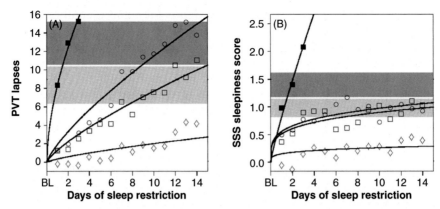

FIGURE 3.3 (A) PVT lapses and (B) sleepiness score after acute sleep deprivation and chronic sleep deprivation. Each panel displays group averages for subjects in the 8 h (◊), 6 h (□), and 4 h (○) chronic sleep period conditions across 14 days, and in the 0 h (■) sleep condition across 3 days.[61] *BL*, Baseline

chronic sleep deprivation could not reliably predict the degree of cognitive impairment on sleepiness alone. Based on the neurocognitive performances gathered across groups for the duration of the experiment, the authors estimated that the minimal 24 h requirement to attain a stable cognitive equilibrium is 3.8 h of continuous sleep. Below this critical threshold, the model diverges from a stable equilibrium and performance continues to degrade as seen in acute total sleep deprivation. This study demonstrates how relative resilience to sleepiness and stabilization of cognitive performance at a shifted new baseline may provide a false sense of coping with sleep deprivation.

Recently, functional neuroimaging techniques, including fMRI and positron emission tomography (PET), have been utilized to better understand the localization of cognitive dysfunction in cases of acute sleep deprivation.[67] Those studies investigate either increase or decrease of blood oxygen level dependent (BOLD) measurements in fMRI or oxygen/glucose consumption in PET. The prefrontal cortex has been reported to be extremely vulnerable, showing decreased function in acute sleep deprivation.[68] fMRI studies have demonstrated reduced activity in the prefrontal cortex after sleep deprivation in a serial subtraction task.[69]

However, in a verbal learning task, the prefrontal cortex and parietal cortex were more activated after acute sleep deprivation.[70] Interestingly, activation of the prefrontal cortex was positively correlated with the degree of subjective sleepiness, whereas activation in the parietal cortex was positively associated with the preservation of near-normal verbal learning. These patterns of increase and decrease in cerebral activation most likely represent compensatory adaptations (CA). In CA, more activation in the prefrontal cortex with acute sleep deprivation indicates a compensation process in response to the increased homeostatic drive for sleep. The prefrontal cortex is involved in working memory, attention and executive function,[71] functions known to be vulnerable to sleep deprivation. On the other hand, activation of the parietal lobe with sleep deprivation indicates an adaptive process to support the decreased function of other areas of the cortex. These patterns of CA have task-specific differences[69,72] and indicate that sleep deprivation can result in a combination of dysfunction and compensatory hyperfunction in the brain.

Level of difficulty modulate the cerebral response,[73] and interindividual differences can contribute to the difficulty of using neuroimaging studies as a generalized parameter of sleepiness.

It has been demonstrated in fMRI studies that activation of the frontoparietal region was more robustin participants found to be less vulnerable to the effects of sleep deprivation.[74] This suggests that more activated CA may correlate with less vulnerability to sleep deprivation.

As mentioned, several aspects of executive function are affected by sleep deprivation. Sleep deprivation results in difficulty determining the scope of a problem in cases of distracting information and impairs divergent thinking and originality as measured on the Torrance Tests of Creative Thinking.[75,76] It also affects temporal memory, as shown by lower performance on tasks of recency (inability to recall the timing for recent events), even in conditions of preserved alertness with use of caffeine.[40] In a working memory task under sleep deprivation, fMRI studies demonstrated decreased activation in the parietal region, and increased activation in prefrontal and thalamic regions in more complex tasks.[77] This is another example of CA.

Another function of sleep is to preserve flexible decision-making. Experiments show that acute and chronic sleep deprivation results in more rigid thinking and perseverative errors with poor appreciation of an updated situation despite intact critical reasoning,[78,79] loss of focus to relevant cues,[80,81] and increased risk taking,[82] possibly due to reduced functioning of the ventro-medial PFC.[39,83]

ACCIDENTS AND SAFETY IN ACUTE AND CHRONIC SLEEP DEPRIVATION

Performance in lane variability of driving, or accuracy and reaction time after more than 16 h of sleep deprivation, has been reported to be similar to the performance in those with a blood alcohol content of 0.05–0.1%.[84,85] However, when comparing factors of sleep deprivation and alcohol intoxication, subjects who were sleep deprived showed progressive deterioration through driving, whereas subjects who were intoxicated demonstrated stable impairment throughout the entire excercise.[86] It has also been demonstrated that the circadian rhythm affects driving performance and risk of motor vehicle accidents. Connor et al.[87] reported a fivefold higher incidence of motor vehicle accidents between 2 and 5 a.m. compared with other times of the day. Extending these findings, Lenné et al.,[88] in a report on driving performance, especially speed control, showed deterioration in the early morning (2–5 a.m.) as well as between 2 and 5 p.m., corresponding to the mid-afternoon dip. These phenomena are more prominent in low stimulus situations, such as highway and rural driving.[89]

Sleep deprivation has been a major issue in the medical field. One of the most significant changes in modern medical training came about after the Libby Zion's case, with the introduction of the duty-hour regulations that helped to decrease the risk of medical errors by sleep-deprived physicians.[90,91] One study demonstrated that the performance of sleep-deprived, postcall residents in vigilance, sustained attention and driving tasks is just as impaired as those with a blood alcohol concentration of 0.04–0.05%, the equivalent of 1–2 standard drinks.[92]

MEMORY

The successive stages of memory are encoding, consolidation/reconsolidation and retrieval. Encoding and retrieval occur in the awake state, but adequate sleep improves all stages of memory formation, particularly consolidation and reconsolidation. Observation of improved encoding capacity following sleep has been demonstrated in numerous studies,[36,93,94] and accounted for by the "synaptic homeostasis" hypothesis:[95,96] for the brain to adapt to the widespread synaptic long-term potentiation (strengthening of postsynaptic nerve cells responses to stimulations across synapses) resulting from the incessant formation of new memories, sleep-dependent mechanisms may downscale synaptic strengths to levels that are energetically sustainable for the brain. Among all stages of sleep, slow-wave activity (SWA, 0.5–4 Hz) may contribute the most to this process.[97,98]

Sleep plays an active role in memory formation because it could provide a favorable neurochemical environment with sleep-specific neurophysiological events such as SWS, sleep spindles and hippocampal ripples in NREM sleep, and hippocampal theta activity in REM sleep. According to the dual process hypothesis,[99] NREM[100–102] and REM[103–106] sleep have differential involvement in declarative (episodic and semantic) and nondeclarative memory (procedural and emotional memory) consolidation, respectively.

Although reports are conflicting in older adults,[107,108] SWS in young adults is positively correlated with declarative memory consolidation.[108] Experimental enhancement of SWA via transcranial application of a 0.75 Hz oscillating potential is associated with improved declarative memory for word-pairs[109] in young adults when compared to a sham stimulation group, indicating the specific involvement of SWA in declarative memory consolidation. Reactivation of hippocampal memory during SWS has been observed in both animal and human studies.[110,111] This process involves synchronization between the hippocampus and cortex, ultimately leading to synaptic consolidation. Studies in rats demonstrate a hippocampal chronological "replay" during SWS following an experience of environment exploration (spatial learning)[112] that are associated with hippocampal "ripples" (100–200 Hz) and sharp wave activity.[113,114] In addition, coupling of hippocampal sharp-wave/ripple complexes with neocortical delta (1–4 Hz) slow oscillation patterns and spindles were demonstrated in rats during SWS.[115] This dialog between the hippocampus and neocortex, mediated by thalamocortical sleep spindles, could facilitate transfer from the hippocampus to neocortical long-term storage sites.[116] The specific implication of sleep spindles in memory consolidation and reconsolidation is supported by studies demonstrating a positive correlation between memory performance and spindle density, independent of the sleep stage.[117–119]

REM sleep may also enhance nondeclarative (procedural, priming) memory[103–106] and the emotional modulation of memory.[120,121] Procedural memory involves the acquisition of motor skills. Most common experimental designs have used tasks of finger-sequence tapping, mirror tracing or other forms of visuomotor learning. Procedural memory involves a wide set of structures, including the motor cortex (supplementary motor area), basal ganglia (striatum), thalamus and cerebellum. The hippocampus[122] is involved to a lesser extent but may be implicated in more complex motor learning tasks.[123] While selective NREM sleep deprivation is difficult without totally depriving an individual from sleep, selective REM sleep deprivation studies have supported the view that REM sleep may be involved in the formation of procedural memory.[124] This has been supported by the finding of decreased performance on a learned finger-sequence tapping task following REM-suppressed sleep via acetylcholine receptor blockade.[125] On the other hand, procedural memory is not affected by REM-suppressant antidepressant medications in both depressed and nondepressed patients.[126–128] In addition, several studies have demonstrated that NREM sleep events such as SWA[129–131] and spindles[132–134] are associated with learning of even simple motor tasks. Finally, similar to the reactivation studies conducted with procedural memory, two studies reexposing participants to parts of a learned melody during SWS resulted in performance benefit.[135,136]

Memories associated with high emotional content are also more strongly associated with REM sleep.[120,121,137] Activation of limbic areas and reward circuits are observed during REM sleep and dreams, and the amygdala may be involved in regulating hippocampal memory-encoding processes.[138] Memories of high emotional valence may benefit from heightened stabilization during sleep. In normal conditions, emotional memory traces may be reactivated and remodeled each night during REM sleep. This process, repeated over time, results in strengthening the nonaffective "core," or facts of the memory and, concomitantly, weakening the affective "layer" initially associated with it.[139] This model, condensed by Walker et al.[139] in the formula "sleep to remember and sleep to forget," is supported by neuroimaging evidence of decreased amygdala activity during the repeated viewing of emotional pictures after an interval filled with sleep.[35] NREM sleep may also play a role in the consolidation of emotional memory as pharmacological augmentation of sleep spindles with GHB or zolpidem may improve recall of highly arousing and negative stimuli.[140] Further evidence stems from the observation of odor-induced reactivation of fear memories in NREM sleep both in animals[141] and humans.[142] Despite this, there is lack of general agreement that NREM sleep, including sleep spindles, may be implicated in both declarative and nondeclarative memory, and REM sleep may be more specifically associated with emotional memory.

INSIGHT AND CREATIVITY, GIST MEMORY

Processes occurring during sleep can extract meaning and result in new insight, as suggested by historical anecdotes such as the discovery of the chemical structure of benzene or the periodic table of elements, which were reportedly sleep-induced breakthroughs.[143] Some have hypothesized that memory consolidation during sleep could facilitate access to general rules or shortcuts, potentially leading to the solution of problems encountered during the wake state. Wagner et al.[144] reported that 3 times more participants accessed a hidden rule for a number reduction task after sleep.

Some mechanisms during memory consolidation may also promote its integration into a metalevel of cognition. This is illustrated by experiments showing that participants tend to create "false" memories of gist words that are not present on a word list task prior to sleep, but which pertain to the list of associated words (e.g., for white coat, stethoscope, hospital, nurse, thermometer, the gist word might be doctor).[145,146] In addition, REM sleep may have a lasting effect on the ability to solve complex anagrams, as demonstrated by a 32% advantage on this task in participants awakened from REM sleep.[147] This effect could be related to the prolongation of a REM-related "hyperassociative state," a state promoting novel thinking and creative problem-solving skills. Conversely, the use of creative thinking has been associated with increased REM sleep during the following sleep period.[148]

IMMUNE FUNCTION

Sleep and circadian rhythm have a notable influence on the regulation of the immune system. Differentiated immune cells with immediate effector functions (natural killer cells and cytotoxic T lymphocytes) peak during the wake period, while naive and central memory T cells, involved in the slowly evolving adaptive immune response, peak during sleep.[149] SWS-related GH and prolactin (PRL) secretion and decreased antiinflammatory catecholamine, and cortisol levels provide a favorable endocrine milieu for antigen presenting T cell interactions, with a shift toward T helper cell (Th) 1 cytokines, increase in Th cell proliferation, and migration of naive T cells to lymph nodes.[149] Sleep restriction to 4 h results in increased plasma levels of interleukin 6 (IL-6)[150] and C-reactive protein (CRP)[151] within 10 days, and even mild sleep restriction to 6 h over 8 days leads to elevated levels of inflammatory cytokines.[18] Elevations have been shown to be mild but consistent, resulting in a state of low-grade systemic inflammation within ranges increasing cardiovascular risk.[152,153] Chronic sleep loss results in an increased risk of viral upper respiratory infections[154] and a higher rate of infections in shift workers.[155] Reports of decreased immune response to influenza virus vaccination in individuals suffering from chronic sleep loss[156] further supports the view that, although associated with elevated markers of inflammation, sleep loss results in a functional state of immunodeficiency.

There is little and conflicting data regarding quantity of sleep and cancer risk, and no evidence of an association with neurological tumors. Kakizaki et al.[157] found an increased risk of breast cancer (hazard ratio = 1.62) associated with 6 h (compared to 7 h) of sleep; however, several large prospective cohort and case-control studies have failed to replicate this association. One prospective cohort study found sleep of less than 7 h to be associated with reduced ovarian cancer risk.[158] Less than 6 h of sleep may be associated with colorectal adenoma,[159] and both extreme short (≤5 h) and long (≥9 h) sleep durations may be associated with increased risk of colorectal cancer.[160]

CONCLUSIONS

In addition to its critical function in endocrine, metabolic, and immune regulation, sleep appears to play an active and important role in maintaining health, emotional well-being and neurocognitive performance during wakefulness. Lifestyle changes have been emphasized in recent years to prevent a wide range of diseases and promote health in general. We hope that in the future a similar effort will be made to stress the importance of sleep-related lifestyle changes that can improve both sleep quality and sleep quantity, not only for our patients but also for society at large.

References

1. Watson NF, Badr MS, Belenky G. Joint consensus statement of the American Academy of Sleep Medicine and Sleep Research Society on the recommended amount of sleep for a healthy adult: methodology and discussion. *J Clin Sleep Med*. 2015;11(8):931–952.
2. Hirshkowitz M, Whiton K, Albert SM. National Sleep Foundation's updated sleep duration recommendations: final report. *Sleep Heal*. 2015;1(4):233–243.
3. Liu Y, Wheaton AG, Chapman DP, Cunningham TJ, Croft JB. Prevalence of healthy sleep duration among adults—United States, 2014. *MMWR Morb Mortal Wkly Rep*. 2016;65(6):137–141.
4. Suratt PM, Barth JT, Diamond R. Reduced time in bed and obstructive sleep-disordered breathing in children are associated with cognitive impairment. *Pediatrics*. 2007;119(2):320–329.
5. Dang-Vu TT, Schabus M, Desseilles M, Sterpenich V, Bonjean M, Maquet P. Functional neuroimaging insights into the physiology of human sleep. *Sleep*. 2010;33(12):1589–1603. www.pubmedcentral.nih.gov/articlerender.fcgi?artid=2982729&tool=pmcentrez&rendertype=abstract..

6. Maquet P. Functional neuroimaging of normal human sleep by positron emission tomography. *J Sleep Res.* 2000;9(3):207–231.

7. Spiegel K, Leproult R, Van Cauter E. Impact of sleep debt on metabolic and endocrine function. *Lancet.* 1999;354(9188):1435–1439.

8. Späth-Schwalbe E, Schöller T, Kern W, Fehm HL, Born J. Nocturnal adrenocorticotropin and cortisol secretion depends on sleep duration and decreases in association with spontaneous awakening in the morning. *J Clin Endocrinol Metab.* 1992;75(6):1431–1435.

9. Winsky-Sommerer R. Interaction between the corticotropin-releasing factor system and hypocretins (orexins): a novel circuit mediating stress response. *J Neurosci.* 2004;24(50):11439–11448.

10. Maquet P. Sleep function(s) and cerebral metabolism. *Behav Brain Res.* 1995;69(1–2):75–83.

11. Simon C, Gronfier C, Schlienger JL, Brandenberger G. Circadian and ultradian variations of leptin in normal man under continuous enteral nutrition: relationship to sleep and body temperature. *J Clin Endocrinol Metab.* 1998;83(6):1893–1899.

12. Mullington JM, Chan JL, Van Dongen HPA. Sleep loss reduces diurnal rhythm amplitude of leptin in healthy men. *J Neuroendocrinol.* 2003;15(9):851–854.

13. Taheri S, Lin L, Austin D, Young T, Mignot E. Short sleep duration is associated with reduced leptin, elevated ghrelin, and increased body mass index. *PLoS Med.* 2004;1(3):210–217.

14. Spiegel K, Tasali E, Penev P, Van Cauter E. Brief communication: Sleep curtailment in healthy young men is associated with decreased leptin levels, elevated ghrelin levels, and increased hunger and appetite. *Ann Intern Med.* 2004;141(11):846–850.

15. van Drongelen A, Boot CRL, Merkus SL, Smid T, van der Beek AJ. The effects of shift work on body weight change—a systematic review of longitudinal studies. *Scand J Work Environ Heal.* 2011;37(4):263–275.

16. Cappuccio FP, Taggart FM, Kandala NB. Meta-analysis of short sleep duration and obesity in children and adults. *Sleep.* 2008;31(5):619–626.

17. Hotamisligil GS. Inflammatory pathways and insulin action. *Int J Obes Relat Metab Disord.* 2003;27(suppl 3): S53–S55.

18. Vgontzas AN, Zoumakis E, Bixler EO. Adverse effects of modest sleep restriction on sleepiness, performance, and inflammatory cytokines. *J Clin Endocrinol Metab.* 2004;89(5):2119–2126.

19. Guerre-Millo M. Adiponect: an, update. *Diabetes Metab.* 2008;34(1):12–18.

20. Kotani K, Sakane N, Saiga K. Serum adiponectin levels and lifestyle factors in Japanese men. *Hear Vessel.* 2007;22(5):291–296.

21. Gronfier C, Luthringer R, Follenius M. A quantitative evaluation of the relationships between growth hormone secretion and delta wave electroencephalographic activity during normal sleep and after enrichment in delta waves. *Sleep.* 1996;19(10):817–824.

22. Van Cauter E, Plat L, Scharf MB. Simultaneous stimulation of slow-wave sleep and growth hormone secretion by gamma-hydroxybutyrate in normal young men. *J Clin Invest.* 1997;100(3):745–753.

23. Van Cauter E, Aschoff J. Endocrine and other biological rhythms. In: DeGroot LJ, Jameson JL, eds. *Endocrinology.* Philadelphia, PA: Elseviers; 2006:341–372.

24. Brabant G, Prank K, Ranft U. Physiological regulation of circadian and pulsatile thyrotropin secretion in normal man and woman. *J Clin Endocrinol Metab.* 1990;70(2):403–409.

25. Luboshitzky R, Herer P, Levi M, Shen-Orr Z, Lavie P. Relationship between rapid eye movement sleep and testosterone secretion in normal men. *J Androl.* 1999;20(6):731–737. www.ncbi.nlm.nih.gov/pubmed/10591612.

26. Axelsson J, Ingre M, Åkerstedt T, Holmbäck U. Effects of acutely displaced sleep on testosterone. *J Clin Endocrinol Metab.* 2005;90(8):4530–4535.

27. Luboshitzky R, Zabari Z, Shen-Orr Z, Herer P, Lavie P. Disruption of the nocturnal testosterone rhythm by sleep fragmentation in normal men. *J Clin Endocrinol Metab.* 2001;86(3):1134–1139.

28. Filicori M, Santoro N, Merriam GR, Crowley WF. Characterization of the physiological pattern of episodic gonadotropin secretion throughout the human menstrual cycle. *J Clin Endocrinol Metab.* 1986;62(6):1136–1144.

29. Schmitt K, Holsboer-Trachsler E, Eckert A. BDNF in sleep, insomnia, and sleep deprivation. *Ann Med.* 2016;3890:1–10.

30. Davies SK, Ang JE, Revell VL. Effect of sleep deprivation on the human metabolome. *Proc Natl Acad Sci USA.* 2014;111:10761–10766.

31. Danilenko KV, Cajochen C, Wirz-Justice A. Is sleep per se a zeitgeber in humans?. *J Biol Rhythms.* 2003;18(2):170–178.

32. Hasler BP, Buysse DJ, Kupfer DJ, Germain A. Phase relationships between core body temperature, melatonin, and sleep are associated with depression severity: further evidence for circadian misalignment in non-seasonal depression. *Psychiatry Res.* 2010;178(1):205–207.

33. Gujar N, McDonald SA, Nishida M, Walker MP. A role for REM sleep in recalibrating the sensitivity of the human brain to specific emotions. *Cereb Cortex.* 2011;21(1):115–123.

34. Pace-Schott EF, Shepherd E, Spencer RMC. Napping promotes inter-session habituation to emotional stimuli. *Neurobiol Learn Mem.* 2011;95(1):24–36.

35. Van Der Helm E, Yao J, Dutt S, Rao V, Saletin JM, Walker MP. REM sleep depotentiates amygdala activity to previous emotional experiences. *Curr Biol.* 2011;21(23):2029–2032.

36. Yoo S-S, Hu PT, Gujar N, Jolesz Fa, Walker MP. A deficit in the ability to form new human memories without sleep. *Nat Neurosci.* 2007;10(3):385–392.

37. Gujar N, Yoo S-S, Hu P, Walker MP. Sleep deprivation amplifies reactivity of brain reward, networks, biasing the appraisal of positive emotional experiences. *J Neurosci.* 2011;31(12):4466–4474.

38. Wallis JD. Orbitofrontal cortex and its contribution to decision-making. *Annu Rev Neurosci.* 2007;30:31–56.

39. Venkatraman V, Chuah YML, Huettel Sa, Chee MWL. Sleep deprivation elevates expectation of gains and attenuates response to losses following risky decisions. *Sleep.* 2007;30(5):603–609.

40. Harrison Y, Horne JA. The impact of sleep deprivation on decision making: a review. *J Exp Psychol Appl.* 2000;6(3):236–249.

41. Ekman P. An argument for basic emotions. *Cogn Emot.* 1992;6(3):169–200.

42. Messinger D, Dondi M, Nelson-Goens GC, Beghi A, Fogel AA, Simion F. How sleeping neonates smile. *Dev Sci.* 2002;5(1):48–54.

43. Dondi M, Messinger D, Colle M. A New perspective on neonatal smiling: differences between the judgments of expert coders and naive observers. *Infancy.* 2007;12(3):235–255.

44. Oudiette D, Dealberto MJ, Uguccioni G. Dreaming without REM sleep. *Conscious Cogn.* 2012;21(3):1129–1140.

45. Fosse R, Stickgold R, Hobson JA. Brain-mind states: reciprocal variation in thoughts and hallucinations. *Psychol Sci.* 2001;12(1):30–36.

46. Freud S. *The Interpretation of Dreams.* New York: The Macmillan Company; 1913. doi:10.1037/10561-004.

47. Jung CG. *Memories, Dreams, Reflections.* Jaffé A, ed. New York, NY: Vintage Books; 1989.

48. Valli K, Revonsuo A. The threat simulation theory in light of recent empirical evidence: a review. *Am J Psychol.* 2009;122(1):17–38. www.ncbi.nlm.nih.gov/pubmed/19353929.

49. Perogamvros L, Schwartz S. The roles of the reward system in sleep and dreaming. *Neurosci Biobehav Rev.* 2012;36(8):1934–1951.

50. Andersen ML, Alvarenga TF, Mazaro-Costa R, Hachul HC, Tufik S. The association of testosterone, sleep, and sexual function in men and women. *Brain Res.* 2011;1416:80–104.

51. Hamilton Na, Catley D, Karlson C. Sleep and the affective response to stress and pain. *Health Psychol.* 2007;26(3):288–295.

52. Buxton OM, Hopcia K, Sembajwe G. Relationship of sleep deficiency to perceived pain and functional limitations in hospital patient care workers. *J Occup Environ Med.* 2012;54(7):851–858.

53. Edwards RR, Almeida DM, Klick B, Haythornthwaite JA, Smith MT. Duration of sleep contributes to next-day pain report in the general population. *Pain.* 2008;137(1):202–207.

54. Roehrs Ta, Harris E, Randall S, Roth T. Pain sensitivity and recovery from mild chronic sleep loss. *Sleep.* 2012;35(12):1667–1672.

55. Dorrian J, Rogers NL, Dinges DF. Psychomotor vigilance performance: neurocognitive assay sensitive to sleep loss. In: Kushida C, ed. *Sleep Deprivation.* New York: Marcel Dekker; 2005:39–70.

56. Lim J, Dinges DF. Sleep deprivation and vigilant attention. *Ann NY Acad Sci.* 2008;1129:305–322.

57. Goel N, Rao H, Durmer JS, Dinges DF. Neurocognitive consequences of sleep deprivation. *Semin Neurol.* 2009;29(4):320–339.

58. Bjerner B. Alpha depression and lowered pulse rate during delayed actions in a serial reaction test: a study of sleep deprivation. *Acta Physiol Scand.* 1949;19(suppl 65):1–93.

59. Wilkinson RT. Sleep deprivation: performance tests for partial and selective sleep deprivation. In: Abt L, ed. *Progress in Clinical Psychology.* New York, NY: Grune; 1969:28–43.

60. Belenky G, Wesensten NJ, Thorne DR. Patterns of performance degradation and restoration during sleep restriction and subsequent recovery: a sleep dose-response study. *J Sleep Res.* 2003;12(1):1–12.

61. Van Dongen HP, Maislin G, Mullington JM, Dinges DF. The cumulative cost of additional wakefulness: dose-response effects on neurobehavioral functions and sleep physiology from chronic sleep restriction and total sleep deprivation. *Sleep*. 2003;26(2):117–126.

62. Dinges DF, Kribbs NB. Performing while sleepy: effects of experimentally-induced sleepiness. In: Monk TH, ed. *Sleep, Sleepiness and Performance*. Chichester: John Wiley & Sons; 1991:97–128.

63. Baranski JV, Pigeau RA. Self-monitoring cognitive performance during sleep deprivation: effects of modafinil, *d*-amphetamine and placebo. *J Sleep Res*. 1997;6(2):84–91.

64. Bard EG, Sotillo C, Anderson AH, Thompson HS, Taylor MM. The DCIEM map task corpus: spontaneous dialogue under sleep deprivation and drug treatment. *Speech Commun*. 1996;20(1–2):71–84.

65. Wechsler D. *Manual for the Wechsler Adult Intelligence Scale—Revised*. New York: Psychological Corporation; 1981.

66. Thorne DR, Genser SG, Sing HC, Hegge FW. The Walter Reed performance assessment battery. *NeurobehavToxicolTeratol*. 1985;7(4):415–418.

67. Dang-Vu TT, Desseilles M, Petit D, Mazza S, Montplaisir J, Maquet P. Neuroimaging in sleep medicine. *Sleep Med*. 2007;8(4):349–372.

68. Horne JA. Human sleep, sleep loss and behaviour. Implications for the prefrontal cortex and psychiatric disorder. *Br J Psychiatry*. 1993;162:413–419.

69. Drummond SP, Brown GG, Stricker JL, Buxton RB, Wong EC, Gillin JC. Sleep deprivation-induced reduction in cortical functional response to serial subtraction. *Neuroreport*. 1999;10(18):3745–3748.

70. Drummond SP, Brown GG, Gillin JC, Stricker JL, Wong EC, Buxton RB. Altered brain response to verbal learning following sleep deprivation. *Nature*. 2000;403(6770):655–657.

71. Jones K, Harrison Y. Frontal lobe function, sleep loss and fragmented sleep. *Sleep Med Rev*. 2001;5(6):463–475.

72. Mu Q, Nahas Z, Johnson KA. Decreased cortical response to verbal working memory following sleep deprivation. *Sleep*. 2005;28:55–67.

73. Drummond SPA, Brown GG, Salamat JS, Gillin JC. Increasing task difficulty facilitates the cerebral compensatory response to total sleep deprivation. *Sleep*. 2004;27:445–451.

74. Mu Q, Mishory A, Johnson KA. i in. Decreased brain activation during a working memory task at rested baseline is associated with vulnerability to sleep deprivation. *Sleep*. 2005;28:433–446.

75. Wimmer F, Hoffmann R, Bonato R, Moffitt A. The effects of sleep deprivation on divergent thinking and attention processes. *J Sleep Res*. 1992;1(4):223–230.

76. Horne JA. Sleep loss and "divergent" thinking ability. *Sleep*. 1988;11(6):528–536.

77. Chee MWL, Choo WC. Functional imaging of working memory after 24 hr of total sleep deprivation. *J Neurosci*. 2004;24:4560–4567.

78. Harrison Y, Horne JA. One night of sleep loss impairs innovative thinking and flexible decision making. *Organ Behav Hum Decis Process*. 1999;78(2):128–145.

79. Herscovitch J, Stuss D, Broughton R. Changes in cognitive processing following short-term cumulative partial sleep deprivation and recovery oversleeping. *J Clin Neuropsychol*. 1980;2(4):301–319.

80. Norton R. The effects of acute sleep deprivation on selective attention. *Br J Psychol*. 1970;61(2):157–161.

81. Blagrove M, Alexander C, Horne Ja. The effects of chronic sleep reduction on the performance of cognitive tasks sensitive to sleep deprivation. *Appl Cogn Psychol*. 1995;9:21–40.

82. Harrison Y, Horne JA. Sleep loss affects risk-taking. *J Sleep Res*. 1998;7(suppl 2):113.

83. Womack SD, Hook JN, Reyna SH, Ramos M. Sleep loss and risk-taking behavior: a review of the literature. *Behav Sleep Med*. 2013;11(5):343–359.

84. Williamson AM, Feyer AM. Moderate sleep deprivation produces impairments in cognitive and motor performance equivalent to legally prescribed levels of alcohol intoxication. *Occup Environ Med*. 2000;57(10):649–655.

85. Dawson D, Reid K. Fatigue, alcohol and performance impairment. *Nature*. 1997;388(6639):235.

86. Hack MA, Choi SJ, Vijayapalan P, Davies RJO, Stradling JR. Comparison of the effects of sleep deprivation, alcohol and obstructive sleep apnoea (OSA) on simulated steering performance. *Respir Med*. 2001;95(7): 594–601. www.sciencedirect.com/science/article/B6WWS-45V29JK-9/2/7dceb05ba44e7d4ed40c1dcebf9067d0.

87. Connor J, Norton R, Ameratunga S. Driver sleepiness and risk of serious injury to car occupants: population based case control study. *BMJ*. 2002;324(7346):1125.

88. Lenné MG, Triggs TJ, Redman JR. Time of day variations in driving performance. *Accid Anal Prev*. 1997;29(4):431–437.

89. Reimer B, D'Ambrosio LA, Coughlin JF. Secondary analysis of time of day on simulated driving performance. *J Safety Res*. 2007;38(5):563–570.

90. The Libby Zion case. *Ann Intern Med*. 1991;115(12):985–986.

91. Asch DA, Parker RM. The Libby Zion case. One step forward or two steps backward?. *N Engl J Med*. 1988;318(12):771–775.

92. Arnedt JT, Owens J, Crouch M, Stahl J, Carskadon Ma. Neurobehavioral performance of residents after heavy night call vs after alcohol ingestion. *JAMA*. 2005;294(9):1025–1033.

93. McDermott CM, LaHoste GJ, Chen C, Musto A, Bazan NG, Magee JC. Sleep deprivation causes behavioral, synaptic, and membrane excitability alterations in hippocampal neurons. *J Neurosci*. 2003;23(29):9687–9695.

94. Mander BA, Santhanam S, Saletin JM, Walker MP. Wake deterioration and sleep restoration of human learning. *Curr Biol*. 2011;21(5).

95. Tononi G, Cirelli C. Sleep and synaptic homeostasis: a hypothesis. *Brain Res Bull*. 2003;62(2):143–150.

96. Tononi G, Cirelli C. Sleep and the price of plasticity: from synaptic and cellular homeostasis to memory consolidation and integration. *Neuron*. 2014;81(1):12–34.

97. Van Der Werf YD, Altena E, Schoonheim MM. Sleep benefits subsequent hippocampal functioning. *Nat Neurosci*. 2009;12(2):122–123.

98. Massimini M, Tononi G, Huber R. Slow waves, synaptic plasticity and information processing: insights from transcranial magnetic stimulation and high-density EEG experiments. *Eur J Neurosci*. 2009;29(9):1761–1770.

99. Ackermann S, Rasch B. Differential effects of non-REM and REM sleep on memory consolidation?. *Curr Neurol Neurosci Rep*. 2014;14(2).

100. Yaroush R, Sullivan MJ, Ekstrand BR. Effect of sleep on memory. II. Differential effect of the first and second half of the night. *J Exp Psychol*. 1971;88(3):361–366.

101. Marshall L, Born J. The contribution of sleep to hippocampus-dependent memory consolidation. *Trends Cogn Sci*. 2007;11(10):442–450.

102. Gais S, Born J. Declarative memory consolidation: mechanisms acting during human sleep. *Learn Mem*. 2004;11(6):679–685.

103. Plihal W, Born J. Effects of early and late nocturnal sleep on declarative and procedural memory. *J Cogn Neurosci*. 1997;9(4):534–547.

104. Wagner U, Gais S, Born J. Emotional memory formation is enhanced across sleep intervals with high amounts of rapid eye movement sleep. *Learn Mem*. 2001;8(2):112–119.

105. Plihal W, Born J. Effects of early and late nocturnal sleep on priming and spatial memory. *Psychophysiology*. 1999;36(5):571–582.

106. Verleger R, Schuknecht SV, Jaskowski P, Wagner U. Changes in processing of masked stimuli across early- and late-night sleep: a study on behavior and brain potentials. *Brain Cogn*. 2008;68(2):180–192.

107. Mander BA, Rao V, Lu B. Prefrontal atrophy, disrupted NREM slow waves and impaired hippocampal-dependent memory in aging. *Nat Neurosci*. 2013;16(3):357–364.

108. Scullin MK. Sleep, memory, and aging: the link between slow-wave sleep and episodic memory changes from younger to older adults. *Psychol Aging*. 2013;28(1):105–114.

109. Marshall L, Helgadóttir H, Mölle M, Born J. Boosting slow oscillations during sleep potentiates memory. *Nature*. 2006;444(7119):610–613.

110. O'Neill J, Pleydell-Bouverie B, Dupret D, Csicsvari J. Play it again: reactivation of waking experience and memory. *Trends Neurosci*. 2010;33(5):220–229.

111. Peigneux P, Laureys S, Fuchs S. Are spatial memories strengthened in the human hippocampus during slow wave sleep?. *Neuron*. 2004;44(3):535–545.

112. Wilson MA, McNaughton BL. Reactivation of hippocampal ensemble memories during sleep. *Science*. 1994;265(5172):676–679.

113. Buzsáki G. Hippocampal sharp waves: their origin and significance. *Brain Res*. 1986;398(2):242–252.

114. Nádasdy Z, Hirase H, Czurkó A, Csicsvari J, Buzsáki G. Replay and time compression of recurring spike sequences in the hippocampus. *J Neurosci*. 1999;19(21):9497–9507.

115. Sirota A, Csicsvari J, Buhl D, Buzsáki G. Communication between neocortex and hippocampus during sleep in rodents. *Proc Natl Acad Sci USA*. 2003;100(4):2065–2069.

116. Mölle M, Born J. Slow oscillations orchestrating fast oscillations memory consolidation. *Prog Brain Res*. 2011;193:93–110.

117. Schabus M, Gruber G, Parapatics S. Sleep spindles and their significance for declarative memory consolidation. *Sleep*. 2004;27(8):1479–1485.

118. Eschenko O, Mölle M, Born J, Sara SJ. Elevated sleep spindle density after learning or after retrieval in rats. *J Neurosci*. 2006;26(50):12914–12920.

119. Piosczyk H, Holz J, Feige B. The effect of sleep-specific brain activity versus reduced stimulus interference on declarative memory consolidation. *J Sleep Res*. 2013;22(4):406–413.

120. Wagner U, Fischer S, Born J. Changes in emotional responses to aversive pictures across periods rich in slow-wave sleep versus rapid eye movement sleep. *Psychosom Med*. 2002;64:627–634.

121. Groch S, Wilhelm I, Diekelmann S, Born J. The role of REM sleep in the processing of emotional memories: Evidence from behavior and event-related potentials. *Neurobiol Learn Mem*. 2013;99:1–9.

122. Albouy G, King BR, Maquet P, Doyon J. Hippocampus and striatum: Dynamics and interaction during acquisition and sleep-related motor sequence memory consolidation. *Hippocampus*. 2013;23(11):985–1004.

123. Peigneux P, Laureys S, Delbeuck X, Maquet P. Sleeping brain, learning brain. The role of sleep for memory systems. *Neuroreport*. 2001;12(18):A111–A124.

124. Smith C. Sleep states and memory processes in humans: procedural versus declarative memory systems. *Sleep Med Rev*. 2001;5(6):491–506.

125. Rasch B, Gais S, Born J. Impaired off-line consolidation of motor memories after combined blockade of cholinergic receptors during REM sleep-rich sleep. *Neuropsychopharmacology*. 2009;34(7):1843–1853.

126. Vertes RP, Eastman KE. The case against memory consolidation in REM sleep. *Behav Brain Sci*. 2000;23(6):867–876.

127. Siegel JM. The REM sleep: memory consolidation hypothesis. *Science*. 2001;294(5544):1058–1063.

128. Rasch B, Pommer J, Diekelmann S, Born J. Pharmacological REM sleep suppression paradoxically improves rather than impairs skill memory. *Nat Neurosci*. 2009;12(4):396–397.

129. Huber R, Felice Ghilardi M, Massimini M, Tononi G. Local sleep and learning. *Nature*. 2004;430(6995):78–81.

130. Aeschbach D, Cutler AJ, Ronda JM. A role for non-rapid-eye-movement sleep homeostasis in perceptual learning. *J Neurosci*. 2008;28(11):2766–2772.

131. Landsness EC, Crupi D, Hulse BK. Sleep-dependent improvement in visuomotor learning: a causal role for slow waves. *Sleep*. 2009;32(10):1273–1284.

132. Walker MP, Brakefield T, Morgan A, Hobson JA, Stickgold R. Practice with sleep makes perfect: sleep-dependent motor skill learning. *Neuron*. 2002;35(1):205–211.

133. Nishida M, Walker MP. Daytime naps, motor memory consolidation and regionally specific sleep spindles. *PLoS One*. 2007;2(4).

134. Fogel SM, Smith CT, Cote KA. Dissociable learning-dependent changes in REM and non-REM sleep in declarative and procedural memory systems. *Behav Brain Res*. 2007;180(1):48–61.

135. Schönauer M, Geisler T, Gais S. Strengthening procedural memories by reactivation in sleep. *J Cogn Neurosci*. 2013;26(1):143–153.

136. Antony JW, Gobel EW, O'Hare JK, Reber PJ, Paller KA. Cued memory reactivation during sleep influences skill learning. *Nat Neurosci*. 2012;15(8):1114–1116.

137. Nishida M, Pearsall J, Buckner RL, Walker MP. REM sleep, prefrontal theta, and the consolidation of human emotional memory. *Cereb Cortex*. 2009;19(5):1158–1166.

138. McGaugh JL. The amygdala modulates the consolidation of memories of emotionally arousing experiences. *Annu Rev Neurosci*. 2004;27:1–28.

139. Walker MP, van der Helm E. Overnight therapy? The role of sleep in emotional brain processing. *Psychol Bull*. 2009;135(5):731–748.

140. Kaestner EJ, Wixted JT, Mednick SC. Pharmacologically increasing sleep spindles enhances recognition for negative and high-arousal memories. *J Cogn Neurosci*. 2013;25(10):1597–1610.

141. Rolls A, Makam M, Kroeger D, Colas D, de Lecea L, Heller HC. Sleep to forget: interference of fear memories during sleep. *Mol Psychiatry*. 2013;18(11):1166–1170.

142. Hauner KK, Howard JD, Zelano C, Gottfried JA. Stimulus-specific enhancement of fear extinction during slow-wave sleep. *Nat Neurosci*. 2013;16(11):1553–1555.

143. Stickgold R, Walker M. To sleep, perchance to gain creative insight?. *Trends Cogn Sci*. 2004;8(5):191–192.

144. Wagner U, Gais S, Haider H, Verleger R, Born J. Sleep inspires insight. *Nature*. 2004;427:352–355.

145. Payne JD, Schacter DL, Propper RE. The role of sleep in false memory formation. *Neurobiol Learn Mem*. 2009;92(3):327–334.

146. Diekelmann S, Born J, Wagner U. Sleep enhances false memories depending on general memory performance. *Behav Brain Res.* 2010;208(2):425–429.

147. Walker MP, Liston C, Hobson JA, Stickgold R. Cognitive flexibility across the sleep-wake cycle: REM-sleep enhancement of anagram problem solving. *Cogn Brain Res.* 2002;14(3):317–324.

148. Lewin I, Gombosh D. Increase in REM time as a function of the need for divergent thinking. In: Levin PKW, ed. *Sleep: Physiology, Biochemistry, Psychology, Pharmacology, Clinical Implications.* Basel: Karger; 1973:399–403.

149. Besedovsky L, Lange T, Born J. Sleep and immune function. *Pflugers Arch Eur J Physiol.* 2012;463(1):121–137.

150. Haack M, Sanchez E, Mullington JM. Elevated inflammatory markers in response to prolonged sleep restriction are associated with increased pain experience in healthy volunteers. *Sleep.* 2007;30(9):1145–1152. www.pubmed-central.nih.gov/articlerender.fcgi?artid=1978405&tool=pmcentrez&rendertype=abstract.

151. Meier-Ewert HK, Ridker PM, Rifai N. Effect of sleep loss on C-reactive protein, an inflammatory marker of cardiovascular risk. *J Am Coll Cardiol.* 2004;43(4):678–683.

152. Mullington JM, Simpson NS, Meier-Ewert HK, Haack M. Sleep loss and inflammation. *Best Pract Res Clin Endocrinol Metab.* 2010;24(5):775–784.

153. Mullington JM, Haack M, Toth M, Serrador JM, Meier-Ewert HK. Cardiovascular, inflammatory, and metabolic consequences of sleep deprivation. *Prog Cardiovasc Dis.* 2009;51(4):294–302.

154. Cohen S, Doyle WJ, Alper CM, Janicki-Deverts D, Turner RB. Sleep habits and susceptibility to the common cold. *Arch Intern Med.* 2009;169(1):62–67.

155. Mohren DCL, Jansen NWH, Kant IJ, Galama J, van den Brandt PA, Swaen GMH. Prevalence of common infections among employees in different work schedules. *J Occup Environ Med.* 2002;44(11):1003–1011.

156. Spiegel K, Sheridan JF, Van Cauter E. Effect of sleep deprivation on response to immunization. *JAMA.* 2002;288(12):1471–1472.

157. Kakizaki M, Kuriyama S, Sone T. Sleep duration and the risk of breast cancer: the Ohsaki Cohort Study. *BR J Cancer.* 2008;99(9):1502–1505.

158. Weiderpas E, Sandin S, Inoue M. Risk factors for epithelial ovarian cancer in Japan—results from the Japan Public Health Center-based Prospective Study cohort. *Int J Oncol.* 2012;40(1):21–30.

159. Thompson CL, Larkin EK, Patel S, Berger NA, Redline S, Li L. Short duration of sleep increases risk of colorectal adenoma. *Cancer.* 2011;117(4):841–847.

160. Jiao L, Duan Z, Sangi-Haghpeykar H, Hale L, White DL, El-Serag HB. Sleep duration and incidence of colorectal cancer in postmenopausal women. *Br J Cancer.* 2013;108(1):213–221.

Sleep and Cognitive Impairment

B.R. Peters*, S.J. Sha**, K. Yaffe†

*Pulmonary and Sleep Associates of Marin, Novato, CA, USA
**Department of Neurology and Neurological Sciences, Palo Alto, CA, USA
†University of California, San Francisco, CA, USA

SLEEP CHANGES IN NORMAL AGING

From early brain development to senescence, sleep is a dynamic and evolving state that changes dramatically as we age. In order to better understand the relationship of cognitive impairment and sleep, it is important to first review the changes that occur during the normal aging process. In reviewing the basic sleep stages that constitute a night of sleep, the vast majority of age-dependent changes in sleep architecture occur well before the age of 60.[1] For example, in late adolescence and early adulthood, slow-wave sleep declines, while in middle age and beyond the proportion of stage 1 nonrapid eye movement (NREM) sleep may

Sleep and Neurologic Disease. http://dx.doi.org/10.1016/B978-0-12-804074-4.00004-2

increase slightly, while rapid eye movement (REM) sleep decreases. Among the elderly, there are few significant changes in sleep architecture, with only slight increases in stage 1 NREM sleep and more wakefulness occurring after sleep onset.

Insomnia and excessive daytime sleepiness are common in elderly individuals. The Established Populations for Epidemiologic Studies of the Elderly (EPESE) evaluated 9000 participants and found that 29% of those older than age 65 had difficulty maintaining sleep.[2] It is speculated that a reduced homeostatic sleep drive may contribute to more frequent awakenings, especially from stage 2 NREM sleep, and lead to more fragmented sleep.[3] In addition, obstructive sleep apnea (OSA) may lead to both nocturnal awakenings and excessive daytime sleepiness, and increases with age.[4]

Aging may also impact the circadian rhythm, a complementary alerting signal that directly affects the timing of optimal sleep and wakefulness. There is evidence that aging contributes to decreased amplitude of the sleep–wake rhythm and core body temperature.[5] The endogenous system may become less responsive to phase shifting due to impairment of the suprachiasmatic nucleus (SCN), the body's circadian pacemaker. In addition, visual impairment, such as that caused by cataracts or macular degeneration, may affect the retinal ganglia cells that project to the SCN, further exacerbating the problem.[6] In one study, elderly subjects with impaired vision were 30–60% more likely to have disturbed night time sleep when compared to age-matched controls.[7] Without light input to reset the innate circadian rhythm, misalignment may occur and contribute to both insomnia and daytime sleepiness.

When considering poor sleep among the elderly, there is an extensive list of potential causes, including primary sleep disorders, comorbid medical and psychiatric illnesses, medications, lifestyle choices, and environmental considerations (Table 4.1).

Regardless of the cause, poor sleep can have significant impacts on quality of life and overall health. Both depression and anxiety are associated with poor sleep, and poor sleep is a risk factor for depression among the elderly.[8–10] In particular, there seems to be a negative association between poor sleep and cognitive function. Women may be particularly susceptible to impaired cognition in the context of poor sleep. In the Study of Osteoporotic Fractures, older women scored lower on the Mini Mental State Examination (MMSE) and took more time to complete Trail Making Tests as sleep efficiency decreased and sleep onset latency increased.[11] Cognitive decline over 15 years was more likely to occur when sleep efficiency was less than 70% in the same population.[12] Another cohort, the participants in the Nurses' Health Study, demonstrated worse cognitive performance, including memory and attention, with sleep durations of less than 5 h.[13] Moreover, sleep-disordered breathing appears to contribute to neurocognitive decline, especially among women.[14,15]

SLEEP IN DEMENTIA

Dementia is defined as a disorder that includes cognitive or behavioral impairment that represents a decline from prior functioning and causes impairment in occupational or social functioning. The impairment must not be better explained by delirium, substance or toxin exposure, or another medical or psychiatric condition.[16] The global prevalence of dementia is estimated to be more than 24 million, mostly constituting Alzheimer's disease (AD), and is expected to double every 20 years through 2040.[17] The criteria used to define these disorders

TABLE 4.1   Causes of Poor Sleep Among the Elderly

Sleep disorders
- Insomnia
- Circadian rhythm disorders
- Obstructive sleep apnea
- Restless legs syndrome
- Periodic limb movements of sleep
- Parasomnias
- Narcolepsy

Medical conditions
- Medication side effects
- Chronic pain (e.g., arthritis, headache, back pain)
- Cardiovascular disease (e.g., CAD, congestive heart failure)
- Stroke
- Gastroesophageal reflux disease (GERD)
- Nocturia
- Mobility limitations
- Visual impairment
- Chronic obstructive pulmonary disease (COPD)

Psychiatric conditions
- Depression
- Anxiety
- Posttraumatic stress disorder (PTSD)
- Bipolar disorder
- Seasonal affective disorder (SAD)
- Psychological/psychosocial stressors (job loss, divorce, bereavement)

Lifestyle factors
- Poor sleep hygiene
- Lack of regular exercise
- Substance use (alcohol, smoking)
- Sleep environment (institutionalization)

continue to evolve as the scientific understanding of the various etiologies advances. This section will review the sleep characteristics associated with eight of the major subtypes of dementia, some of which are associated with other medical disorders (Table 4.2).

Alzheimer's Disease

AD is a progressive neurodegenerative disorder characterized by decline in episodic memory, language, visuospatial function, and executive function.[18] The pathology of the disease is

TABLE 4.2   Dementia Subtypes

- Alzheimer's disease
- Frontotemporal dementia
- Parkinson's disease dementia
- Lewy body dementia
- Progressive supranuclear palsy
- Vascular dementia
- Huntington's disease
- Creutzfeldt-Jakob disease

characterized by the deposition of neurofibrillary tangles and neuritic plaques, the accumulation of which leads to the dysfunction and loss of healthy neurons. Ultimately, widespread structural degeneration can include the entorhinal cortex, hippocampus, amygdala, nucleus basalis of Meynert, suprachiasmatic nucleus, intralaminar nuclei of the thalamus, locus coeruleus, raphe nuclei, central autonomic regulators, and finally cortex.[19] These abnormalities can have profound effects on sleep.

In mild to moderate AD, sleep disturbances occur in 25% of patients; in advanced stages, more than 50% are impacted.[20] Symptoms include excessive daytime sleepiness (often manifesting as naps), insomnia, frequent nocturnal awakenings and early morning awakenings, and may gradually lead to greater functional impairment.[21] There are specific sleep state changes noted on polysomnography (PSG) and characteristic electroencephalographic (EEG) abnormalities in wakefulness with AD. The proportion of stage 1 NREM increases with more frequent awakenings, with a corresponding decrease in slow-wave sleep.[22] The K-complexes of stage 2 sleep become less frequent and poorly formed, of lower amplitude, shorter duration, and lower frequency. The density of delta waves in stage NREM 3 sleep is also reduced.[23] Episodes of REM sleep are of shorter duration, perhaps due to degeneration of the cholinergic nucleus basalis of Meynert. Normally this structure inhibits the nucleus reticularis of the thalamus, which generates NREM sleep.[24,25] With the loss of this inhibition, NREM sleep constitutes more of the total sleep time. In wakefulness, slowing of the dominant occipital rhythm occurs with increased theta and delta intrusion.[26,27] REM sleep behavior disorder (RBD) rarely occurs in Alzheimer's disease and if present may suggest an alternate or comorbid diagnosis, such as dementia with Lewy bodies (DLB).

Structural changes of the hippocampus occur with disrupted sleep. As measured by volumetric MRI, reduced sleep efficiency (time asleep/time in bed) leads to reduced hippocampal volumes over time.[28] Reduced sleep efficiency may occur in the setting of insomnia and OSA as well. Growing evidence suggests treatment of the latter condition with continuous positive airway pressure (CPAP) may help reverse these morphological changes.[29]

There appears to be a bidirectional relationship between amyloid-β(Aβ) plaque formation and compromised sleep. Levels of Aβ are regulated by neuronal synaptic activity and the sleep–wake cycle. Sleep deprivation increases Aβ deposition, while sleep recovery decreases Aβ. This process can be potentiated by orexin, a wake-promoting neurotransmitter.[30] Amyloid deposition in cognitively normal individuals, a preclinical stage of Alzheimer's disease, is associated with poor sleep quality.[31] Following plaque formation, the sleep–wake cycle degenerates in mice models and the normal diurnal fluctuation is lost. Artificial dissolution of these plaques normalizes the sleep–wake cycles.[32] The circadian rhythm disturbances often occur early in the course of the disease and impairment of circadian clocks at the cellular level may have a role in the neurodegenerative process by contributing to oxidative stress (Fig. 4.1).[33]

These neuropathological observations have important clinical implications. Chronic sleep deprivation, due to either restricted quantity with failure to meet sleep needs, or undermined quality due to other sleep disorders like sleep apnea, are risk factors that may contribute to these pathological changes. Moreover, loss of normal sleep–wake cycles may lead to the type of circadian disruption and sleep irregularities frequently seen in dementia.

Beyond the electrical, imaging, and pathological changes, important clinical consequences occur as AD progresses. The most disruptive phenomenon may be the circadian

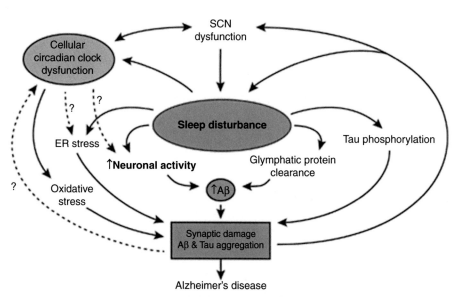

FIGURE 4.1 **Proposed mechanisms linking sleep deprivation, circadian dysfunction, and Alzheimer's disease (AD).** *Dotted arrows* represent hypothetical links. *Source: Adapted from Musiek, ES, et al. Sleep, circadian rhythms, and the pathogenesis of Alzheimer disease.* Exp Mol Med. 2015;47:e148.

occurrence of sundowning, a delirium-like state that includes nocturnal agitation and wandering behaviors.[34] The nucleus basalis of Meynert also modulates the SCN and its degeneration may contribute to altered sleep–wake rhythms and melatonin production[34,35] One manifestation of these circadian changes is the development of an irregular sleep–wake rhythm disorder (ISWRD). This condition is characterized by chronic patterns of irregular sleep and wake episodes throughout the 24-h period, with insomnia at night and excessive daytime sleepiness (with napping) during the daytime.[36] Instead of a major period of sleep lasting 6 or more hours and occurring overnight, multiple irregular sleep bouts occur throughout the day and night.[37,38] The longest period of sleep typically lasts less than 4 h. Despite the disorganization and variability in the timing of sleep, the total amount of sleep obtained may be normal for age. ISWRD may be exacerbated by institutionalization, with a lack of exposure to synchronizing factors such as natural light and regular social schedules. Beyond the safety concerns associated with wandering and falls, the condition is highly disruptive to the sleep of caregivers and this may lead to higher rates of placement in assisted living.[39]

Despite the high incidence of insomnia, the use of hypnotics as first-line treatment among this population is discouraged. The use of sedative-hypnotics may exacerbate the development of delirium, worsen cognition, and may be associated with a higher incidence of falls and hip fractures. Over-the-counter agents containing diphenhydramine are associated with an exceptionally high risk of delirium in this patient population due to anticholinergic side effects.[40] Though there may be modest improvements in total sleep time with the use of zolpidem or zaleplon (averaging about 25 min), research indicates that the associated risks for altered cognition (odds ratio 4.78) and daytime fatigue or sleepiness (odds ratio 3.82) are substantial.[41] One retrospective case-control study found the associated risk of a hip fracture doubled with the use of zolpidem in the elderly population.[42] It should be noted that insomnia itself, especially when occurring in the context of nocturia and daytime sleepiness, seems to be independently associated with an increased likelihood of falling.[43]

Alternatively, efforts to reinforce the circadian rhythm through preservation of *zeitgebers* (time-givers), such as properly timed light exposure and efforts to minimize napping, may be helpful. It is highly important to educate caregivers and to engage them actively in the intervention. Although 15–30 min of natural sunlight exposure upon awakening is preferred, the use of a light box may be considered. Research has shown that sleep duration improves by about 15 min with both morning and all-day exposures to 2500 lux of artificial light.[44] By combining light exposure with efforts to limit time in bed during the day, increased daytime physical activity, and reduced noise and light at night, dramatic reductions in daytime sleep have been noted in nursing home patients.[45] There was no significant change in hours of night time sleep or number of awakenings, however. The exposure to light may have further impacts on mood and cognition.[46] There is little evidence that the use of oral melatonin improves on these effects.[47]

Frontotemporal Dementia

Frontotemporal dementia (FTD) accounts for 5–15% of all dementias.[17] FTD is characterized by early personality changes, executive dysfunction, and language difficulties.[48,49] Behavioral changes may manifest as disinhibition, loss of insight, hyperorality, lack of social awareness, and apathy. It is estimated that up to 75% of patients with FTD have sleep disturbances. The sleep–wake rhythm is disturbed, but with less marked changes compared to AD. These rhythms may be highly fragmented with phase advancement of the overnight sleep epoch, resulting in an earlier onset and offset of sleep relative to conventional sleep timing.[50] EEG may demonstrate generalized slowing with increased theta and delta activity in the frontotemporal regions, corresponding with the areas of degeneration.[51,52]

Parkinson's Disease

Parkinson's disease (PD) is a neurodegenerative disease with distinctive motor features of rigidity, bradykinesia, and tremor. Cognitively, patients may have mild cognitive impairment affecting executive and visuospatial function. About 80% of patients with PD will go on to develop dementia within 8 years.[53] Pathologically, Lewy bodies, eosinophilic spherical cytoplasmic inclusions with dense cores and peripheral halo, are present and may be indistinguishable from those of Lewy body disease. Dopaminergic medications used to treat PD may exacerbate confusion and hallucinations, which may actually represent the intrusion of REM elements into wakefulness.[54,55] In one study that reviewed the EEG findings of patients with PD, electrocortical slowing was noted in up to one third of patients, predominately in the temporal, occipital and frontal regions.[56] In addition, patients with RBD have been observed to have EEG slowing of the dominant occipital frequency, along with higher theta power during wakefulness in all regions of the brain.[57]

There is substantial evidence that RBD is an early manifestation of neurodegenerative disorders of alpha-synucleinopathy deposition [PD, DLB, and multiple system atrophy (MSA)], sometimes preceding these diagnoses by decades.[58–61] Clinically, RBD can interrupt sleep for the patient and caregiver, leading to impaired cognition and fatigue, and may result in injury to the patient or bed partner. For a more thorough review of this topic, the reader is directed to the movement disorders chapter of this text.

Lewy Body Dementia

DLB is second only to AD in terms of prevalence, affecting an estimated 1–5% of the population.[62] Compared to patients with AD, patients with DLB more frequently complain of sleep disturbances, abnormal nocturnal movements, and excessive sleepiness, with approximately 50% reporting elevated Epworth sleepiness scales.[63] PSG may demonstrate reduced sleep efficiency, respiratory disturbances consistent with obstructive sleep apnea in 88%, and periodic limb movements of sleep (PLMs) associated with arousals in 74%.[64] The latter finding may represent an increased incidence of restless legs syndrome in this population, which by itself can contribute to sleep fragmentation, nocturnal awakenings, and insomnia.[65] One study found that 75% of patients who underwent PSG had arousals unrelated to movement or breathing abnormalities, and that these arousals were not entirely accounted for by a primary sleep disorder.[66] This same study found that 96% of patients reported RBD, and this was confirmed by PSG in over 80%. Another study found 12 of 15 patients with RBD and a neurodegenerative disease had limbic or neocortical Lewy body disease at autopsy; consequently, RBD is a supportive feature in the diagnostic criteria.[67–69]

The awake EEG may demonstrate generalized slowing with loss of alpha and increased theta activity, correlating with the severity of the disease.[70] Transient sharp waves may be noted in the temporal regions.[71] Frontal intermittent rhythmic delta activity (FIRDA) has also been reported.[72] The degree of fluctuation in cognition seems to be reflected by variability in the mean EEG power.[73] Reducing fluctuations in arousal with cholinesterase inhibitors not only improves cognition but may also treat the visual hallucinations that sometimes occur.[74,75] The reduction in the psychiatric, sleep, and cognitive aspects of the disease can greatly improve the quality of life for these patients.

Progressive Supranuclear Palsy

Progressive supranuclear palsy (PSP) is a dementia associated with degeneration of the frontosubcortical neural networks, often resulting in apathy, decreased executive function, or impulsivity.[76] It is characterized by progressive axial rigidity, postural instability, and supranuclear gaze palsy, corresponding with impaired downward gaze. Pathologically, tau tangles are present along with gliosis and neuronal loss in the brainstem and basal ganglia.[77]

Patients with PSP complain of excessive daytime sleepiness and are noted to have a dramatic reduction in REM sleep.[78] Cognitive decline is reflected by EEG slowing that predominates in the frontal regions during wakefulness. There is also evidence that CSF hypocretin 1 (orexin A) levels may be reduced in some patients, a finding common in narcolepsy, correlating with the duration of disease.[79] REM without atonia has been reported, indicating a possible association with RBD.[80] One study estimated that 28% of patients with PSP have loss of atonia and 13% have clinical RBD, confirmed by PSG.[81]

Vascular Dementia

Vascular dementia is characterized by progressive cognitive impairment secondary to recurrent vascular injury.[82] Sleep in vascular dementia is characterized by more disruption of the sleep–wake cycle and decreased sleep quality.[83] The EEG demonstrates slowing of the dominant occipital rhythm, increased theta activity, and increased delta activity.[84,85] In addition, there is more alpha activity noted anteriorly, which may correlate to diminished vigilance and fluctuating cognition.[86]

Huntington's Disease

Huntington's disease (HD) is an autosomal dominant neurodegenerative disorder characterized by caudate atrophy. The pathophysiology involves expansion of a CAG trinucleotide repeat in the *IT15* gene located on the short arm of chromosome four.[87] Symptoms include choreiform movements, cognitive impairment with early executive dysfunction, and psychiatric manifestations such as depression.[17] Dementia can occur as the disease progresses.

Degeneration affects the SCN, motor cortex, and striatum and the loss of SCN function leads to disruption of the circadian rhythm.[88] As a result, sleep latency can increase, sleep efficiency and slow-wave sleep can decrease, and nocturnal awakenings occur more frequently.[89,90] When HD progresses in severity, PLMs increase, although this may not necessarily translate to increased daytime sleepiness. Moreover, gradual slowing and reduction of the wake EEG amplitude is noted with increased theta and decreased alpha activities.[91]

Creutzfeldt-Jakob Disease

Creutzfeldt-Jakob disease (CJD) is a rapidly progressive dementia resulting from accumulation of misfolded prion proteins.[92] Pathological findings include gliosis and vacuolation. CJD is associated with myoclonus and ataxia. Associated diagnostic findings include EEG sharp wave abnormalities, elevated CSF tau levels, and prion induction via real-time quaking-induced conversion (RT-QuIC). The disease is rapidly progressive, with a mean survival of 4–8 months, although 5–10% may survive 2 years or more.[17]

More than half of patients experience sleep disturbances, mostly insomnia, and up to 15% present initially with sleep complaints.[93] Aggressive behaviors and dream-reality confusion, in which it is hard to distinguish what occurs as part of dream mentation from real life events, may occur in sleep.[94] Central or obstructive sleep apnea events are significantly elevated, although the exact prevalence of these conditions is not well studied.[95]

PSG studies in CJD demonstrate disorganized sleep patterns with abrupt transitions between sleep stages, few sleep spindles and K-complexes, and a lower amount of REM sleep.[96] During wakefulness, the EEG demonstrates hallmark periodic sharp wave complexes—biphasic, triphasic, or mixed spikes or slow waves—with a generalized slow background.[97,98]

SLEEP DISORDERS AS A RISK FACTOR FOR COGNITIVE IMPAIRMENT

There is a growing body of literature that suggests that sleep deprivation and sleep disorders can independently contribute to the development of cognitive impairment and dementia. While the role of sleep in humans in not completely understood, it is clear that it serves at least several important restorative functions. Researchers in 2012 described the gliovascular clearance system, or "glymphatics," a network of CSF-filled paravascular spaces that flush interneuronal debris from the brain parenchyma.[99] Sleep is, at least in part, a process by which these metabolic waste products—including adenosine (a signal for sleepiness)—are removed from the brain. Slow-wave sleep seems to enhance the activity of the glymphatic system by approximately 60%.[100]

As mentioned earlier, there may be a bidirectional relationship between Aβ plaque deposition and sleep changes. Mouse models demonstrate that plaque formation may lead to more fragmented sleep with a loss of the natural diurnal variation of Aβ, while dissolution of these plaques normalizes sleep–wake cycles.[101] It is possible that disorders that disturb sleep quantity and quality may exacerbate neurodegeneration.

OSA has long been implicated as a potential contributing factor in the development of dementia.[102] Sleep-disordered breathing occurs more frequently in Alzheimer's disease than in age-matched controls, with prevalence estimates ranging from 40% to 50%, and its severity is correlated with the degree of cognitive impairment.[103] Though a causal relationship is not yet established, it is hypothesized that both sleep fragmentation and intermittent hypoxia may contribute to neurodegenerative changes. Chronic and recurrent hypoxia may affect highly oxygen-dependent regions of the brain, including the hippocampus and neocortex, while secondary hypertension may exacerbate cognitive decline.

Elevated blood pressure during sleep may also play a direct role in Aβ deposition. The normal reduction in blood pressure that occurs during sleep is referred to as "dipping." It has been demonstrated that "nondippers," or those who do not exhibit this normal blood pressure reduction, have higher levels of Aβ deposition in the posterior cingulate region when compared to normal "dippers."[104] It has also been demonstrated that patients with both OSA and insomnia have a greater risk of nondipping during sleep. In addition, hypoxia can lead to endothelial dysfunction, inflammation, and oxidative stress in vulnerable cell populations.[105] This cellular stress may become sufficient to impair cell–cell interactions, synaptic function, and neural circuitry. OSA also results in extensive sleep fragmentation, undermining sleep quality, which may further potentiate Aβ deposition. Individuals with Down syndrome serve as an interesting example as they have an increased incidence of both OSA, due to midface hypoplasia, and early-onset AD.[106] Based on the available evidence, it is reasonable to assume that untreated OSA may lead to more rapid progression of dementia, and treatment with CPAP may delay the onset of the disease.[107] Normalization of sleep should be a priority in managing those with dementia to optimize daily function and to protect remaining cognitive reserve.

Insomnia is very common in patients with dementia, and many of these patients take medication to help them fall asleep. Although there is some data to suggest that those taking over-the-counter and prescription sleeping pills have a greater risk of developing dementia, a direct link has not been established and there are likely several confounding variables at play. Anticholinergics[108] and benzodiazepines[109] have been demonstrated to worsen cognitive impairment. Limited research suggests that the most popular nonbenzodiazepine sleep aid, zolpidem, may increase the risk of dementia, although again this has not been firmly established and many confounding variables may be contributing.[110] Interestingly, nocturnal awakenings and resulting insomnia commonly occur in untreated OSA. This might explain the increased association of dementia among the elderly population with comorbid hypertension, diabetes, and stroke—all conditions that are provoked by OSA. This might also explain the association between dementia and proton pump inhibitors, as OSA frequently incites symptoms of reflux due to the increase of negative intrathoracic pressure during apneic events.[111] These medications may be used to treat symptoms that are actually the result of untreated OSA, chronically damaging vulnerable brain cell populations.

TREATMENT

Insomnia

Difficulty initiating or maintaining sleep may occur as part of insomnia and contribute to fragmented sleep and agitated behaviors. Insomnia may be exacerbated by irregular sleep habits, excessive daytime sleeping, pain, medication effects, poor sleep environment, and comorbid depression.[112] Proper sleep hygiene includes a regular bedtime routine, avoidance of caffeine and alcohol in the hours preceding sleep, and increased activity during the day. As much as possible, benzodiazepine and anticholinergic medications should be avoided as they may contribute to confusion and cognitive decline.

Circadian Disorders

Degeneration of the circadian rhythm via disruption of the SCN and pineal gland, or inadequate or improperly timed light exposure may contribute to sleep irregularities, resulting in insomnia and daytime sleepiness.[113,114] Although some of these changes are irreparable due to the disease process itself, a regular sleep schedule and properly timed light exposure may still be helpful.[115-117] In addition, low doses of melatonin may enhance the circadian signal and reduce sundowning phenomena without significant adverse effects, although the benefits may be modest compared to light exposure.[118,119]

Obstructive Sleep Apnea

As discussed, there is growing evidence that OSA may worsen cognitive decline and contribute to dementia.[120] Untreated, the condition is associated with snoring, excessive daytime sleepiness, nocturia, nocturnal awakenings, bruxism, and mood complaints. A small percentage of dementia patients with comorbid OSA may experience cognitive benefit with CPAP treatment.[121] Clinical experience demonstrates that CPAP is surprisingly well tolerated by those with dementia and caregivers may likewise benefit from improved sleep.

Restless Legs Syndrome and Periodic Limb Movements of Sleep

Restless legs syndrome (RLS), or Willis-Ekbom disease, is characterized by an uncomfortable feeling, often in the legs, associated with an urge to move, relieved by movement, that occurs more often in the evening and during periods of prolonged immobilization. It may trigger nocturnal wandering and be associated with periodic limb movements during sleep that lead to sleep fragmentation. The exact prevalence of these conditions among patients with dementia is difficult to determine and remains unknown. RLS does increase in incidence among older individuals, with a peak prevalence in older women (16.3% among those age 60–69) and late middle-aged men (7.8% among those age 50–59).[122] Iron replacement when serum ferritin levels fall below 50 may be helpful. Dopamine agonists may also be an effective therapeutic option, though these medications may exacerbate hallucinations and psychosis among susceptible individuals. Therefore, they should be used with caution or not at all among patients with DLB. For a complete review of RLS including treatment options, the reader is directed to the movement disorders chapter in this text.

REM Behavior Disorder and Parasomnias

Dream-enactment behaviors such as kicking, hitting, and jumping out of bed may occur as part of RBD. These can be potentially harmful to both the patient and his or her bed partner. Therefore, safety precautions and protective measures are recommended with the removal of dangerous objects, clearing away bedside furniture, and lowering the mattress to the floor. Clonazepam and high-dose melatonin are effective in reducing these behaviors with few side effects.[123,124] RBD and its association with neurodegenerative disease is also covered in detail in the movements disorders section of this text.

CONCLUSIONS

Sleep disorders are extremely common in patients with neurodegenerative disease. As the population ages, this will become an even greater issue. The interplay between sleep and cognition is complex and it is only recently that efforts have been made to systematically evaluate these connections; thus, much research is still needed. Sleep is an often-overlooked component of patients' health, and the evaluation of any patient with cognitive symptoms should include a review of their sleep history. Treatment can be difficult, but if successful may slow progression of disease and lead to greater quality of life for both patients and caregivers.

References

1. Ohayon MM, Carskadon MA, Guilleminault C, et al. Meta-analysis of quantitative sleep parameters from childhood to old age in healthy individuals: developing normative sleep values across the human lifespan. *Sleep.* 2004;27:1255–1273.
2. Foley DJ, Monjan AA, Brown SL, et al. Sleep complaints among elderly persons: an epidemiologic study of three communities. *Sleep.* 1995;18:425–432.
3. Dijk DJ, Duffy JF, Czeisler CA. Age-related increase in awakenings impaired consolidation of non-REM sleep at all circadian phases. *Sleep.* 2001;24:565–577.

4. Punjabi NM. The epidemiology of adult obstructive sleep apnea. *Proc Am Thorac Soc.* 2008 Feb 15;5(2):136–143.

5. Richardson GS, Carskadon MA, Orav EJ. Circadian variation of sleep tendency in elderly and young adult subjects. *Sleep.* 1982;5(suppl. 2):S82–S94.

6. Nygard M, Hill RH, Wikstrom MA, et al. Age-related changes in electrophysiological properties of the mouse suprachiasmatic nucleus in vitro. *Brain Res Bull.* 2005;65:149–154.

7. Asplund R. Sleep, health, and visual impairment in the elderly. *Arch Gerontol Geriatr.* 2000;30:7–15.

8. Paudel ML, Taylor BC, Diem SJ, et al. Associations between depressive symptoms and sleep disturbances in community-dwelling older men. *J Am Geriatr Soc.* 2008;56:1228–1235.

9. Spira A, Stone K, Beaudreau SA, et al. Anxiety symptoms and objectively measured sleep quality in older women. *Am J Geriatric Psychiatry.* 2009;17:136–143.

10. Perlis ML, Smith LJ, Lyness JM, et al. Insomnia as a risk factor for onset of depression in the elderly. *Behav Sleep Med.* 2006;4:104–113.

11. Blackwell T, Yaffe K, Ancoli-Israel S, et al. Poor sleep is associated with impaired cognitive function in older women: the study of osteoporotic fractures. *J Gerontol A Biol Sci Med Sci.* 2006;61:405–410.

12. Yaffe K, Blackwell T, Barnes DE, et al. Preclinical cognitive decline and subsequent sleep disturbance in older women. *Neurology.* 2007;69:237–242.

13. Tworoger SS, Lee S, Schernhammer ES, et al. The association of self-reported sleep duration, difficulty sleeping, and snoring with cognitive function in older women. *Alzheimer Dis Assoc Disord.* 2006;20:41–81.

14. Spira AP, Blackwell T, Stone KL, et al. Sleep-disordered breathing and cognition in older women. *J Am Geriatr Soc.* 2008;56:45–50.

15. Cohen-Zion M, Stepnowsky C, Marler M, et al. Changes in cognitive function associated with sleep disordered breathing in older people. *J Am Geriatr Soc.* 2001;49:1622–1627.

16. Tay L, Lim WS, Chan M, et al. New DSM-V neurocognitive disorders criteria and their impact on diagnostic classifications of mild cognitive impairment and dementia in a memory clinic setting. *Am J Geriatr Psychiatry.* 2015;23(8):768–779.

17. Bradley WG, Daroff RB, Fenichel GM, et al. *The Dementias. Neurology in Clinical Practice. Expert Consult.* 5th ed. Philadelphia, PA: Elsevier Saunders; 2011:1855-1907.

18. Reitz C, Brayne C, Mayeux R. Epidemiology of Alzheimer disease. *Nat Rev Neurol.* 2011 Mar;7(3):137–152.

19. Petit D, Montplaisir J, Boeve BF. Alzheimer's disease and other dementias. In: Kryger MH, Roth T, Dement WC, eds. *Principles and Practices of Sleep Medicine.* St Louis (MO): Elsevier Saunders; 2011:1038.

20. McKhann G, Drachman D, Folstein M, et al. Clinical diagnosis of Alzheimer's disease: report of Health and Human Services Task Force on Alzheimer's Disease. *Neurology.* 1984;34:939–944.

21. Lee JH, Bliwise DL, Ansari FP, et al. Daytime sleepiness and functional impairment in Alzheimer disease. *Am J Geriatr Psychiatry.* 2007;15:620–626.

22. Prinz PN, Vitaliano PP, Vitiello MV, et al. Sleep, EEG and mental function changes in senile dementia of the Alzheimer's type. *Neurobiol Aging.* 1982;3:361–370.

23. Ktonas PY, Golemati S, Xanthopoulos P, et al. Potential dementia biomarkers based on the time-varying microstructure of sleep EEG spindles. *Conf Proc IEEE Eng Med Biol Soc.* 2007;2007:2464–2467.

24. Montplaisir J, Petit D, Lorrain D, et al. Sleep in Alzheimer's disease: further considerations on the role of brainstem and forebrain cholinergic populations in sleep-wake mechanisms. *Sleep.* 1995;18:145–148.

25. Petit D, Montplaisir J, Lorrain D, et al. Spectral analysis of the rapid eye movement sleep electroencephalogram in right and left temporal regions: a biological marker of Alzheimer's disease. *Ann Neurol.* 1992;32:172–176.

26. Crowley K, Sullivan EV, Adalsteinsson E, et al. Differentiating pathologic delta from healthy physiologic delta in patients with Alzheimer disease. *Sleep.* 2005;28:865–870.

27. Hassania F, Petit D, Nielsen T, et al. Quantitative EEG and statistical mapping of wakefulness and REM sleep in the evaluation of mild to moderate Alzheimer's disease. *Eur Neurol.* 1997;37:219–224.

28. Elcombe EL, Lagopoulos J, Duffy SL, Lewis SJ, Norrie L, Hickie IB, Naismith SL. Hippocampal volume in older adults at risk of cognitive decline: the role of sleep, vascular risk, and depression. *J Alzheimers Dis.* 2015;44(4):1279–1290.

29. Ferini-Strambi L, Marelli S, Galbiati A, Castronovo C. Effects of continuous positive airway pressure on cognition and neuroimaging data in sleep apnea. *Int J Psychophysiol.* 2013;89(2):203–212.

30. Roh JH, Jiang H, Finn MB, et al. Potential role of orexin and sleep modulation in the pathogenesis of Alzheimer's disease. *J Exp Med.* 2014;211(13):2487–2496.

31. Ju YE, McLeland JS, Toedebusch CD, et al. Sleep quality and preclinical Alzheimer's disease. *JAMA Neurol.* 2013;70(5):587–593.

32. Roh JH, Huang Y, Bero AW, Kasten T, Stewart FR, Bateman RJ, Holtzman DM. Sleep-wake cycle and diurnal fluctuation of amyloid-β as biomarkers of brain amyloid pathology. *Sci Tranl Med*. 2012 Sep 5;4(150):150ra122.

33. Musiek ES, Xiong DD, Holtzman DM. Sleep, circadian rhythms, and the pathogenesis of Alzheimer disease. *Exp Mol Med*. 2015;47:e148.

34. Klaffke S, Staedt J. Sundowning and circadian rhythm disorders in dementia. *Acta Neurol Belg*. 2006;106:168–175.

35. Wu YH, Swaab DF. The human pineal gland and melatonin in aging and Alzheimer's disease. *J Pineal Res*. 2005;38:145–152.

36. American Academy of Sleep Medicine*International Classification of Sleep Disorders*. 3rd ed. Darien, IL: American Academy of Sleep Medicine; 2014.

37. Wittin W, Kwa I, Eikelenboom P, Mirmiran M, Swaab D. Alterations in the circadian rest-activity rhythm in aging and Alzheimer's disease. *Biol Psychiatry*. 1990;27:563–572.

38. Okawa M, Mishima K, Hishikawa Y, Hozumi S, Hori H, Takahashi K. Circadian rhythm disorders in sleep-waking and body temperature in elderly patients with dementia and their treatment. *Sleep*. 1991;14:478–485.

39. Pollack C, Stokes P. Circadian rest-activity rhythms in demented and non-demented older community residents and their caregivers. *J Am Geriatr Soc*. 1997;45:446–452.

40. Agostini JV, Leo-Summers LS, Inouye SK. Cognitive and other adverse effects of diphenhydramine use in hospitalized patients. *Arch Inter Med*. 2001;161:2091–2097.

41. Glass J, Lanctot KL, Herrmann N, et al. Sedative hypnotics in older people with insomnia: meta-analysis of risks and benefits. *BMJ*. 2005;331(7526):1169.

42. Wang PS, Bohn RL, Glynn RJ, et al. Zolpidem use and hip fractures in older people. *J Am Geriatr Soc*. 2001;49:1685–1690.

43. Avidan AY, Bries BE, James ML, Szafara KL, Wright GT, Chervin RD. Insomnia and hypnotic use, recorded in the minimum data set as predictors of falls and hip fractures in Michigan nursing homes. *J Am Geriatr Soc*. 2005;53:955–962.

44. Sloane PD, Williams CS, Mitchell M, et al. High-intensity environmental light in dementia: effect on sleep and activity. *J Am Geriatr Soc*. 2007;55:1524–1533.

45. Alessi CA, Martin JL, Webber AP, et al. Randomized, controlled trial of a nonpharmacological intervention to improve abnormal sleep/wake patterns in nursing home residents. *J Am Geriatr Soc*. 2005;53:803–810.

46. Riemersma-van der Lek RF, Swaab DF, Twisk J, et al. Effect of bright light and melatonin on cognitive and noncognitive function in elderly residents of group care facilities: a randomized controlled trial. *JAMA*. 2008;299:2642–2655.

47. Singer C, Tractenberg RE, Kaye J, et al. A multicenter, placebo-controlled trial of melatonin for sleep disturbance in Alzheimer's disease. *Sleep*. 2003;26:893–901.

48. McKhann GM, Albert MS, Grossman M, et al. Clinical and pathological diagnosis of frontotemporal dementia: report of the Work Group on Frontotemporal Dementia and Pick's Disease. *Arch Neurol*. 2001;58:1803–1809.

49. Rascovsky K, Hodges JR, Knopman D, et al. Sensitivity of revised diagnostic criteria for the behavioral variant of frontotemporal dementia. *Brain*. 2011;134(Pt 9):2456–2477.

50. Harper DG, Stopa EG, McKee AC, et al. Differential circadian rhythm disturbances in men with Alzheimer disease and frontotemporal degeneration. *Arch Gen Psychiatry*. 2001;58:353–360.

51. Besthorn C, Sattel H, Hentschel F, et al. Quantitative EEG in frontal lobe dementia. *J Neural Transm Suppl*. 1996;47:169–181.

52. Yener GG, Leuchter AF, Jenden D, et al. Quantitative EEG in frontotemporal dementia. *Clin Electroencephalogr*. 1996;27:61–68.

53. Aarsland D, Andersen K, Larsen JP, et al. Prevalence and characteristics of dementia in Parkinson disease: an 8-year prospective study. *Arch Neurol*. 2003;60:387–392.

54. Emre M. Dementia associated with Parkinson's disease. *Lancet*. 2003;2:229–237.

55. Onofrj M, Thomas A, D'Andreamatteo G, et al. Incidence of RBD and hallucinations in patients affected by Parkinson's disease: 8-year follow-up. *Neurol Sci*. 2002;23:S91–S94.

56. Soikkeli R, Partanen J, Soininen H, et al. Slowing of EEG in Parkinson's disease. *Electroencephalogr Clin Neurophysiol*. 1991;79:159–165.

57. Fantini ML, Gagnon J-F, Petit D, et al. Slowing of EEG in idiopathic REM sleep behavior disorder. *Ann Neurol*. 2003;53:774–780.

58. Boeve B, Silber M, Ferman T, et al. Association of REM sleep behavior disorder and neurodegenerative disease may reflect an underlying synucleinopathy. *Mov Disord*. 2001;16:622–630.

59. Boeve B, Silber M, Ferman T, et al. REM sleep behavior disorder in Parkinson's disease, dementia with Lewy bodies, and multiple system atrophy. In: Bedard M, Agid Y, Chouinard S, eds. *Mental and Behavioral Dysfunction in Movement Disorders*. Totowa, NJ: Humana Press; 2003:383–397.

60. Gagnon JF, Bédard MA, Fantini ML, et al. REM sleep behavior disorder and REM sleep without atonia in Parkinson's disease. *Neurology*. 2002;59:585–589.

61. Boeve BF, Silber MH, Ferman TJ, et al. Clinicopathologic correlations in 172 cases of rapid eye movement sleep behavior disorder with or without a coexisting neurologic disorder. *Sleep Med*. 2013;14(8):754–762.

62. McKeith IG, Dickson DW, Lowe J, et al. Dementia with Lewy bodies: diagnosis and management: third report of the DLB Consortium. *Neurology*. 2005;65:1863–1872.

63. Grace JB, Walker MP, McKeith IG. A comparison of sleep profiles in patients with dementia with Lewy bodies and Alzheimer's disease. *Int J Geriatr Psychiatry*. 2000;15:1028–1033.

64. Boeve BF, Ferman TJ, Silber MH, et al. Sleep disturbances in dementia with Lewy bodies involve more than REM sleep behavior disorder. *Neurology*. 2003;60:A79.

65. Boeve BF, Silber MH, Ferman TJ. Current management of sleep disturbances in dementia. *Curr Neurol Neurosci Rep*. 2002;2:169–177.

66. Pao WC, Boeve BF, Ferman TJ, et al. Polysomnographic findings in dementia with Lewy bodies. *Neurologist*. 2013;19(1):1–6.

67. Boeve BF, Silber MH, Ferman TJ, et al. REM sleep behavior disorder and degenerative dementia: an association likely reflecting Lewy body disease. *Neurology*. 1998;51:363–370.

68. Boeve BF, Silber MH, Parisi JE, et al. Synucleinopathy pathology and REM sleep behavior disorder plus dementia or parkinsonism. *Neurology*. 2003;61:40–45.

69. Turner RS. Idiopathic rapid eye movement sleep behavior disorder is a harbinger of dementia with Lewy bodies. *J Geriatr Psychiatr Neurol*. 2002;15:195–199.

70. Briel RC, McKeith IG, Barker WA, et al. EEG findings in dementia with Lewy bodies and Alzheimer's disease. *J Neurol Neurosurg Psychiatr*. 1999;66:401–403.

71. Barber PA, Varma AR, Lloyd JJ, et al. The electroencephalogram in dementia with Lewy bodies. *Acta Neurol Scand*. 2000;101:53–56.

72. Calzetti S, Bortone F, Negrotti A, et al. Frontal intermittent rhythmic delta activity (FIRDA) in patients with dementia with Lewy bodies: a diagnostic tool?. *Neurol Sci*. 2002;23:S65–S66.

73. Walker MP, Ayre GA, Cummings JL, et al. Quantifying fluctuation in dementia with Lewy bodies, Alzheimer's disease, and vascular dementia. *Neurology*. 2000;54:1616–1624.

74. Mori E, Ikeda M, Kosaka K. Donepezil-DLB Study Investigators. Donepezil for dementia with Lewy bodies: a randomized, placebo-controlled trial. *Ann Neurol*. 2012;72:41–52.

75. McKeith I, Del Ser T, Spano P, Emre M, Wesnes K, Anand R, Cicin-Sain A, Ferrara R, Spiegel R. Efficacy of rivastigmine in dementia with Lewy bodies: a randomised, double-blind, placebo-controlled international study. *Lancet*. 2000;356:2031–2036.

76. Litvan I, Agid Y, Calne D, et al. Clinical research criteria for the diagnosis of progressive supranuclear palsy (Steele-Richardson-Olszewski syndrome):report of the NINDS-SPSP international workshop. *Neurology*. 1996;44:1–9.

77. Dickson DW, Rademakers R, Hutton ML. Progressive supranuclear palsy: pathology and genetics. *Brain Pathol*. 2007;17:74–82.

78. Montplaisir J, Petit D, Décary A, et al. Sleep and quantitative EEG in patients with progressive supranuclear palsy. *Neurology*. 1997;49:999–1003.

79. Yasui K, Inoue Y, Kanbayashi T, et al. CSF orexin levels of Parkinson's disease, dementia with Lewy bodies, progressive supranuclear palsy and corticobasal degeneration. *J Neurol Sci*. 2006;250:120–123.

80. Aldrich MS, Foster NL, White RF, et al. Sleep abnormalities in progressive supranuclear palsy. *Neurology*. 1989;25:577–581.

81. Arnulf I, Merino-Andreu M, Bloch F, et al. REM sleep behavior disorder and REM sleep without atonia in patients with progressive supranuclear palsy. *Sleep*. 2005;28:349–354.

82. Erkinjuntti T, Partinen M, Sulkava R, et al. Sleep apnea in multi-infarct dementia and Alzheimer's disease. *Sleep*. 1987;10:419–425.

83. Aharon-Peretz J, Masiah A, Pillar T, et al. Sleep-wake cycles in multi-infarct dementia and dementia of the Alzheimer type. *Neurology*. 1991;41:1616–1619.

84. Sato K, Kamiya S, Okawa M, et al. On the EEG component waves of multi-infarct dementia seniles. *Int J Neurosci.* 1996;86:95–109.

85. Signorino M, Pucci E, Belardinelli N, et al. EEG spectral analysis in vascular and Alzheimer dementia. *Electroencephalogr Clin Neurophysiol.* 1995;94:313–332.

86. Tsuno N, Shigeta M, Hyoki K, et al. Fluctuations of source locations of EEG activity during transition from alertness to sleep in Alzheimer's disease and vascular dementia. *Neuropsychobiology.* 2004;50:267–272.

87. The Huntington's Disease Collaborative Research GroupA novel gene containing a trinucleotide repeat this is expanded and unstable on Huntington's disease chromosomes. *Cell.* 1993;72:971–983.

88. Morton AJ, Wood NI, Hastings MH, et al. Disintegration of the sleep-wake cycle and circadian timing in Huntington's disease. *J Neurosci.* 2005;25:157–163.

89. Hansotia P, Wall R, Berendes J. Sleep disturbances and severity of Huntington's disease. *J Neurosci.* 2007;27:7869–7878.

90. Wiegand M, Moller AA, Lauer CJ, et al. Nocturnal sleep in Huntington's disease. *J Neurol.* 1991;238:203–208.

91. Streletz LJ, Reyes PF, Zalewska M, et al. Computer analysis of EEG activity in dementia of the Alzheimer's type and Huntington's disease. *Neurobiol Aging.* 1990;11:15–20.

92. Creutzfeldt-Jakob disease. *WHO Recommended Surveillance Standards.* 2nd ed. WHO: Geneva; 1999:35–38.

93. Landolt HP, Glatzel M, Blättler T, et al. Sleep-wake disturbances in sporadic Creutzfeldt-Jakob disease. *Neurology.* 2006;66:1418–1424.

94. Wall CA, Rummans TA, Aksamit AJ, et al. Psychiatric manifestations of Creutzfeldt-Jakob disease: a 25-year analysis. *J Neuropsychiatry Clin Neurosci.* 2005;17:489–495.

95. Cohen OS, Chapman J, Korczyn AD, Warman-Alaluf N, Orlev Y, Givaty G, Nitsan Z, Appel S, Rosenmann H, Kahana E, Shechter-Amir D. Characterization of sleep disorders in patients with E200K familial Creutzfeldt-Jakob disease. *J Neurol.* 2015;262(2):443–450.

96. Donnet A, Famarier G, Gambarelli D, et al. Sleep electroencephalogram at the early stage of Creutzfeldt-Jakob disease. *Clin Electroencephalogr.* 1992;23:118–125.

97. World Health OrganizationConsensus on criteria for diagnosis of sporadic CJD. *Weekly Epidemiol Rec.* 1998;73:361–365.

98. Steinhoff BJ, Zerr I, Glatting M, et al. Diagnostic value of periodic complexes in Creutzfeldt-Jakob disease. *Ann Neurol.* 2004;56:702–708.

99. Iliff JJ, Wang M, Liao Y, Plogg BA, Peng W, Gundersen GA, Benveniste H, Vates GE, Deane R, Goldman SA, Nagelhus EA, Nedergaard M. A paravascular pathway facilitates CSF flow through the brain parenchyma and the clearance of interstitial solutes, including amyloid β. *Sci Trans Med.* 2012;4(147):147ra111.

100. Xie L, Kang H, Xu Q, Chen MJ, Lioa Y, Thiyagarajan M, O'Donne J, Christensen DJ, Nicholson C, Iliff JJ, Takano T, Deane R, Nedergaard M. Sleep drives metabolite clearance from the adult brain. *Science.* 2013;342(6156):373–377.

101. Roh JH, Huang Y, Bero AW, Kasten T, Stewart FR, Bateman RJ, Holtzman DM. Disruption of the sleep-wake cycle and diurnal fluctuation of amyloid-β in mice with Alzheimer's disease pathology. *Sci Transl Med.* 2012;5:150ra122.

102. Pan W, Kastin AJ. Can sleep apnea cause Alzheimer's disease?. *Neurosci Biobehav Rev.* 2014;47:656–669.

103. Janssens JP, Pautex S, Hilleret H, Michel JP. Sleep disordered breathing in the elderly. *Aging Clin Exp Res.* 2000;12(6):417–429.

104. Tarumi T, Harris TS, Hill C, German Z, Riley J, Turner M, Womack KB, Kerwin DR, Monson NL, Stowe AM, Mathews D, Cullum CM, Zhang R. Amyloid burden and sleep blood pressure in amnestic mild cognitive impairment. *Neurology.* 2015;85(22):1922–1929.

105. Daulatzai MA. Evidence of neurodegeneration in obstructive sleep apnea: relationship between obstructive sleep apnea and cognitive dysfunction in the elderly. *J Neurosci Res.* 2015;93(12):1778–1794.

106. Wilcock DM, Schmitt FA, Head E. Cerebrovascular contributions to aging and Alzheimer's disease in Down syndrome. *Biochim Biophys Acta.* 2015;1862(5):909–914.

107. Osorio RS, Gumb T, Pirraglia E, Varga AW, Lu SE, Lim J, Wohlleber ME, Ducca EL, Koushyk V, Glodzik L, Mosconi L, Ayappa I, Rapoport DM, de Leon MJ. Sleep-disordered breathing advances cognitive decline in the elderly. *Neurology.* 2015;84(19):1964–1971.

108. Gray SL, Anderson ML, Dublin S, Hanlon JT, Hubbard R, Walker R, Yu O, Crane PK, Larson EB. Cumulative use of strong anticholinergics and incident dementia. *JAMA Intern Med.* 2015;175(3):401–407.

109. Biollioti de Gage S, Moride Y, Ducruet T, Kurth T, Verdoux H, Tournier M, Pariente A, Begaud B. Benzodiazepine use and risk of Alzheimer's disease: case-control study. *BMJ*. 2014;349:g5205.

110. Shih HI, Lin CC, Tu YF, Chiang CM, Hsu HC, Chi CH, Kao CH. An increased risk of reversible dementia may occur after zolpidem derivative use in the elderly population: a population-based case-control study. *Medicine (Baltimore)*. 2015;94(17):e809.

111. Gomm W, von Hold K, Thome F, Broich K, Maier W, Fink A, Doblhammer G, Haenisch B. Association of proton pump inhibitors with risk of dementia: a pharmacoepidemiological claims data analysis. *JAMA Neurol*. 2016;10:1001.

112. Dauvilliers Y. Insomnia in patients with neurodegenerative conditions. *Sleep Med*. 2007;8:S27–S34.

113. Ferrari E, Arcaini A, Gornati R, et al. Pineal and pituitary-adrenocortical function in physiological aging and in senile dementia. *Exp Gerontol*. 2000;35:1239–1250.

114. Wu YH, Fischer DF, Kalsbeek A, et al. Pineal clock gene oscillation is disturbed in Alzheimer's disease, due to functional disconnection from the "master clock". *FASEB J*. 2006;20:1874–1876.

115. Lyketsos CG, Lindell Veiel L, Baker A, et al. A randomized controlled trial of bright light therapy for agitated behaviors in dementia patients residing in long-term care. *Intl J Geriatr Psychiatry*. 1999;14:520–525.

116. Van Someren E, Kessler A, Mirmiran M, et al. Indirect bright light improves circadian rest-activity rhythm disturbances in demented patients. *Biol Psychiatry*. 1997;41:955–963.

117. Ancoli-Israel S, Gehrman P, Martin JL, et al. Increased light exposure consolidates sleep and strengthens circadian rhythms in severe Alzheimer's disease patients. *Behav Sleep Med*. 2003;1:22–36.

118. Srinivasan V, Pandi-Perumal SR, Cardinali DP, et al. Melatonin in Alzheimer's disease and other neurodegenerative disorders. *Behav Brain Functions*. 2006;2:15–37.

119. Wang JZ, Wang ZF. Role of melatonin in Alzheimer-like neurodegeneration. *Acta Pharmacol Sin*. 2006;27:41–49.

120. Osorio RS, Gumb T, Pirraglia E, et al. Sleep-disordered breathing advances cognitive decline in the elderly. *Neurology*. 2015;84(19):1964–1971.

121. Cooke JR, Ayalon L, Palmer BW, et al. Sustained use of CPAP slows deterioration of cognition, sleep, and mood in patients with Alzheimer's disease and obstructive sleep apnea: a preliminary study. *J Clin Sleep Med*. 2009;5:305–309.

122. Hogl B, Kiechl S, Willeit J, et al. Restless legs syndrome: a community-based study of prevalence, severity and risk factors. *Neurology*. 2005;64:1920–1924.

123. Aurora RN, Zak RS, Maganti RK, et al. Best practice guide for the treatment of REM sleep behavior disorder (RBD). *J Clin Sleep Med*. 2010;6:85–95.

124. Kunz D, Mahlberg R. A two-part, double-blind, placebo-controlled trial of exogenous melatonin in REM sleep behavior disorder. *J Sleep Res*. 2010;19(4):591–596.

Sleep and Movement Disorders

L. Ashbrook*, E.H. During*,**

*The Stanford Center for Sleep Sciences and Medicine, Redwood City, CA, USA,
**Division of Sleep Medicine, Stanford University, Redwood City, CA, USA

OUTLINE

Sleep and Neurologic Disease. http://dx.doi.org/10.1016/B978-0-12-804074-4.00005-4

TABLE 5.1   Sleep History: Important Questions

- What time do you go to bed and wake up?
- How long does it take you to fall asleep?
- How often do you awaken overnight?
- Do you snore?
- Do you feel sleepy during the daytime?
- Do you experience any disturbing movement overnight including
  - An urge to move the legs relieved by moving them? (RLS)
  - Trouble moving or rolling over in bed? (rigidity)
  - Abnormal postures? (dystonia)
- Do you act out your dreams overnight?

INTRODUCTION

Sleep disruption is a common nonmotor symptom of many movement disorders. Unfortunately, a good sleep history is not always obtained during clinical appointments, and the patient may not necessarily volunteer this information. As a result there are many missed opportunities to improve these patients' quality of life.[1] A good sleep history should include assessment of a bedtime and wake time; the amount of time it takes to fall asleep and the amount of time spent awake at night; snoring or other abnormal breathing sounds; excessive daytime sleepiness (EDS) and fatigue; nocturnal motor and sensory symptoms, including rigidity, dystonia and restless legs, and nocturnal behaviors such as dream enactment. Table 5.1 A caregiver or bed partner is often essential to obtain a complete history, especially regarding questions about dream enactment. The timing of symptoms to medication dosing should be addressed, particularly sedating medications such as dopamine agonists. The treatment of these sleep disorders can lead to not only improved sleep for both patient and caregiver, but also improved daytime functioning.

The α-synucleinopathies are a group of neurodegenerative proteinopathies that are characterized by the deposition of the protein α-synuclein in distinct neuroanatomical areas of the brainstem and cortex. These diseases include Parkinson's disease (PD), multiple system atrophy (MSA), dementia with Lewy bodies (DLB), and pure autonomic failure (PAF). In PD and DLB, α-synuclein is deposited in the form of Lewy bodies. In MSA, these deposits take the form of glial cytoplasmic inclusions. PAF is a predominantly peripheral disease of Lewy body deposition. Due to the lack of motor symptoms, it will not be discussed in this section, and is discussed in greater detail in the chapter on autonomic disorders.

PARKINSON'S DISEASE

PD is associated with more sleep complaints than any other movement disorder. The motor manifestations of PD include tremor, bradykinesia, and rigidity; however, many times the nonmotor symptoms can be just as disabling. Sleep impairment is one of the most bothersome nonmotor symptoms, and patients may experience any combination of EDS, insomnia, circadian rhythm disruption, sleep apnea, REM sleep behavior disorder (RBD), and restless legs syndrome (RLS) (Table 5.2).

TABLE 5.2   Common Sleep Disturbances in α-Synucleinopathies, Treatments, and Pitfalls

Sleep complaint	Treatment	Look out for
Hypersomnia	Modafinil (limited data)[2,3] Consider reducing dopaminergic medication[4] May consider other stimulants or agents to improve nocturnal sleep[5,6]	Sleep attacks associated with dopaminergic medication[7]
Insomnia	Long acting dopaminergic medication overnight[8,9] CBTI[10,11] Melatonin, doxepin, and eszopiclone all improve subjective sleep but not objective sleep[10,12,13]	Unclear link between hypnotics and development of Parkinson's disease[14]
Circadian rhythm disruption	Melatonin[12] Bright light therapy[11]	Possible circadian shifting effects of dopaminergic medication[15,16]
OSA	CPAP[17]	CPAP use in the subset of MSA patients with stridor and floppy epiglottis[18,19]
RBD	Melatonin[20,21] Clonazepam[22]	Injurious behavior to self or bed partner[22]
RLS	Ensure adequate iron stores First line: α-2-delta voltage-dependent calcium channels agonists[23–25] First line: dopamine agonists[26,27] Second line: opioids[28]	Augmentation with dopamine agonist medications[29]

Excessive daytime sleepiness affects approximately 15–50% of patients with PD, compared to 2–20% of age-matched controls, depending on the study.[30–34] Sleepiness in PD is more common in male patients and those with longer duration of illness, greater severity of disease, subjective reports of pain, mood disorders, and autonomic dysfunction.[4,33,35] The EDS in PD is commonly associated with medication side effects, especially dopaminergics medications.[2] Sleepiness may be present prior to treatment and typically increases as the disease progresses.[2] There is not always a clear correlation between the patient's symptoms and objective measures of sleepiness such as polysomnography (PSG) or multiple sleep latency (MSLT) testing in the sleep lab. When present, objective correlates of sleepiness on these studies have been associated with patients who are obese and those who have comorbid obstructive sleep apnea (OSA).[35] The presence of OSA is likely unrelated to α-synuclein deposition.

Treatment

Despite their widespread use, there is limited data to support the use of stimulants in this patient population. In a small trial, modafinil at a dose of 200 mg improved subjective daytime sleepiness; however, there was no improvement in the results of the maintenance of wakefulness test (a standardized test whereby patients are kept in a dark room in the sleep lab and asked to maintain wakefulness).[3,36] Caffeine at a dose of 200 mg twice daily had a nonsignificant benefit on reports of subjective sleepiness in one small randomized trial, although it did improve some objective motor measures with a reduction in Unified Parkinson's Disease Rating Scale (UPDRS) scores by an average of 4.69 points.[5] Surprisingly, clonazepam led to less subjective daytime sleepiness and fewer periodic limb movements (PLMs) in a small group of patients who underwent PSG and MSLT testing.[6] Sodium oxybate was evaluated in an open label trial for EDS in PD patients with sleep complaints and Epworth sleepiness scale (ESS) scores ≥ 10 (a score ≥10 is indicative of EDS).[7] The 27 subjects who completed the 6-week-trial took a mean dose of 7.8 g and reported improvement in subjective sleepiness, with a decrease in Epworth sleepiness scores from 15.6–9, and significantly increased slow wave sleep duration, from 41–78 min. Adverse events included dizziness, enuresis, daytime sleepiness, and reduced alertness in 8/30 participants.[7]

While more research is needed to determine the most effective treatment in this population, limited data suggest that a treatment that improves sleep fragmentation may lead to an improvement in daytime sleepiness. Although it is possible that clonazepam and sodium oxybate accomplish this task, these medications may not be ideal in the PD population because they can also worsen cognitive impairment and worsen gait instability, and are therefore generally avoided. In the meantime, modafinil may be considered for EDS, although high-quality evidence is lacking.

Sleep attacks are another concern in patients with PD, especially those on dopaminergic medications. Sleep attacks are sudden, irresistible bouts of sleep and may occur during active behaviors, such as during conversation or while eating. Sleep attacks are estimated to occur in 1–6% of patients.[37] They have been attributed to multiple dopamine agonist medications, including case reports of three patients having sleep attacks at the wheel while taking bromocriptine, pergolide, or piribedil.[38] Attacks occur predominantly in those with pre-existing EDS, although all patients may be susceptible. In one cohort of 2,952 German patients, 6% experienced sleep attacks and 18% of those with sleep attacks had an ESS <10, suggesting that not all patients with sleep attacks have EDS.[39] If a patient has sleep attacks, he or she should be advised not to drive until this symptom is effectively treated.

Insomnia affects 27–80% of patients with PD.[40–42] Insomnia is more likely to occur in females, those with a longer duration of disease, comorbid depression or anxiety, or cognitive complaints.[1,40,43] Insomnia has also been found to correlate with symptoms of autonomic dysfunction including thermoregulatory, pupillomotor, and cardiovascular symptoms such as orthostasic intolerance.[44]

Sleep maintenance insomnia is more common than sleep initiation insomnia, and those with PD do not differ from controls in subjective sleep latency.[40] One exception is those patients with LRRK2 mutations, who have significantly more sleep initiation insomnia than those with idiopathic PD.[8] Nocturnal awakenings and difficulty returning to sleep are attributed to a range of problems, including dystonia, akinesia, restless legs, painful cramps, tremor, difficulty turning in bed, circadian clock disruption, nocturia, and urinary incontinence.[9]

Treatment

Treatment of insomnia should begin with optimizing the treatment of motor symptoms of PD. Long acting dopaminergic medications such as controlled-release carbidopa/levodopa or rotigotine transdermal patches have been shown to significantly improve subjective sleep quality, onset, and maintenance.[45,46] Similarly, deep brain stimulation can improve subjective sleep quality in those eligible for this intervention.[10] In patients with neurogenic bladder, anticholinergic medications or botulinum toxin treatment may be a good option to reduce nocturia, although anticholinergics should be use with caution in those with cognitive impairment.[11] Cognitive behavioral therapy for insomnia (CBTI) and light therapy have also been studied and shown to be effective in one small study.[12,13]

From the limited data available, it appears that pharmacologic interventions tend to improve sleep subjectively but not objectively. In one small randomized control trial of 18 patients, melatonin 3 mg at bedtime resulted in subjective improvement in the Pittsburgh Sleep Quality Index scores, although PSG sleep latency and sleep efficiency did not change.[14] Doxepin 10 mg has also been demonstrated to improve sleep subjectively by the Insomnia Severity Index, SCOPA-night scale (a PD specific scale to evaluate nocturnal sleep and daytime drowsiness), Pittsburgh Sleep Quality Index-sleep disturbances subscale, and fatigue severity scale.[12] Eszopiclone was tested against placebo in 30 patients with PD and insomnia, and while subjective sleep quality did improve, total sleep time as measured by PSG did not.[47] Similarly, while some case reports describe symptomatic improvement with zolpidem, there is no clear data to support its use as a sleep aid in PD patients. There is open-label trial evidence that quetiapine at low doses (average dose 31.9 mg) may be helpful in treating insomnia in PD patients;[14] however, this medication is generally avoided given its side effect profile in elderly patients.

There is some concern that hypnotic use may be linked to neurodegenerative disease. Zolpidem as a treatment for insomnia was correlated with a greater incidence of PD in a large cohort study of 59,548 Taiwanese patients (Fig. 5.1).[48] While causation is not clear, there was a 30–40% increased risk of developing PD in the group treated with zolpidem when compared to age-matched controls over a 5-year period.

The circadian rhythm is an endogenous clock of approximately 24 h that is influenced by environmental cues and regulated largely by the suprachiasmatic nucleus (SCN) in the hypothalamus.[49,50] The SCN promotes wake during the day and sleep at night by stimulating the release of melatonin from the pineal gland, which peaks in the middle of the night in most individuals. PD patients have been demonstrated to have blunted melatonin curves with peak melatonin levels only 25% as high as those of control subjects.[1,51] The severity of melatonin impairment correlates with daytime sleepiness, and patients with EDS have been shown to have 50% lower peak levels than patients without EDS.

The etiology of reduced melatonin levels is hypothesized to be related to Lewy body deposition and degeneration of central pacemaker cells; however, there is also some evidence pointing to the role of dopaminergic medications.[1] Fertl and colleagues found that peak melatonin levels were similar in PD and control patients; however, those on dopaminergic treatment had peak values at earlier times in the evening.[15,16] Patients not taking dopaminergic medication had melatonin trends similar to controls. This leads to concern that treatment may cause phase shifting and thereby contribute to sleep disturbance. The exact cause for these conflicting findings is not clear but does demonstrate that there is far more to learn about the role of the circadian clock in this disease.

Some components of circadian dysfunction may be treatable. For example, melatonin at a dose of 3 mg has been shown to improve subjective sleep quality in a small study, although it did not improve abnormalities seen on PSG.[14] Bright light therapy with 10,000 Lux for at least 30 min each morning may also be beneficial.[13]

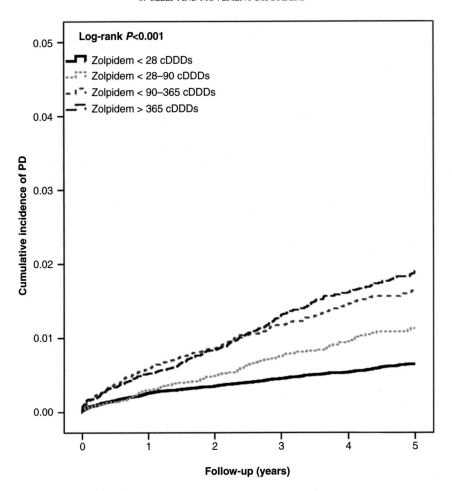

FIGURE 5.1 **Cumulative incidence of Parkinson's disease by cumulative defined daily doses (*cDDDs*) of zol-pidem use during the follow-up period in the cohort with sleep disturbance.** *Source: Reprinted with permission from Yang Y-W et al. Zolpidem and the risk of Parkinson's disease: a nationwide population-based study. J Psychiatr Res. 2014;58:84–88.*

OSA is characterized by recurrent collapse of the upper airway completely (apnea) or par-tially (hypopnea) causing hypoxemia and/or arousals. The incidence of OSA in PD ranges from 20–60%, varying by study population and event scoring criteria.[52] While this incidence is similar to that of the general population, OSA in PD patients may be less correlated with BMI.[53] Patients with PD also spend more time sleeping in the supine position, which increases their risk of airway obstruction.[54]

Treatment

Treatment of OSA in patients with PD is similar to treatment in the general population. Continuous positive airway pressure (CPAP) has demonstrated efficacy in reducing AHI, improving the percentage of time spent in stage 3 non-REM (NREM) sleep compared to stage 2 NREM sleep, and improving nocturnal oxygenation in PD patients.[17] There are no data on the efficacy of mandibular advancement devices in this population. Long acting dopaminergic treatment, such as extended release carbidopa/levodopa, may also be considered in the treatment of OSA in PD patients based on the results of an observational study that suggested dopaminergic therapy may reduce collapse of upper airway musculature in the latter half of the night, however, this has not been firmly established.[55]

Sleep benefit describes a phenomenon in PD whereby patients experience a transient improvement in movement upon waking, and can report feeling in an "on" state prior to taking their morning dopaminergic medication dose. The duration of sleep benefit can range from 3–300 min and occurs in about 50% of patients.[56,57] This phenomenon can even allow some patients to delay or skip their initial medication dose. Sleep benefit is a subjective benefit and objective measures of dexterity testing have not been demonstrated to reliably improve.[58]

MULTIPLE SYSTEM ATROPHY

MSA is a neurodegenerative disease characterized by the deposition of α-synuclein in the form of glial cytoplasmic inclusions in cortical, subcortical, and spinal cord regions. Patients with predominantly parkinsonian symptoms are classified as having MSA-P, while patients with predominantly cerebellar symptoms are classified as MSA-C. Sleep and autonomic dysfunction is common. In adults over the age of 50, the prevalence of MSA is estimated at 3 per 100,000.[59] Like patients with PD, patients with MSA experience frequent insomnia and EDS. Unlike PD, those with MSA are at increased risk of sleep disordered breathing and stridor, which can prove deadly. RBD is also more common in this population.

Hypersomnia and Insomnia

The prevalence of EDS is similar in MSA and PD, with about 25% of patients scoring >10 on the ESS.[34] Both MSA-C and MSA-P patients are equally sleepy and the degree of their EDS does not seem to correlate with comorbid sleep disordered breathing.[34,60] Patients with MSA may have more sleepiness in response to dopaminergic medication compared to patients with PD. In a small study of 17 patients with MSA-P and 23 patients with PD, 200 mg of levodopa plus 50 mg of benserazide resulted in a significant increase in sleepiness in the MSA group but not the PD group.[61] Sleep attacks have also been reported in MSA following dopaminergic medication.[62] Insomnia is more common than EDS in MSA, and affects up to 70% of patients.[63] On PSG, MSA patients have been demonstrated to have reduced total sleep time and reduced sleep efficiency, although the macro and micro structures of sleep were not different from the controls.[64]

Sleep Disordered Breathing and Stridor

Nearly all patients with MSA have sleep disordered breathing. In one study of 25 patients, 96% had sleep disordered breathing.[60] These patients are more likely to have both central and obstructive sleep apnea, due to degeneration of brainstem chemoreceptors.[65] Even more concerning is the risk of stridor. Stridor is a high-pitched, inspiratory tone distinct from snoring, easily noticed by the bed partner. It is thought to be secondary to laryngeal abductor weakness in the upper airway.[66] Prevalence has been estimated at 35–40%, based on two series of 40 and 42 MSA patients.[67,68] Stridor is very serious and has been associated with sudden nocturnal death; as a result, prognosis is worse for MSA patients with stridor.[68] CPAP is the treatment of choice and may improve survival.[67] If positive airway pressure is ineffective or if stridor occurs during the day, tracheostomy may be considered.[69] There are reports of patients with MSA and stridor experiencing sudden death despite CPAP treatment, which may be related to a floppy epiglottis, making CPAP a relative contraindication in these patients.[18,70] Floppy epiglottis can some times be detected with laryngoscopy, although this association has not been firmly established. Despite this association, the standard of care for treatment of sleep disordered breathing remains positive airway pressure therapy.[18]

DEMENTIA WITH LEWY BODIES

The cardinal features of DLB include fluctuating levels of alertness, parkinsonism and visual hallucinations. The need to distinguish DLB from other dementias such as Alzheimer's dementia (AD) arises commonly, and comorbid sleep disorders such as RBD may be used to aid in the diagnosis.

Hypersomnia and Insomnia

A core feature of DLB is fluctuations in mental status. Approximately 40–60% of patients report EDS.[19,71] On MSLT, patients with DLB are sleepier than those with AD, despite similar nocturnal total sleep times.[72] Mean sleep latency during MSLT was 6.4 minutes on four naps in 81 DLB patients, significantly shorter than patients with AD, and not correlated with dopaminergic medication.[72] Sleep efficiency is reduced; in one sample of 78 patients, 72% had a sleep efficiency of less than 80%, and 25% had a sleep efficiency less than 60%.[73] Insomnia was noted in 53% of patients in one cohort.[71]

Sleep Apnea

The prevalence of sleep apnea is similar in PD and DLB, according to a study of 29 DLB patients and 29 matched PD patients without dementia.[74] However, another study of 78 DLB patients demonstrated that sleep apnea was present in 55% of patients. Eighteen percent of patients had central events, and one met criteria for the diagnosis of central sleep apnea syndrome.[73]

REM SLEEP BEHAVIOR DISORDER

One defining component of REM sleep is atonia of the skeletal musculature. Experiments by Jouvet and colleagues in 1965 demonstrated that lesions near the locus coeruleus in the brains of cats could eliminate atonia during active sleep, and cats with this lesion were observed to have complex motor activity during sleep.[75] The clinical correlate in humans was described by Schenck and colleagues in 1986 in the cases of five patients with a variety of neurologic illnesses, each acting out their dreams with loss of REM atonia demonstrated on PSG.[76]

Low EMG tone is a defining feature of REM. A state in which other features of REM are met but atonia is absent may be considered a dissociated state, with features of wake and features of REM sleep occurring simultaneously. Thus RBD may be thought of as a dissociated state.

RBD is defined by dream enactment and the loss of REM atonia on PSG. This dream enactment often manifests as yelling, kicking, punching, sitting, grabbing, jumping from bed or violent thrashing. Dreams are commonly violent, even though individuals experiencing the dream may not have violent tendencies. Patients can typically describe the dreams they were experiencing in detail. Many patients with RBD describe an attempt to save a loved one or themselves, which can lead to self-injury or injury of their bed partners. Of 93 patients in an early Mayo clinic series, 32% had injured themselves and 64% had injured their spouses.[77]

Epidemiology

RBD is present in 0.5% of the general population and 2% of those over 60-years-old.[22,78,79] Patients who come to clinical attention are overwhelmingly male, with estimates ranging from 80–90%.[80–82] However, by screening questionnaire in a population of 107 patients, a similar prevalence of probable RBD was found in both men and women (43 and 31%, respectively).[83] Men did have significantly more violent and vigorous movements. This has lead to the postulation that the greater prevalence in men may be at least partly explained by the greater clinical attention paid when movements are violent.

RBD has been associated with a wide array of neurologic conditions, the most prominent being the disorders of α-synuclein deposition. RBD is also seen to a lesser extent in narcolepsy. There is recent recognition that patients with Wilson's disease have a high prevalence of both RBD and REM without atonia (RWA), the elevation of EMG tone during PSG but without symptoms of dream enactment.[84] The list of conditions that have been associated with RBD is long and includes progressive supranuclear palsy, frontotemporal dementia, amyotrophic lateral sclerosis, multiple sclerosis, brainstem infarction, subarachnoid hemorrhage, Guillain–Barre syndrome, myotonic dystrophy type 2, and voltage-gated potassium channel antibody-associated limbic encephalitis.[77,85–89] Many of these result from lesions in the medial and tegmental pons.[90] RBD can also be secondary to medications, most commonly SSRIs and SNRIs, as discussed in the following.

α-SYNUCLEINOPATHIES AND REM SLEEP BEHAVIOR DISORDER

Disorders of α-synuclein deposition are the most common cause of RBD, occurring in 15–44% of patients with PD, 30–83% of patients with DLB, and 85–100% of patients with MSA.[71–73,91–94] What was once considered idiopathic RBD (iRBD) is now thought to be a premotor manifestation of a yet undiagnosed α-synucleinopathy. In one series, 80% of 26 patients diagnosed with iRBD had developed parkinsonism at a 16-year follow-up.[95] The rates of disease-free survival based on a cohort of 44 patients by Iranzo and colleagues was estimated to be 65.2% at 5 years, 26.6% at 10 years, and only 7.5% at 14 years.[96] Of the remaining disease-free patients who underwent further testing, all four had dopamine transporter (DAT) positive scans, suggesting subclinical disease. This prodromal stage, which can precede motor symptoms by up to several decades, may prove very important clinically as a window for intervention when disease modifying therapies become available.[97] RBD is less frequent and less severe in PD patients with the LRRK2-PD genetic mutation, suggesting that RBD may not be a prodromal state of LRRK2-PD as it is in idiopathic PD.[8] In those with iRBD, excessive sleepiness may predict an accelerated progression to parkinsonism and dementia.[98]

The strong relationship between iRBD and α-synucleinopathies is further supported by multiple imaging studies, demonstrating that when patients with iRBD are compared to both control patients and PD patients, the findings are more similar to PD than to the controls. This has been demonstrated with [123]I-2β-carbomethoxy-3β-(4-iodophenyl)-N-(3-fluoropropyl)-nortropane (^{123}I-FP-CIT) SPECT binding in the striatum, diffusion tensor imaging, and transcranial Doppler imaging of the substantia nigra.[99–101]

NARCOLEPSY AND REM SLEEP BEHAVIOR DISORDER

Narcolepsy is the second most common cause of RBD.[82] Symptoms of RBD typically emerge early in the course of narcolepsy, either at onset or within the first few years.[82] Narcolepsy accounts for 8% of cases of neurologically attributable RBD by one estimation, and 36–54% of those with narcolepsy complain of RBD symptoms.[77,102,103] In patients who have both RBD and narcolepsy, the SSRIs used to treat cataplexy may exacerbate RBD. Unlike other groups of RBD patients, those with narcolepsy are equally likely to be male or female, and the average age in one cohort of 55 narcoleptic patients was 41.[102]

Medications

Antidepressants, including selective serotonin reuptake inhibitors (SSRIs), serotonin and norepinephrine reuptake inhibitors (SNRIs), tricyclic antidepressants (TCAs), and monoamine oxidase inhibitors (MAOIs) have been associated with increased risk of RBD.[104] Given the fact that all of these medications interact with the serotonin system, one might hypothesize that this system is somehow involved in the mechanism of RBD. It is possible that the raphe nucleus, involved in REM on/off circuitry, is affected and that motor control during REM is altered.[90] Antidepressants may trigger RBD in up to 6% of patients based on 1235 patients attending an outpatient psychiatric clinic in Hong Kong.[105] However, many patients who have RBD attributed to antidepressant use also have early markers of neurodegenerative disease, such as changes in motor behavior and olfactory discrimination.[106] This suggests that in some patients the use of antidepressants may unmask RBD rather than trigger it independently. Therefore, patients diagnosed with RBD who are on antidepressants should still be clinically evaluated for other signs and symptoms of parkinsonism.

Diagnosis

The sleep history in older adults visiting a neurology or sleep clinic should always include questions about dream enactment behavior. If present, the details of the events should be

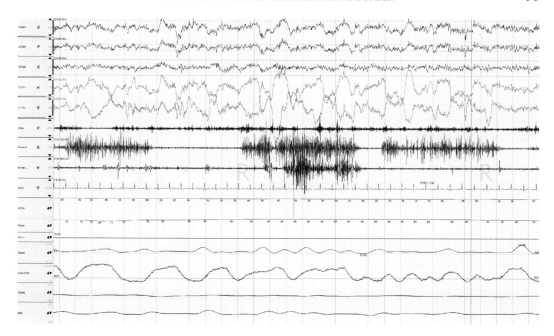

FIGURE 5.2 **Polysomnography of REM sleep behavior disorder.** PSG in a 65-year-old man with a 20-year history of dream enactment. During epoch pictured, video showed the patient laying supine with eyes closed, arms raised and gesturing, with lips moving as if speaking.

elicited, including both the observations of the bed partner and the recall of the patient. The diagnosis of RBD requires both the clinical history of dream enactment and the presence of REM without atonia on PSG (Fig. 5.2). This manifests as both elevated muscle tone and frequent muscle twitching during REM. The lack of atonia may not be evident if only the chin EMG lead is used, as is traditionally done on PSG. Arms and legs may move independently, therefore monitoring each limb is recommended.[107] If there is excess tone on EMG during PSG but no clinical history of dream enactment, the diagnosis can be made if excessive limb movement is seen by video during the PSG recording. If no excessive movement is seen during PSG, and the patient does not endorse a history of dream enactment, the diagnosis of RBD cannot be made.

Differential Diagnosis

The differential diagnoses of RBD commonly includes NREM parasomnias and nocturnal seizures. While nocturnal seizures can occur at all ages, NREM parasomnias are more frequent in childhood and often resolve in teenage years or early adulthood; however, they can also occur de-novo in adults. Other diagnoses to consider include periodic limb movement disorder and dissociative disorder.

NREM parasomnias are the consequence of a sudden and incomplete arousal from sleep, typically from stage 3 NREM slow-wave sleep, causing a range of manifestations that can be placed on a scale of increasing complexity, including confusional arousals, sleep terrors, and somnambulism (sleepwalking). The simplest form, a confusional arousal, is characterized by rudimentary, clumsy, and disorganized movements often associated with unintelligible speech, generally lasting a few minutes, with spontaneous return to sleep in most cases. "Sexsomnia" is a particular form of confusional arousal associated with forceful sexual behaviors out of character for the patient. An acute stress response with sympathetic activation in reaction to a perceived threat is the hallmark of a sleep terror. These should be distinguished from nightmares, which emerge from REM sleep and are the recollection of a frightening image or situation. The central feature of somnambulism is ambulation during sleep with minimal awareness of the environment; the individual moves slowly, avoiding obstacles, often with eyes open and with a blank expression and may carry on some purposeless activities, such as moving objects or rearranging the furniture. More severe cases have led to more complex and dangerous behaviors such as sleep driving. Sleep-related eating disorder occurs when sleepwalking is accompanied by a compulsion to eat, typically food of poor nutritional value and high calorie content. This condition is highly associated with RLS and can be triggered by hypnotics, particularly benzodiazepine receptor agonists.[108,109] It has also been associated with OSA.[110]

Several features present in NREM parasomnias help distinguish them from RBD: (1) their propensity to occur in the first half of the night when there is a higher proportion of slow-wave sleep; (2) sleep inertia leading to difficulty awakening with residual confusion; and (3) amnesia, which can be partial. In the case of sleep terrors causing a patient to escape a threat by jolting out of bed and occasionally running, the differential diagnosis can be challenging because individuals often have a good recollection of the frightening situation or object, such as animals chasing them or a ceiling falling on them. However, the scene remembered is primitive and not as elaborate as when arising from REM sleep. Confusion dissipates gradually over seconds to minutes when awakening from NREM sleep, as opposed to the rapid orientation that occurs when awakening from REM.

Nocturnal frontal lobe epilepsy (NFLE), the most classic form being sleep-related hypermotor epilepsy (SHE), is commonly mistaken for RBD. First described in 1977, SHE seizures commonly arise from NREM sleep (most commonly stage 1 and 2 NREM sleep) and result in complex, bizarre, hypermotor behaviors with full or partial preservation of consciousness, even with bilateral motor involvement. The supplementary motor area is commonly activated and there is often minimal or no postictal confusion or amnesia.[111,112] The electrographic diagnosis of NFLE can be challenging due to movement-related muscle artifacts and poor sensitivity of scalp EEG for deep or mesial frontal foci.[113] Nocturnal seizures can be distinguished from NREM parasomnias by their stereotyped presentation, the occasional presence of an aura, their tendency to cluster in the same night, and occurrence during naps.

Although RBD can result in leg kicking during sleep, it should be differentiated from periodic limb movements (PLMs) of sleep. PLMs are repeated twitching or flexing of one or both legs occurring at intervals of about 20–40 s, often associated with RLS and obstructive sleep apnea. PLMs may be an epiphenomenon of sleep instability. As opposed to RBD, PLMs are stereotyped and do not correlate with a dream.

Finally, the differential diagnosis of RBD includes sleep-related dissociative disorder, a rare psychiatric disorder which manifests with violent spells during which the individual may re-enact traumatic experiences or unconscious fantasies, usually followed by complete amnesia of the episode.[114] In contrast with parasomnias, nocturnal seizures, and sleep-related movement disorders, the dissociative episodes do not arise from sleep but from wake. They should however be considered in the differential diagnosis of RBD in the setting of psychiatric comorbidities, or when secondary gain is suspected. Dissociative disorders are often associated with post-traumatic stress disorder and borderline personality disorder spectrum.

Clinical Course

The average age at RBD diagnosis is between 52 and 63, and cases arising during childhood have been reported.[77,115–117] Symptoms may begin insidiously with sleep talking, then escalate to increasingly active behavior, finally presenting to medical attention once injurious behavior occurs. There is an estimated 7-year delay to diagnosis.[81] Symptoms can wax and wane in severity, and in a case series of 89 patients, only 3 patients had RBD at all three time points at 0, 4, and 8 years.[91]

Treatment

The most important component of treatment is the need to safeguard the environment. Patients should remove breakable objects from their bedside and ensure that sharp corners near the bed are padded. If they are near a window they should place thick drapes over the window and ensure that dangerous objects, such as guns and knives, are locked and inaccessible. It may be necessary to place the mattress on the ground if the patient is at risk of falling out of bed, and bed partners may need to sleep in a separate room.

A safe sleep environment is essential, but pharmacologic treatment is often necessary. Treatment may be as simple as withdrawing or decreasing the offending medication, such as an SSRI, if possible. There are few placebo-controlled trials of medications for treatment of RBD. Clonazepam was the first recognized treatment and is completely or partially successful in 87% of the patients in one series of 93 patients.[77] In another series of 82 patients, 81% had self-injurious behavior prior to treatment, and only 6% with treatment.[118] The starting dose is typically 0.5 mg, and it can be titrated to 2 mg as needed. This medication should be increased slowly owing to its potential side effects of transient cognitive impairment and gait imbalance. For this reason, melatonin has been used more frequently in recent years as the first line drug of choice for RBD. Doses of 3–12 mg have been shown to reduce dream enactment behavior, with an efficacy similar to clonazepam.[20,119]

If neither melatonin nor clonazepam is effective, then there are mostly anecdotal reports that rivastigmine, donepezil, pramipexole, levodopa, and imipramine may be effective.[21,120,121] While SSRI, SNRIs and TCAs are thought to exacerbate RBD, there are some data to suggest that in refractory cases they may be helpful, possibly due to REM suppression.[122]

RESTLESS LEG SYNDROME

RLS, or Willis–Ekbom disease, is defined as an urge to move the legs that is at least partially relieved by the act of movement. It is typically preceded by an unpleasant sensation and either begins or significantly worsens during the evening hours and during periods of immobility. RLS can be defined as intermittent if it occurs less than twice weekly for the past year with at least five lifetime events, or chronic if it occurs at least twice weekly.[123]

Epidemiology

RLS affects 5–10% of the adult population of European origin; however, it is estimated that only 2–3% come to clinical attention.[124,125] In a cohort of 16,202 patients, 337 (2.2%) reported that they discussed symptoms with their primary care doctor, however, only 21 (6.2% of those with symptoms) were diagnosed with RLS, suggesting that this diagnosis is frequently over-looked.[126] While the differential includes arthritic pain, leg cramping, positional discomfort, peripheral neuropathy, and myalgic pain, the key features of RLS are an urge to move, its circadian rhythmicity, and its relief with movement. RLS is more common in those with renal failure, iron deficiency, neuropathy, multiple sclerosis, spinal cord injury and pregnancy.

Patients with PD are more likely to have RLS than those in the general population, with greater incidence as the disease progresses.[126] In a four-year study of RLS in PD, 109 patients free of dopaminergic medication were evaluated for symptoms of RLS. Symptoms were present in 4.6% of the patients at study onset ($n = 5$), which is similar to general population prevalence estimates. However, this percentage had increased to 6.5% ($n = 7$) at 2 years and 16.3% at 4 years ($n = 16$).[126] It is possible that long-term dopaminergic treatment, rather than the disease itself, may contribute to higher rates in the PD population, essentially unmasking pre-clinical RLS.[127] Despite similar initial rates of RLS, motor leg restlessness is over twice as common in the early stages of the disease, prior to dopaminergic treatment. Motor leg restlessness is defined as an urge to move the legs but not meeting full criteria for RLS. In a cohort of 200 newly diagnosed PD patients, motor leg restlessness was present in 40%, compared to 17.9% of age-matched controls.[128] This is thought to relate to the akathisia of PD and may further argue against a link between PD and RLS.

There is frequently a genetic component to RLS, with 63–92% of patients reporting a family history.[129,130] The genetic subtypes are not well understood, however genes involved in iron homeostasis have been associated with increased risk of future disease.[131,132] Nocturnal eating may be part of the spectrum of RLS, and this behavior can occur in up to 31–61% of patients.[108,133] Antidepressants such as SSRIs and SNRIs and antihistamines are known to ex-acerbate RLS in some individuals.

Etiology

RLS has been associated with reduced central iron stores. This is supported by evidence from autopsy studies revealing cellular iron deficiency in the substantia nigra in the RLS patients studied.[134] Cerebrospinal fluid iron levels are lower and transferrin levels are higher in RLS patients when compared to controls, while serum levels are typically normal.[135] Imaging studies using gradient echo MRI and transcranial ultrasound of the midbrain have demon-strated lower iron stores in patients with RLS.[136,137] Iron is a cofactor for tyrosine hydroxylase, the catalyst for the rate limiting step of dopamine synthesis, which may explain why dopa-mine replacement is an effective treatment for RLS, although there has been no convincing evidence that RLS patients are deficient in dopamine.[138] In summary, the etiology of this con-dition remains elusive and is likely multifactorial in nature.

Treatment: Nonpharmacologic

Simple nonpharmacologic approaches include withdrawing offending medications, if clinically acceptable. Medications that have been reported to exacerbate RLS symptoms include antihistamines, neuroleptic agents, dopamine-blocking antiemetics and most antidepressants. Bupropion, which increases dopamine levels, is an alternate antidepressant to consider because it is not associated with RLS and instead may actually improve symptoms.[139] Many individuals find that a warm bath or shower relieves symptoms. The impact of caffeine on symptoms is disputed, but a trial of caffeine withdrawal is reasonable to determine the impact for each individual. A moderate level of exercise as opposed to sedentary or excessive exercise may also minimize symptoms; however, patients should be advised to limit exercise close to bedtime as this may worsen symptoms. In patients with comorbid OSA, treatment with CPAP may improve RLS symptoms.[140] The Relaxis pad provides vibration to the lower legs or affected area and is an alternative FDA-approved treatment for RLS. The efficacy of this treatment requires further study but could be considered for a select group of patients.

Treatment: Iron Replacement

Screening for low iron with a serum ferritin is important in the evaluation of all RLS patients. The classic approach has been to treat with iron supplements if the ferritin level is lower than 50–75 ng/mL.[141] There is no single treatment regimen for oral iron supplementation, but 325 mg of ferrous sulfate (65 mg of elemental iron) combined with 100–200 mg of vitamin C with each dose to enhance absorption twice daily has traditionally been recommended.[142] An iron panel should then be rechecked in 3–4 months to ensure the ferritin is improving to goal of greater than 75 ng/ml. If oral iron is not adequate, intravenous (IV) iron may be indicated. Low molecular weight iron dextran is far less likely to lead to anaphylaxis than high molecular weight preparations.[143] IV iron has been shown to be effective not just for those RLS patients with iron deficiency anemia but also for those without anemia.[144] Dextran 1000 mg as a one time dose decreased RLS in 76% of patients with iron deficiency anemia and RLS.[145] In a group of 25 patients with RLS but without anemia, 250 mg iron dextran weekly for four doses was used. 68% of patients reported moderate or complete resolution of their symptoms 1–6 weeks after treatment. This treatment effect lasted on average 31 weeks, ranging from 2–97 weeks.[144] In this group, interestingly, neither blood nor CSF iron predicted response to IV iron treatment. Serum ferritin can be elevated as an acute phase reactant and, therefore, may not be accurate in the context of acute inflammation or infection.

Treatment: Intermittent Symptoms

Further treatment is based on how frequently bothersome symptoms occur. For intermittent symptoms occurring less than twice weekly, treatment can include dopaminergic medications, benzodiazepines or opioids on an as needed basis. Carbidopa/levodopa, 25 mg/100 mg (0.5–1 tablet) has good efficacy but should not be used more than 2–3 times weekly given the risk for augmentation. Augmentation is the paradoxical worsening of symptoms such that they occur earlier in the day, are more intense, or extend beyond the legs despite escalating medication dose. Carbidopa/levodopa is a good as-needed option for RLS that occurs occasionally in the context of long car or air trips, theater visits, or on rare bothersome nights.[146] Patients taking carbidopa/levodopa daily, either immediate or controlled release formulations, are at risk not only of augmentation but also of recurrence of symptoms in the morning, or rebound effect.[26] Taking dopaminergic medications no more than 2–3 times weekly minimizes the risk of augmentation and rebound. Benzodiazepines and hypnotics are thought to be useful for RLS; however, few studies are available and the mechanism may be simply that of enhanced sleep initiation, therefore these drugs are not typically chosen as first line treatment. In the right context, zolpidem or zaleplon may be useful for intermittent symptoms at the beginning of the night, while temazepam and eszopiclone may be more useful for intermittent symptoms awakening a patient from sleep, given the half-lives and duration of action of each of these medications.[142] However, sedatives and hypnotics may also increase the risk of amnestic sleep-related eating disorder or sleep walking when given to patients with RLS, and this risk should be discussed with patients if these drugs are considered.[107] Opioids and opioid analogs are also used occasionally for intermittent symptoms, most commonly tramadol.

Treatment: Daily Symptoms

Daily treatment of RLS is reserved for those who experience severe symptoms more than twice weekly. When treating daily symptoms, it is helpful for patients to take their medication before their symptoms begin because once symptoms start a higher dose is typically required to provide relief. There are two classes of medication considered first line for RLS, nonergot dopamine agonists and α-2-delta voltage-dependent calcium channels agonists. The commonly used dopamine agonists include pramipexole, ropinirole and the rotigotine patch.[146,147] The starting doses are typically lower than those used in PD. Recommended starting doses are pramipexole 0.125 mg 2 h before symptoms start, ropinirole 0.25 mg 1.5 h before symptoms start, and rotigotine patch 1 mg applied around the same time every 24 h. Dosing is slowly increased every few days as needed and maintained at the lowest effective dose. Maximum doses are pramipexole 1 mg, ropinirole 2 mg, and rotigotine patch 3 mg.[142,144] Higher dosing is not recommended due to the risk of augmentation.

Augmentation is mostly commonly associated with the dopamine agonists and carbidopa/levodopa, with greater risk at higher doses.[27] After the first year of pramipexole use, annual discontinuation due to augmentation is 9% per year, according to one report of 164 patients treated with pramipexole over a ten-year period at a tertiary care facility.[29] Rates are thought to be similar between the dopamine agonists pramipexole and ropinirole, although they may be slightly lower with transdermal rotigotine given its slow release over a 24 h period.

If augmentation does develop, then attempts should be made to wean the patient off the offending medication. This approach is typically associated with a marked worsening of symptoms, therefore cross-taper with a medication of a different class is recommended. The second category of medication utilized in RLS treatment is the category of α-2-delta voltage-dependent calcium channels agonists. These medications include gabapentin, pregabalin and gabapentin enacarbil. Most have similar efficacy and may even be superior to dopamine agonists without the risk of augmentation.[23,24,148] As a result, clinicians are increasingly choosing α-2-delta ligands over dopamine agonists as their first line treatment.

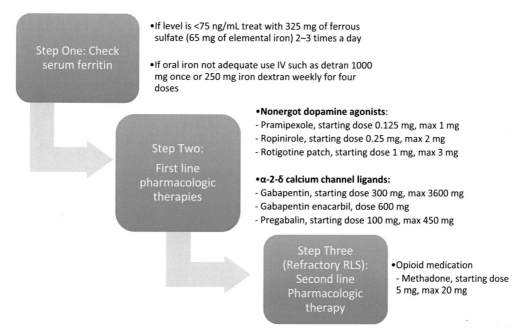

FIGURE 5.3 **Management of chronic persistent RLS.**

For symptoms that remain refractory, opioid medication can prove helpful. Low dose methadone, starting at 5 mg and titrating up to 20 mg daily, is quite effective in treating RLS.[25] Methadone can offer a 75% reduction in symptoms, and a combination of oxycodone and naloxone also significantly improved symptoms over placebo.[25,28] While up to 15% of patients may stop these medications in the first year due to side effects, those who remain on therapy are subsequently able to continue with stable benefit and without augmentation.[29] Side effects that should be taken into account include sleepiness, cognitive changes, nausea, gait imbalance, central sleep apnea and the risk of dependence, although rapid dose escalation is rare (Fig. 5.3).

PLMs occur in up to 80% of patients with RLS. Patients are usually unaware of their occurrence, and PLMs are often confused for RLS. RLS is a sensation that occurs while the patient is awake, while PLMs are movements that occur during sleep. The clinical significance of PLMs is not clear. While they do increase in frequency with age, there is not a clear association with comorbid insomnia or hypersomnia.[149–151] PLMs may be either part of the normal aging process or an epiphenomenon related to other sleep disorders, such as OSA.

HUNTINGTON'S DISEASE

Huntington's disease is an autosomal dominant neurodegenerative disorder caused by a CAG repeat expansion in the huntingtin gene. Classic symptoms include progressive chorea and psychiatric disturbances. Sleep disturbances are common. Patients report significantly more difficulty falling and staying asleep compared to controls, with 80% reporting some form of sleep disturbance.[152] While insomnia does not develop until later in the disease, mild sleep disturbances may be one of the earliest manifestations, often preceding the motor and psychiatric symptoms.[153] By the time that patients develop the motor and neuropsychiatric manifestations of the disease they begin to have shorter, more fragmented sleep and reduced percentages of REM sleep.[154,155] Circadian disturbance is also a common feature of Huntington's disease, and patients with Huntington's disease stage II/III have been shown to have reduced melatonin levels when compared to controls.[156,157] High quality data for treatment of sleep disorders in Huntington's disease are lacking. For circadian disruption, bright light therapy combined with melatonin may be effective.[158]

WILSON'S DISEASE

Wilson's disease is an autosomal recessive disorder of the copper metabolism in which patients have a defective ATP7B protein and cannot properly excrete and store copper, resulting in the accumulation of copper in multiple organs, most prominently the liver and brain.[159] Sleep disruption is common in Wilson's disease. In a sample of 55 patients with Wilson's disease, 80% reported some sleep complaint with more frequent problems in those with primarily neurologic symptoms compared to primarily liver symptoms.[160] Patients with Wilson's disease patients report significantly more EDS and daytime napping compared to controls, as well cataplexy-like episodes.[160] Sleep complaints may be the first symptom of Wilson's disease. In a cohort of 41 patients, 7% reported insomnia as their first symptom, and another 7% reported symptoms of RBD.[84] Given this information, screening for Wilson's disease may be considered in young patients with iRBD, as early treatment may be available in the form of copper chelation.

CONCLUSIONS

Sleep complaints can be a disabling manifestation of many movement disorders, which significantly impact patients' quality of life. Patients with PD, MSA, and DLB have a higher prevalence of insomnia, hypersomnia, circadian disruption, RBD and RLS. Ensuring optimal treatment of nocturnal motor symptoms is very important, and there is emerging data that reduced melatonin levels may be quite common in these patients. MSA patients are at risk of stridor, which can be deadly and is therefore important to screen for. RBD is quite common in all three of these conditions, and can be used as a clinical factor to aid in the diagnosis of DLB when other dementing illnesses are also being considered, such as AD. Idiopathic RBD is now considered an α-synucleinopathy precursor and is best treated with melatonin or clonazepam. RLS is also likely linked to dopamine metabolism, though its etiology remains unclear. Effective treatments include iron supplementation, dopamine agonists, α-2-delta voltage-dependent calcium channels agonists, and opioids. Data continues to emerge on sleep changes and their significance in other movement disorders such as Huntington's and Wilson's disease.

Acknowledgments

The authors would like to thank Drs. Mark Mahowald and Mitchell Miglis for their aid in reviewing this manuscript.

References

1. Breen DP, Vuono R, Nawarathna U, Fisher K, Shneerson JM, Reddy AB, Barker RA. Sleep and circadian rhythm regulation in early Parkinson disease. *JAMA Neurol.* 2014;71(5):589–595.

2. Tholfsen LK, Larsen JP, Schulz J, Tysnes O-B, Gjerstad MD. Development of excessive daytime sleepiness in early Parkinson disease. *Neurology.* 2015;85(2):162–168.

3. Adler CH, Caviness JN, Hentz JG, Lind M, Tiede J. Randomized trial of modafinil for treating subjective daytime sleepiness in patients with Parkinson's disease. *Mov Disord.* 2003;18(3):287–293.

4. Simuni T, Caspell-Garcia C, Coffey C, Chahine LM, Lasch S, Oertel WH, Mayer G, Högl B, Postuma R, Videnovic A, Amara AW, Marek K. Correlates of excessive daytime sleepiness in de novo Parkinson's disease: a case control study. *Mov Disord.* 2015;30(10):1371–1381.

5. Postuma RB, Lang AE, Munhoz RP, Charland K, Pelletier A, Moscovich M, Filla L, Zanatta D, Rios Romenets S, Altman R, Chuang R, Shah B. Caffeine for treatment of Parkinson disease: a randomized controlled trial. *Neurology.* 2012;79(7):651–658.

6. Shpirer I, Miniovitz A, Klein C, Goldstein R, Prokhorov T, Theitler J, Pollak L, Rabey JM. Excessive daytime sleepiness in patients with Parkinson's disease: a polysomnography study. *Mov Disord.* 2006;21(9): 1432–1438.

7. Ondo WG, Perkins T, Swick T, Hull KL, Jimenez JE, Garris TS, Pardi D. Sodium oxybate for excessive daytime sleepiness in Parkinson disease: an open-label polysomnographic study. *Arch Neurol.* 2008;65(10): 1337–1340.

8. Pont-Sunyer C, Iranzo A, Gaig C, Fernández-Arcos A, Vilas D, Valldeoriola F, Compta Y, Fernández-Santiago R, Fernández M, Bayés A, Calopa M, Casquero P, de Fàbregues O, Jaumà S, Puente V, Salamero M, José Martí M, Santamaría J, Tolosa E. Sleep Disorders in Parkinsonian and Nonparkinsonian LRRK2 Mutation Carriers. *PLoS One.* 2015;10(7):e0132368.

9. Barone P, Amboni M, Vitale C, Bonavita V. Treatment of nocturnal disturbances and excessive daytime sleepiness in Parkinson's disease. *Neurology.* 2004;63(8 Suppl 3):S35–S38.

10. Breen DP, Low HL, Misbahuddin A. The impact of deep brain stimulation on sleep and olfactory function in Parkinson's disease. *Open Neurol J.* 2015;9:70–72.

11. Anderson RU, Orenberg EK, Glowe P. OnabotulinumtoxinA office treatment for neurogenic bladder incontinence in Parkinson's disease. *Urology.* 2014;83(1):22–27.

12. Rios Romenets S, Creti L, Fichten C, Bailes S, Libman E, Pelletier A, Postuma RB. Doxepin and cognitive behavioural therapy for insomnia in patients with Parkinson's disease—a randomized study. *Parkinsonism Relat Disord.* 2013;19(7):670–675.

13. Rutten S, Vriend C, van den Heuvel OA, Smit JH, Berendse HW, van der Werf YD. Bright light therapy in Parkinson's disease: an overview of the background and evidence. *Parkinsons Dis.* 2012;2012:767105.

14. Medeiros CAM, Carvalhedo de Bruin PF, Lopes LA, Magalhães MC, de Lourdes Seabra M, de Bruin VMS. Effect of exogenous melatonin on sleep and motor dysfunction in Parkinson's disease. A randomized, double blind, placebo-controlled study. *J Neurol.* 2007;254(4):459–464.

15. Fertl E, Auff E, Doppelbauer A, Waldhauser F. Circadian secretion pattern of melatonin in Parkinson's disease. *J Neural Transm Park Dis Dement Sect.* 1991;3(1):41–47.

16. Fertl E, Auff E, Doppelbauer A, Waldhauser F. Circadian secretion pattern of melatonin in de novo parkinsonian patients: evidence for phase-shifting properties of l-dopa. *J Neural Transm Park Dis Dement Sect.* 1993;5(3):227–234.

17. Neikrug AB, Liu L, Avanzino JA, Maglione JE, Natarajan L, Bradley L, Maugeri A, Corey-Bloom J, Palmer BW, Loredo JS, Ancoli-Israel S. Continuous positive airway pressure improves sleep and daytime sleepiness in patients with Parkinson disease and sleep apnea. *Sleep.* 2014;37(1):177–185.

18. Shimohata T, Tomita M, Nakayama H, Aizawa N, Ozawa T, Nishizawa M. Floppy epiglottis as a contraindication of CPAP in patients with multiple system atrophy. *Neurology.* 2011;76(21):1841–1842.

19. Elder GJ, Colloby SJ, Lett DJ, O'Brien JT, Anderson KN, Burn DJ, McKeith IG, Taylor J-P. Depressive symptoms are associated with daytime sleepiness and subjective sleep quality in dementia with Lewy bodies. *Int J Geriatr Psychiatry.* 2016;31(7):765–770.

20. Boeve BF, Silber MH, Ferman TJ. Melatonin for treatment of REM sleep behavior disorder in neurologic disorders: results in 14 patients. *Sleep Med.* 2003;4(4):281–284.

21. Di Giacopo R, Fasano A, Quaranta D, Della Marca G, Bove F, Bentivoglio AR. Rivastigmine as alternative treatment for refractory REM behavior disorder in Parkinson's disease. *Mov Disord.* 2012;27(4):559–561.

22. Ohayon MM, Caulet M, Priest RG. Violent behavior during sleep. *J Clin Psychiatry*. 1997;58(8):369–376.
23. Garcia-Borreguero D, Patrick J, DuBrava S, Becker PM, Lankford A, Chen C, Miceli J, Knapp L, Allen RP. Pregabalin versus pramipexole: effects on sleep disturbance in restless legs syndrome. *Sleep*. 2014;37(4):635–643.
24. Allen RP, Chen C, Garcia-Borreguero D, Polo O, DuBrava S, Miceli J, Knapp L, Winkelman JW. Comparison of pregabalin with pramipexole for restless legs syndrome. *N Engl J Med*. 2014;370(7):621–631.
25. Ondo WG. Methadone for refractory restless legs syndrome. *Mov Disord*. 2005;20(3):345–348.
26. Guilleminault C, Cetel M, Philip P. Dopaminergic treatment of restless legs and rebound phenomenon. *Neurology*. 1993;43(2):445.
27. Hening WA, Allen RP, Earley CJ, Picchietti DL, Silber MH. An update on the dopaminergic treatment of restless legs syndrome and periodic limb movement disorder. *Sleep*. 2004;27(3):560–583.
28. Frampton JE, Oxycodone/Naloxone PR. A review in severe refractory restless legs syndrome. *CNS Drugs*. 2015;29(6):511–518.
29. Silver N, Allen RP, Senerth J, Earley CJ. A 10-year, longitudinal assessment of dopamine agonists and methadone in the treatment of restless legs syndrome. *Sleep Med*. 2011;12(5):440–444.
30. Young TB. Epidemiology of daytime sleepiness: definitions, symptomatology, and prevalence. *J Clin Psychiatry*. 2004;65(suppl 1):12–16.
31. Hobson DE, Lang AE, Martin WRW, Razmy A, Rivest J, Fleming J. Excessive daytime sleepiness and sudden-onset sleep in Parkinson disease: a survey by the Canadian Movement Disorders Group. *JAMA*. 2002;287(4):455–463.
32. Gjerstad MD, Alves G, Wentzel-Larsen T, Aarsland D, Larsen JP. Excessive daytime sleepiness in Parkinson disease: is it the drugs or the disease?. *Neurology*. 2006;67(5):853–858.
33. Ondo WG, Dat Vuong K, Khan H, Atassi F, Kwak C, Jankovic J. Daytime sleepiness and other sleep disorders in Parkinson's disease. *Neurology*. 2001;57(8):1392–1396.
34. Moreno-López C, Santamaría J, Salamero M, Del Sorbo F, Albanese A, Pellecchia MT, Barone P, Overeem S, Bloem B, Aarden W, Canesi M, Antonini A, Duerr S, Wenning GK, Poewe W, Rubino A, Meco G, Schneider SA, Bhatia KP, Djaldetti R, Coelho M, Sampaio C, Cochen V, Hellriegel H, Deuschl G, Colosimo C, Marsili L, Gasser T, Tolosa E. Excessive daytime sleepiness in multiple system atrophy (SLEEMSA study). *Arch Neurol*. 2011;68(2):223–230.
35. Cochen De Cock V, Bayard S, Jaussent I, Charif M, Grini M, Langenier MC, Yu H, Lopez R, Geny C, Carlander B, Dauvilliers Y. Daytime sleepiness in Parkinson's disease: a reappraisal. *PLoS One*. 2014;9(9):e107278.
36. Högl B, Saletu M, Brandauer E, Glatzl S, Frauscher B, Seppi K, Ulmer H, Wenning G, Poewe W. Modafinil for the treatment of daytime sleepiness in Parkinson's disease: a double-blind, randomized, crossover, placebo-controlled polygraphic trial. *Sleep*. 2002;25(8):905–909.
37. Arnulf I, Leu-Semenescu S. Sleepiness in Parkinson's disease. *Parkinsonism Relat Disord*. 2009;15(suppl 3):S101–S104.
38. Ferreira JJ, Galitzky M, Montastruc JL, Rascol O. Sleep attacks and Parkinson's disease treatment. *Lancet (London, England)*. 2000;355(9212):1333–1334.
39. Paus S, Brecht HM, Köster J, Seeger G, Klockgether T, Wüllner U. Sleep attacks, daytime sleepiness, and dopamine agonists in Parkinson's disease. *Mov Disord*. 2003;18(6):659–667.
40. Ratti P-L, Nègre-Pagès L, Pérez-Lloret S, Manni R, Damier P, Tison F, Destée A, Rascol O. Subjective sleep dysfunction and insomnia symptoms in Parkinson's disease: insights from a cross-sectional evaluation of the French CoPark cohort. *Parkinsonism Relat Disord*. 2015;21(11):1323–1329.
41. Gjerstad MD, Wentzel-Larsen T, Aarsland D, Larsen JP. Insomnia in Parkinson's disease: frequency and progression over time. *J Neurol Neurosurg Psychiatry*. 2007;78(5):476–479.
42. Caap-Ahlgren M, Dehlin O. Insomnia and depressive symptoms in patients with Parkinson's disease. Relationship to health-related quality of life. An interview study of patients living at home. *Arch Gerontol Geriatr*. 2001;32(1):23–33.
43. Chung S, Bohnen NI, Albin RL, Frey KA, Müller MLTM, Chervin RD. Insomnia and sleepiness in Parkinson disease: associations with symptoms and comorbidities. *J Clin Sleep Med*. 2013;9(11):1131–1137.
44. Kurtis MM, Rodriguez-Blazquez C, Martinez-Martin P. Relationship between sleep disorders and other non-motor symptoms in Parkinson's disease. *Parkinsonism Relat Disord*. 2013;19(12):1152–1155.
45. Trenkwalder C, Kies B, Rudzinska M, Fine J, Nikl J, Honczarenko K, Dioszeghy P, Hill D, Anderson T, Myllyla V, Kassubek J, Steiger M, Zucconi M, Tolosa E, Poewe W, Surmann E, Whitesides J, Boroojerdi B, Chaudhuri KR.

Rotigotine effects on early morning motor function and sleep in Parkinson's disease: a double-blind, randomized, placebo-controlled study (RECOVER). *Mov Disord*. 2011;26(1):90–99.

46. Mark MH, Sage JI. Controlled-release carbidopa-levodopa (Sinemet) in combination with standard Sinemet in advanced Parkinson's disease. *Ann Clin Lab Sci*. 1989;19(2):101–106.

47. Menza M, Dobkin RD, Marin H, Gara M, Bienfait K, Dicke A, Comella CL, Cantor C, Hyer L. Treatment of insomnia in Parkinson's disease: a controlled trial of eszopiclone and placebo. *Mov Disord*. 2010;25(11):1708–1714.

48. Yang Y-W, Hsieh T-F, Yu C-H, Huang Y-S, Lee C-C, Tsai T-H. Zolpidem and the risk of Parkinson's disease: a nationwide population-based study. *J Psychiatr Res*. 2014;58:84–88.

49. Czeisler CA. Stability, precision, and near-24-hour period of the human circadian pacemaker. *Science (80-)*. 1999;284(5423):2177–2181.

50. Duffy JF, Cain SW, Chang A-M, Phillips AJK, Münch MY, Gronfier C, Wyatt JK, Dijk D-J, Wright KP, Czeisler CA. Sex difference in the near-24-hour intrinsic period of the human circadian timing system. *Proc Natl Acad Sci USA*. 2011;108(suppl 3):15602–15608.

51. Videnovic A, Noble C, Reid KJ, Peng J, Turek FW, Marconi A, Rademaker AW, Simuni T, Zadikoff C, Zee PC. Circadian melatonin rhythm and excessive daytime sleepiness in Parkinson disease. *JAMA Neurol*. 2014;71(4): 463–469.

52. Kaminska M, Lafontaine A-L, Kimoff RJ. The interaction between obstructive sleep apnea and Parkinson's disease: possible mechanisms and implications for cognitive function. *Parkinsons Dis*. 2015;2015:849472.

53. Zeng J, Wei M, Li T, Chen W, Feng Y, Shi R, Song Y, Zheng W, Ma W. Risk of obstructive sleep apnea in Parkinson's disease: a meta-analysis. *PLoS One*. 2013;8(12):e82091.

54. Sommerauer M, Werth E, Poryazova R, Gavrilov YV, Hauser S, Valko PO. Bound to supine sleep: Parkinson's disease and the impact of nocturnal immobility. *Parkinsonism Relat Disord*. 2015;21(10):1269–1272.

55. Gros P, Mery VP, Lafontaine A-L, Robinson A, Benedetti A, Kimoff RJ, Kaminska M. Obstructive sleep apnea in Parkinson's disease patients: effect of Sinemet CR taken at bedtime. *Sleep Breath*. 2016;20(1):205–212.

56. Merello M, Hughes A, Colosimo C, Hoffman M, Starkstein S, Leiguarda R. Sleep benefit in Parkinson's disease. *Mov Disord*. 1997;12(4):506–508.

57. van Gilst MM, Bloem BR, Overeem S. Sleep benefit in Parkinson's disease: a systematic review. *Parkinsonism Relat Disord*. 2013;19(7):654–659.

58. van Gilst MM, van Mierlo P, Bloem BR, Overeem S. Quantitative motor performance and sleep benefit in Parkinson disease. *Sleep*. 2015;38(10):1567–1573.

59. Bower JH, Maraganore DM, McDonnell SK, Rocca WA. Incidence of progressive supranuclear palsy and multiple system atrophy in Olmsted County, Minnesota, 1976 to 1990. *Neurology*. 1997;49(5):1284–1288.

60. Shimohata T, Nakayama H, Tomita M, Ozawa T, Nishizawa M. Daytime sleepiness in Japanese patients with multiple system atrophy: prevalence and determinants. *BMC Neurol*. 2012;12:130.

61. Seppi K, Högl B, Diem A, Peralta C, Wenning GK, Poewe W. Levodopa-induced sleepiness in the Parkinson variant of multiple system atrophy. *Mov Disord*. 2006;21(8):1281–1283.

62. Högl B, Seppi K, Brandauer E, Wenning G, Poewe W. Irresistible onset of sleep during acute levodopa challenge in a patient with multiple system atrophy (MSA): placebo-controlled, polysomnographic case report. *Mov Disord*. 2001;16(6):1177–1179.

63. Ghorayeb I, Yekhlef F, Chrysostome V, Balestre E, Bioulac B, Tison F. Sleep disorders and their determinants in multiple system atrophy. *J Neurol Neurosurg Psychiatry*. 2002;72(6):798–800.

64. Nam H, Hong Y-H, Kwon H-M, Cho J. Does multiple system atrophy itself affect sleep structure?. *Neurologist*. 2009;15(5):274–276.

65. Iranzo A. Sleep and breathing in multiple system atrophy. *Curr Treat Options Neurol*. 2007;9(5):347–353.

66. Ozawa T, Sekiya K, Aizawa N, Terajima K, Nishizawa M. Laryngeal stridor in multiple system atrophy: clinicopathological features and causal hypotheses. *J Neurol Sci*. 2016;361:243–249.

67. Iranzo A, Santamaria J, Tolosa E, Vilaseca I, Valldeoriola F, Martí MJ, Muñoz E. Long-term effect of CPAP in the treatment of nocturnal stridor in multiple system atrophy. *Neurology*. 2004;63(5):930–932.

68. Silber MH, Levine S. Stridor and death in multiple system atrophy. *Mov Disord*. 2000;15(4):699–704.

69. Ferini-Strambi L, Marelli S. Sleep dysfunction in multiple system atrophy. *Curr Treat Options Neurol*. 2012;14(5):464–473.

70. Shimohata T, Ozawa T, Nakayama H, Tomita M, Shinoda H, Nishizawa M. Frequency of nocturnal sudden death in patients with multiple system atrophy. *J Neurol*. 2008;255(10):1483–1485.

71. Chwiszczuk L, Breitve M, Hynninen M, Gjerstad MD, Aarsland D, Rongve A. Higher frequency and complexity of sleep disturbances in dementia with lewy bodies as compared to Alzheimer's disease. *Neurodegener Dis.* 2016;16(3–4):152–160.

72. Ferman TJ, Smith GE, Dickson DW, Graff-Radford NR, Lin S-C, Wszolek Z, Van Gerpen JA, Uitti R, Knopman DS, Petersen RC, Parisi JE, Silber MH, Boeve BF. Abnormal daytime sleepiness in dementia with Lewy bodies compared to Alzheimer's disease using the Multiple Sleep Latency Test. *Alzheimers Res Ther.* 2014;6(9):76.

73. Pao WC, Boeve BF, Ferman TJ, Lin S-C, Smith GE, Knopman DS, Graff-Radford NR, Petersen RC, Parisi JE, Dickson DW, Silber MH. Polysomnographic findings in dementia with Lewy bodies. *Neurologist.* 2013;19(1):1–6.

74. Terzaghi M, Arnaldi D, Rizzetti MC, Minafra B, Cremascoli R, Rustioni V, Zangaglia R, Pasotti C, Sinforiani E, Pacchetti C, Manni R. Analysis of video-polysomnographic sleep findings in dementia with Lewy bodies. *Mov Disord.* 2013;28(10):1416–1423.

75. Jouvet M, Delorme F. Locus coeruleus et sommeil paradoxal. *CR Soc Biol.* 1965;159:895–899.

76. Schenck CH, Bundlie SR, Ettinger MG, Mahowald MW. Chronic behavioral disorders of human REM sleep: a new category of parasomnia. *Sleep.* 1986;9(2):293–308.

77. Olson EJ, Boeve BF, Silber MH. Rapid eye movement sleep behaviour disorder: demographic, clinical and laboratory findings in 93 cases. *Brain.* 2000;123(Pt 2):331–339.

78. Chiu HF, Wing YK. REM sleep behaviour disorder: an overview. *Int J Clin Pract.* 1997;51(7):451–454.

79. Kang S-H, Yoon I-Y, Lee SD, Han JW, Kim TH, Kim KW. REM sleep behavior disorder in the Korean elderly population: prevalence and clinical characteristics. *Sleep.* 2013;36(8):1147–1152.

80. Iranzo A, Santamaría J, Rye DB, Valldeoriola F, Martí MJ, Muñoz E, Vilaseca I, Tolosa E. Characteristics of idiopathic REM sleep behavior disorder and that associated with MSA and PD. *Neurology.* 2005;65(2):247–252.

81. Postuma RB, Gagnon JF, Vendette M, Fantini ML, Massicotte-Marquez J, Montplaisir J. Quantifying the risk of neurodegenerative disease in idiopathic REM sleep behavior disorder. *Neurology.* 2009;72(15):1296–1300.

82. Schenck CH, Mahowald MW. REM sleep behavior disorder: clinical, developmental, and neuroscience perspectives 16 years after its formal identification in SLEEP. *Sleep.* 2002;25(2):120–138.

83. Bjørnarå KA, Dietrichs E, Toft M. REM sleep behavior disorder in Parkinson's disease—is there a gender difference?. *Parkinsonism Relat Disord.* 2013;19(1):120–122.

84. Tribl GG, Trindade MC, Bittencourt T, Lorenzi-Filho G, Cardoso Alves R, Ciampi de Andrade D, Fonoff ET, Bor-Seng-Shu E, Machado AA, Schenck CH, Teixeira MJ, Barbosa ER. Wilson's disease with and without rapid eye movement sleep behavior disorder compared to healthy matched controls. *Sleep Med.* 2016;17:179–185.

85. Lo Coco D, Cupidi C, Mattaliano A, Baiamonte V, Realmuto S, Cannizzaro E. REM sleep behavior disorder in a patient with frontotemporal dementia. *Neurol Sci.* 2012;33(2):371–373.

86. Ebben MR, Shahbazi M, Lange DJ, Krieger AC. REM behavior disorder associated with familial amyotrophic lateral sclerosis. *Amyotroph Lateral Scler.* 2012;13(5):473–474.

87. Nomura T, Inoue Y, Takigawa H, Nakashima K. Comparison of REM sleep behaviour disorder variables between patients with progressive supranuclear palsy and those with Parkinson's disease. *Parkinsonism Relat Disord.* 2012;18(4):394–396.

88. Chokroverty S, Bhat S, Rosen D, Farheen A. REM behavior disorder in myotonic dystrophy type 2. *Neurology.* 2012;78(24):2004.

89. Iranzo A, Graus F, Clover L, Morera J, Bruna J, Vilar C, Martínez-Rodriguez JE, Vincent A, Santamaría J. Rapid eye movement sleep behavior disorder and potassium channel antibody-associated limbic encephalitis. *Ann Neurol.* 2006;59(1):178–181.

90. Manni R, Ratti P-L, Terzaghi M. Secondary incidental" REM sleep behavior disorder: do we ever think of it?. *Sleep Med.* 2011;12(suppl 2):S50–S53.

91. Gjerstad MD, Boeve B, Wentzel-Larsen T, Aarsland D, Larsen JP. Occurrence and clinical correlates of REM sleep behaviour disorder in patients with Parkinson's disease over time. *J Neurol Neurosurg Psychiatry.* 2008;79(4):387–391.

92. Plazzi G, Corsini R, Provini F, Pierangeli G, Martinelli P, Montagna P, Lugaresi E, Cortelli P. REM sleep behavior disorders in multiple system atrophy. *Neurology.* 1997;48(4):1094–1097.

93. Gagnon JF, Bédard MA, Fantini ML, Petit D, Panisset M, Rompré S, Carrier J, Montplaisir J. REM sleep behavior disorder and REM sleep without atonia in Parkinson's disease. *Neurology.* 2002;59(4):585–589.

94. Boeve BF, Silber MH, Ferman TJ, Lucas JA, Parisi JE. Association of REM sleep behavior disorder and neurodegenerative disease may reflect an underlying synucleinopathy. *Mov Disord.* 2001;16(4):622–630.

95. Schenck CH, Boeve BF, Mahowald MW. Delayed emergence of a parkinsonian disorder or dementia in 81% of older men initially diagnosed with idiopathic rapid eye movement sleep behavior disorder: a 16-year update on a previously reported series. *Sleep Med*. 2013;14(8):744–748.

96. Iranzo A, Tolosa E, Gelpi E, Molinuevo JL, Valldeoriola F, Serradell M, Sanchez-Valle R, Vilaseca I, Lomeña F, Vilas D, Lladó A, Gaig C, Santamaria J. Neurodegenerative disease status and post-mortem pathology in idiopathic rapid-eye-movement sleep behaviour disorder: an observational cohort study. *Lancet Neurol*. 2013;12(5):443–453.

97. Claassen DO, Josephs KA, Ahlskog JE, Silber MH, Tippmann-Peikert M, Boeve BF. REM sleep behavior disorder preceding other aspects of synucleinopathies by up to half a century. *Neurology*. 2010;75(6):494–499.

98. Arnulf I, Neutel D, Herlin B, Golmard J-L, Leu-Semenescu S, Cochen de Cock V, Vidailhet M. Sleepiness in Idiopathic REM Sleep Behavior Disorder and Parkinson Disease. *Sleep*. 2015;38(10):1529–1535.

99. Iranzo A, Lomeña F, Stockner H, Valldeoriola F, Vilaseca I, Salamero M, Molinuevo JL, Serradell M, Duch J, Pavía J, Gallego J, Seppi K, Högl B, Tolosa E, Poewe W, Santamaria J. Decreased striatal dopamine transporter uptake and substantia nigra hyperechogenicity as risk markers of synucleinopathy in patients with idiopathic rapid-eye-movement sleep behaviour disorder: a prospective study [corrected]. *Lancet Neurol*. 2010;9(11):1070–1077.

100. Unger MM, Belke M, Menzler K, Heverhagen JT, Keil B, Stiasny-Kolster K, Rosenow F, Diederich NJ, Mayer G, Möller JC, Oertel WH, Knake S. Diffusion tensor imaging in idiopathic REM sleep behavior disorder reveals microstructural changes in the brainstem, substantia nigra, olfactory region, and other brain regions. *Sleep*. 2010;33(6):767–773.

101. Iwanami M, Miyamoto T, Miyamoto M, Hirata K, Takada E. Relevance of substantia nigra hyperechogenicity and reduced odor identification in idiopathic REM sleep behavior disorder. *Sleep Med*. 2010;11(4):361–365.

102. Nightingale S, Orgill JC, Ebrahim IO, de Lacy SF, Agrawal S, Williams AJ. The association between narcolepsy and REM behavior disorder (RBD). *Sleep Med*. 2005;6(3):253–258.

103. Luca G, Haba-Rubio J, Dauvilliers Y, Lammers G-J, Overeem S, Donjacour CE, Mayer G, Javidi S, Iranzo A, Santamaria J, Peraita-Adrados R, Hor H, Kutalik Z, Plazzi G, Poli F, Pizza F, Arnulf I, Lecendreux M, Bassetti C, Mathis J, Heinzer R, Jennum P, Knudsen S, Geisler P, Wierzbicka A, Feketeova E, Pfister C, Khatami R, Baumann C, Tafti M. Clinical, polysomnographic and genome-wide association analyses of narcolepsy with cataplexy: a European Narcolepsy Network study. *J Sleep Res*. 2013;22(5):482–495.

104. Hoque R, Chesson AL. Pharmacologically induced/exacerbated restless legs syndrome, periodic limb movements of sleep, and REM behavior disorder/REM sleep without atonia: literature review, qualitative scoring, and comparative analysis. *J Clin Sleep Med*. 2010;6(1):79–83.

105. Lam SP, Fong SYY, Ho CKW, Yu MWM, Wing YK. Parasomnia among psychiatric outpatients: a clinical, epidemiologic, cross-sectional study. *J Clin Psychiatry*. 2008;69(9):1374–1382.

106. Postuma RB, Gagnon J-F, Tuineaig M, Bertrand J-A, Latreille V, Desjardins C, Montplaisir JY. Antidepressants and REM sleep behavior disorder: isolated side effect or neurodegenerative signal?. *Sleep*. 2013;36(11):1579–1585.

107. Frauscher B, Iranzo A, Hogl B. Quantification of electromyographic activity during REM sleep in multiple muscles in REM sleep behavior disorder. *Sleep*. 2008;31(5):724–731.

108. Howell MJ, Schenck CH. Restless nocturnal eating: a common feature of Willis-Ekbom Syndrome (RLS). *J Clin Sleep Med*. 2012;8(4):413–419.

109. Provini F, Antelmi E, Vignatelli L, Zaniboni A, Naldi G, Calandra-Buonaura G, Vetrugno R, Plazzi G, Montagna P. Association of restless legs syndrome with nocturnal eating: a case-control study. *Mov Disord*. 2009;24(6):871–877.

110. Olbrich K, Mühlhans B, Allison KC, Hahn EG, Schahin SP, de Zwaan M. Night eating, binge eating and related features in patients with obstructive sleep apnea syndrome. *Eur Eat Disord Rev*. 2009;17(2):120–127.

111. Pedley TA, Guilleminault C. Episodic nocturnal wanderings responsive to anticonvulsant drug therapy. *Ann Neurol*. 1977;2(1):30–35.

112. Yeh S-B, Schenck CH. Sporadic nocturnal frontal lobe epilepsy: a consecutive series of 8 cases. *Sleep Sci*. 2014;7(3):170–177.

113. Beleza P, Pinho J. Frontal lobe epilepsy. *J Clin Neurosci*. 2011;18(5):593–600.

114. Schenck CH, Milner DM, Hurwitz TD, Bundlie SR, Mahowald MW. A polysomnographic and clinical report on sleep-related injury in 100 adult patients. *Am J Psychiatry*. 1989;146(9):1166–1173.

115. Schenck C, Hurwitz T, Mahowald M. Symposium: normal and abnormal REM sleep regulation: REM sleep behaviour disorder: an update on a series of 96 patients and a review of the world literature. *J Sleep Res*. 1993;2(4):224–231.

116. Ju Y-E, Larson-Prior L, Duntley S. Changing demographics in REM sleep behavior disorder: possible effect of autoimmunity and antidepressants. *Sleep Med*. 2011;12(3):278–283.

117. Hickey MG, Demaerschalk BM, Caselli RJ, Parish JM, Wingerchuk DM. Idiopathic rapid-eye-movement (REM) sleep behavior disorder is associated with future development of neurodegenerative diseases. *Neurologist*. 2007;13(2):98–101.

118. Wing YK, Lam SP, Li SX, Yu MWM, Fong SYY, Tsoh JMY, Ho CKW, Lam VKH. REM sleep behaviour disorder in Hong Kong Chinese: clinical outcome and gender comparison. *J Neurol Neurosurg Psychiatry*. 2008;79(12): 1415–1416.

119. Kunz D, Mahlberg R. A two-part, double-blind, placebo-controlled trial of exogenous melatonin in REM sleep behaviour disorder. *J Sleep Res*. 2010;19(4):591–596.

120. Ringman JM, Simmons JH. Treatment of REM sleep behavior disorder with donepezil: a report of three cases. *Neurology*. 2000;55(6):870–871.

121. Sasai T, Inoue Y, Matsuura M. Effectiveness of pramipexole, a dopamine agonist, on rapid eye movement sleep behavior disorder. *Tohoku J Exp Med*. 2012;226(3):177–181.

122. Aurora RN, Zak RS, Maganti RK, Auerbach SH, Casey KR, Chowdhuri S, Karippot A, Ramar K, Kristo DA, Morgenthaler TI. Best practice guide for the treatment of REM sleep behavior disorder (RBD). *J Clin Sleep Med*. 2010;6(1):85–95.

123. Allen RP, Picchietti DL, Garcia-Borreguero D, Ondo WG, Walters AS, Winkelman JW, Zucconi M, Ferri R, Trenkwalder C, Lee HB. Restless legs syndrome/Willis-Ekbom disease diagnostic criteria: updated International Restless Legs Syndrome Study Group (IRLSSG) consensus criteria—history, rationale, description, and significance. *Sleep Med*. 2014;15(8):860–873.

124. Allen RP, Walters AS, Montplaisir J, Hening W, Myers A, Bell TJ, Ferini-Strambi L. Restless legs syndrome prevalence and impact: REST general population study. *Arch Intern Med*. 2005;165(11):1286–1292.

125. Nichols DA, Allen RP, Grauke JH, Brown JB, Rice ML, Hyde PR, Dement WC, Kushida CA. Restless legs syndrome symptoms in primary care: a prevalence study. *Arch Intern Med*. 2003;163(19):2323–2329.

126. Moccia M, Erro R, Picillo M, Santangelo G, Spina E, Allocca R, Longo K, Amboni M, Palladino R, Assante R, Pappatà S, Pellecchia MT, Barone P, Vitale C. A Four-Year Longitudinal Study on Restless Legs Syndrome in Parkinson Disease. *Sleep*. 2016;39(2):405–412.

127. Lee JE, Shin H-W, Kim KS, Sohn YH. Factors contributing to the development of restless legs syndrome in patients with Parkinson disease. *Mov Disord*. 2009;24(4):579–582.

128. Gjerstad MD, Tysnes OB, Larsen JP. Increased risk of leg motor restlessness but not RLS in early Parkinson disease. *Neurology*. 2011;77(22):1941–1946.

129. Ondo W, Jankovic J. Restless legs syndrome: clinicoetiologic correlates. *Neurology*. 1996;47(6):1435–1441.

130. Montplaisir J, Boucher S, Poirier G, Lavigne G, Lapierre O, Lespérance P. Clinical, polysomnographic, and genetic characteristics of restless legs syndrome: a study of 133 patients diagnosed with new standard criteria. *Mov Disord*. 1997;12(1):61–65.

131. García-Martín E, Jiménez-Jiménez FJ, Alonso-Navarro H, Martínez C, Zurdo M, Turpín-Fenoll L, Millán-Pascual J, Adeva-Bartolomé T, Cubo E, Navacerrada F, Rojo-Sebastián A, Rubio L, Ortega-Cubero S, Pastor P, Calleja M, Plaza-Nieto JF, Pilo-de-la-Fuente B, Arroyo-Solera M, García-Albea E, Agúndez JAG. Heme oxygenase-1 and 2 common genetic variants and risk for restless legs syndrome. *Medicine (Baltimore)*. 2015;94(34):e1448.

132. Catoire H, Dion PA, Xiong L, Amari M, Gaudet R, Girard SL, Noreau A, Gaspar C, Turecki G, Montplaisir JY, Parker JA, Rouleau GA. Restless legs syndrome-associated MEIS1 risk variant influences iron homeostasis. *Ann Neurol*. 2011;70(1):170–175.

133. Antelmi E, Vinai P, Pizza F, Marcatelli M, Speciale M, Provini F. Nocturnal eating is part of the clinical spectrum of restless legs syndrome and an underestimated risk factor for increased body mass index. *Sleep Med*. 2014;15(2):168–172.

134. Connor JR, Wang XS, Patton SM, Menzies SL, Troncoso JC, Earley CJ, Allen RP. Decreased transferrin receptor expression by neuromelanin cells in restless legs syndrome. *Neurology*. 2004;62(9):1563–1567.

135. Earley CJ, Connor JR, Beard JL, Malecki EA, Epstein DK, Allen RP. Abnormalities in CSF concentrations of ferritin and transferrin in restless legs syndrome. *Neurology*. 2000;54(8):1698–1700.

136. Rizzo G, Manners D, Testa C, Tonon C, Vetrugno R, Marconi S, Plazzi G, Pizza F, Provini F, Malucelli E, Gramegna LL, Lodi R. Low brain iron content in idiopathic restless legs syndrome patients detected by phase imaging. *Mov Disord*. 2013;28(13):1886–1890.

137. Godau J, Klose U, Di Santo A, Schweitzer K, Berg D. Multiregional brain iron deficiency in restless legs syndrome. *Mov Disord*. 2008;23(8):1184–1187.

138. Earley CJ, Allen RP, Connor JR, Ferrucci L, Troncoso J. The dopaminergic neurons of the A11 system in RLS autopsy brains appear normal. *Sleep Med*. 2009;10(10):1155–1157.

139. Bayard M, Bailey B, Acharya D, Ambreen F, Duggal S, Kaur T, Rahman ZU, Roller K, Tudiver F. Bupropion and restless legs syndrome: a randomized controlled trial. *J Am Board Fam Med*. 2011;24(4):422–428.

140. Delgado Rodrigues RN, Alvim de Abreu E, Silva Rodrigues AA, Pratesi R, Krieger J. Outcome of restless legs severity after continuous positive air pressure (CPAP) treatment in patients affected by the association of RLS and obstructive sleep apneas. *Sleep Med*. 2006;7(3):235–239.

141. Wang J, O'Reilly B, Venkataraman R, Mysliwiec V, Mysliwiec A. Efficacy of oral iron in patients with restless legs syndrome and a low-normal ferritin: A randomized, double-blind, placebo-controlled study. *Sleep Med*. 2009;10(9):973–975.

142. Silber MH, Becker PM, Earley C, Garcia-Borreguero D, Ondo WG. Willis-Ekbom Disease Foundation revised consensus statement on the management of restless legs syndrome. *Mayo Clin Proc*. 2013;88(9):977–986.

143. Auerbach M, Al Talib K. Low-molecular weight iron dextran and iron sucrose have similar comparative safety profiles in chronic kidney disease. *Kidney Int*. 2008;73(5):528–530.

144. Cho YW, Allen RP, Earley CJ. Lower molecular weight intravenous iron dextran for restless legs syndrome. *Sleep Med*. 2013;14(3):274–277.

145. Mehmood T, Auerbach M, Earley CJ, Allen RP. Response to intravenous iron in patients with iron deficiency anemia (IDA) and restless leg syndrome (Willis-Ekbom disease). *Sleep Med*. 2014;15(12):1473–1476.

146. Aurora RN, Kristo DA, Bista SR, Rowley JA, Zak RS, Casey KR, Lamm CI, Tracy SL, Rosenberg RS. The treatment of restless legs syndrome and periodic limb movement disorder in adults—an update for 2012: practice parameters with an evidence-based systematic review and meta-analyses: an American Academy of Sleep Medicine Clinical Practice Guideline. *Sleep*. 2012;35(8):1039–1062.

147. Scholz H, Trenkwalder C, Kohnen R, Riemann D, Kriston L, Hornyak M. Dopamine agonists for restless legs syndrome. *Cochrane database Syst Rev*. 2011;3:CD006009.

148. Wilt TJ, MacDonald R, Ouellette J, Khawaja IS, Rutks I, Butler M, Fink HA. Pharmacologic therapy for primary restless legs syndrome: a systematic review and meta-analysis. *JAMA Intern Med*. 2013;173(7):496–505.

149. Mendelson W. Are periodic leg movements associated with clinical sleep disturbance?. *Sleep*. 1996;19(3):219–223.

150. Youngstedt SD, Kripke DF, Klauber MR, Sepulveda RS, Mason WJ. Periodic leg movements during sleep and sleep disturbances in elders. *J Gerontol A Biol Sci Med Sci*. 1998;53(5):M391–M394.

151. Carrier J, Frenette S, Montplaisir J, Paquet J, Drapeau C, Morettini J. Effects of periodic leg movements during sleep in middle-aged subjects without sleep complaints. *Mov Disord*. 2005;20(9):1127–1132.

152. Goodman AOG, Morton AJ, Barker RA. Identifying sleep disturbances in Huntington's disease using a simple disease-focused questionnaire. *PLoS Curr*. 2010;2:RRN1189.

153. Hansotia P, Wall R, Berendes J. Sleep disturbances and severity of Huntington's disease. *Neurology*. 1985;35(11):1672–1674.

154. Lazar AS, Panin F, Goodman AOG, Lazic SE, Lazar ZI, Mason SL, Rogers L, Murgatroyd PR, Watson LPE, Singh P, Borowsky B, Shneerson JM, Barker RA. Sleep deficits but no metabolic deficits in premanifest Huntington's disease. *Ann Neurol*. 2015;78(4):630–648.

155. Piano C, Losurdo A, Della Marca G, Solito M, Calandra-Buonaura G, Provini F, Bentivoglio AR, Cortelli P. Polysomnographic findings and clinical correlates in Huntington disease: a cross-sectional cohort study. *Sleep*. 2015;38(9):1489–1495.

156. Morton AJ, Wood NI, Hastings MH, Hurelbrink C, Barker RA, Maywood ES. Disintegration of the sleep-wake cycle and circadian timing in Huntington's disease. *J Neurosci*. 2005;25(1):157–163.

157. Kalliolia E, Silajdžić E, Nambron R, Hill NR, Doshi A, Frost C, Watt H, Hindmarsh P, Björkqvist M, Warner TT. Plasma melatonin is reduced in Huntington's disease. *Mov Disord*. 2014;29(12):1511–1515.

158. van Wamelen DJ, Roos RA, Aziz NA. Therapeutic strategies for circadian rhythm and sleep disturbances in Huntington disease. *Neurodegener Dis Manag*. 2015;5(6):549–559.

159. Rodriguez-Castro KI, Hevia-Urrutia FJ, Sturniolo GC. Wilson's disease: a review of what we have learned. *World J Hepatol*. 2015;7(29):2859–2870.

160. Nevsimalova S, Buskova J, Bruha R, Kemlink D, Sonka K, Vitek L, Marecek Z. Sleep disorders in Wilson's disease. *Eur J Neurol*. 2011;18(1):184–190.

6

Sleep and Stroke

G.J. Meskill*, C. Guilleminault**

*Comprehensive Sleep Medicine Associates, Houston, TX, USA
**The Stanford Center for Sleep Sciences and Medicine, Redwood City, CA, USA

INTRODUCTION

Stroke, defined as a focal neurological deficit of acute onset and vascular origin, is the second leading cause of death worldwide and is a major source of physical, psychological, and monetary hardship. There are more than 15 million cases worldwide annually, and 795,000 in the United States alone. In total, 137,000 stroke patients will die from stroke complications, and more than 50% will have physical and mental impairment.[1] Factoring healthcare services, medications, and productivity loss, strokes cost the United States $34 billion annually.[2]

SLEEP APNEA IS A RISK FACTOR FOR STROKE

Obstructive sleep apnea (OSA) is characterized by recurrent episodes of partial or complete interruption in breathing during sleep due to increased upper airway resistance. OSA has been identified as an independent risk factor for several cardiovascular and cerebrovascular

Sleep and Neurologic Disease. http://dx.doi.org/10.1016/B978-0-12-804074-4.00006-6

morbidities, including hypertension, ischemic heart disease, heart failure, atrial fibrillation, and hypertrophic cardiomyopathy.[3–6] The incidence of OSA varies due to significant differences in the sensitivity of testing sites, testing modalities (e.g., in-laboratory polysomnography vs. home sleep testing), and changes in the American Academy of Sleep Medicine (AASM) scoring rules over time. The severity of OSA is measured by the numbers of apneas and hypopneas per hour of sleep, otherwise termed the Apnea-Hypopnea Index (AHI). A recent source estimates that 1 in 5 adults has at least mild OSA.[4] An epidemiologic study of the citizens of Sao Paulo, Brazil demonstrated that moderate OSA, defined as an AHI > 15, was present in 25% of men and 9% of women.[7]

OSA and stroke have a bidirectional relationship: individuals with OSA have a greater risk of stroke, and stroke survivors have a greater incidence of OSA. Studies have shown that 50–75% of patients with stroke or TIA have OSA.[8–10] Despite this relationship, as many as 80% of all patients with OSA are neither diagnosed nor treated. Barriers to diagnosis include patient resistance, provider awareness, and access to diagnostic facilities.[11]

After adjusting for independent cardiovascular risk factors such as hypertension, atrial fibrillation, diabetes, hyperlipidemia, and smoking status, the hazard ratio of myocardial infarction and stroke in patients with OSA was observed in one study to be 2.0 (AHI > 5), and 3.3 in patients with severe OSA (AHI > 30).[12] Another study demonstrated that in patients with moderate OSA (AHI > 20), regardless of the presence of obesity, the odds ratio of stroke occurrence was 4.3.[13] The Sleep Heart Health Study showed that the increased incidence of stroke in men with OSA compared to men without OSA is 2.26.[14]

Meta-analysis of more than 20 studies has demonstrated that more than 50% of patients with stroke have sleep-disordered breathing (SDB), which includes both OSA and Central Sleep Apnea (CSA), with an AHI > 10.[15] Another metaanalysis evaluating 29 stroke articles (ischemic, hemorrhagic, and TIA) suggested that the incidence of SDB was 72% in stroke patients.[16] While SDB improves in the subacute period after a stroke (6 months after stroke onset), as many as 50% of stroke patients continue to have OSA months to years following stroke.[17] Higher AHI values, particularly in men, are associated with higher risk of stroke.[18] While correlation between OSA and stroke has been clearly demonstrated, to date there have been no randomized controlled trials on this topic.

PATHOPHYSIOLOGY

There are two types of strokes: ischemic (85%) and hemorrhagic (15%). An ischemic stroke is a cerebrovascular event in which an intracranial artery is occluded, leading to cerebral ischemia and cell death. An ischemic event that resolves spontaneously before cellular death occurs is considered a transient ischemic attack (TIA). Previously, TIA was defined as an event whose symptoms resolved in less than 24 h. This definition has been replaced because many such events were found to be associated with lesions on diffusion-weighted magnetic resonance imaging (MRI). A hemorrhagic stroke is a cerebrovascular event in which an intracranial vessel ruptures, leading to brain injury.

OSA occurs when changes in cortical regulation of upper airway musculature lead to recurrent episodes of airway obstruction and disruptions in sleep. Relaxation of pharyngeal musculature and the genioglossus during sleep leads to retroposition of these structures, narrowing the diameter of the upper airway.[18] These changes require increased inspiratory effort to generate negative intrathoracic pressure in order to maintain consistent airflow. The resulting increase in airflow velocity through a narrowed lumen causes turbulence and pharyngeal soft tissue vibration (i.e., snoring). This increased airflow velocity creates a vacuum effect on the upper airway soft tissue, and when combined with intrathoracic negative pressure this can lead to partial or complete collapse of the upper airway. Resolution of this occlusion is achieved by intermittent interruptions in cortical sleep, which leads to increased pharyngeal muscle tone and restored patency of the upper airway. More subtle episodes of obstructive breathing, sometimes termed "upper airway resistance syndrome," may not lead to airway obstruction but the respiratory dynamics and increased negative intrathoracic pressure are the same.[19]

Dynamic fluctuations in intrathoracic pressure contribute to long-term cardiovascular consequences. Obstructive respirations during sleep have been demonstrated to cause a leftward shift of the cardiac interventricular septum and pulsus paradoxus.[20] The increased negative pressure in the thoracic cavity leads to increased venous return on diastole, with right atrial and ventricular distention. Conversely, systole is resisted, leading to decreased ejection into the systemic circulation. This can lead to increased atrial natriuretic peptide (ANP) and brain natriuretic peptide (BNP) secretion, inducing nocturnal polyuria as a defense mechanism to reduce cardiac load.[21] Peripheral vasoconstriction has been observed to occur with SDB.[22,23] In addition, when a patient with OSA experiences an obstructive respiratory event during sleep, pulmonary autonomic afferents are largely inhibited due to the prolonged increase in negative intrathoracic pressure, which is the result of inspiring against a closed or partially closed glottis.[24] As a result, hyperventilation is prevented, baroreceptors are stimulated, and sympathetic vasomotor tone increases, leading to peripheral vasoconstriction.[25]

In the context of observed changes in intrathoracic pressure and their resultant effect on cardiac ejection fraction, these changes in the peripheral vasculature could be a response to maintain perfusion pressure. Conversely, given that cortical sleep is fragmented and airway patency is restored by cortical regulation of upper airway muscle tone, cardiac ejection is also normalized. While the peripheral vasculature remains constricted, transient episodes of capillary hypertension occur, which may contribute to end-organ damage. When obstructive apneas recur and cardiac ejection again becomes impaired, the lag between decreased cardiac output and peripheral vascular constriction may cause brief periods of cerebral hypotension, particularly in the distal vasculature. This is one proposed mechanism for the association between OSA and anterior ischemic optic neuropathy.[26]

As SDB becomes more significant, intermittent periods of hypoxia and hypercapnia occur due to periodic airflow impairment. This hypoxia stimulates sympathetic activation, oxidative stress, metabolic derangement, and systemic inflammation.[27] This has been demonstrated in both obstructive and central sleep apnea, indicating that it is the hypoxemia and not the obstruction itself that activates sympathetic tone. In addition, metabolic and sleep dysfunction increases the propensity for obesity, insulin resistance and diabetes mellitus, and nonalcoholic fatty liver disease.[28] The changes in intrathoracic pressure dynamics, sympathetic activation, inflammatory regulation, and their resultant stress to the cardiovascular system contribute to the increased risk of congestive heart failure, hypertension,[29] and atrial fibrillation,[30] all of which are associated with increased risk of stroke.[31–33] The apnea-induced hypoxia also triggers the activation of the so-called "diving reflex," a protective mechanism in all mammals whereby cardiac vagal tone increases, resulting in transient bradycardia. This mechanism helps preserve blood flow to the heart and brain while limiting cardiac oxygen demand. In susceptible individuals, however, the diving reflex can trigger sinus pauses and bradyarrhythmias, such as AV block.[34] When breathing resumes, cardiac output increases and sympathetic tone remains elevated, predisposing susceptible individuals to tachyarrhythmias. Some animal studies have raised the hypothesis that the vibration induced by snoring might have an impact on the atherosclerotic process in adjacent arteries, such as the carotid arteries.[35–37] However, a recent prospective study showed no acceleration in subclinical change in carotid atherosclerosis.[38]

STROKE INCREASES THE RISK OF SLEEP APNEA

Not only is there a large amount of evidence to suggest that sleep apnea is a risk factor for stroke, there is also a large amount of evidence to suggest that stroke increases the risk of sleep apnea. Lesions of the brainstem or certain hemispheric lesions (bulbar and pseudobulbar palsies) can lead to disturbances of respiratory muscles and trigger obstructive respiratory events.[39,40] Rostrolateral medullary lesions may lead to reduced carbon dioxide sensitivity.[41] Bilateral supratentorial lesions have been linked to Cheyne-Stokes respirations due to carbon dioxide hypersensitivity.[42] SDB in stroke patients carries increased neurologic morbidity, including longer hospitalization and earlier neurologic worsening.[43] Chronically, the presence of SDB increases 4-year stroke mortality to 21%.[44] The presence of OSA following stroke worsens 6-month functional outcomes and mortality.[45,46] Stroke patients with OSA performed worse on tests of attention, executive functioning, visual perception, psychomotor ability, and intelligence than those without OSA.[47] It is for all of these reasons that OSA screening should be recommended for all patients with stroke or TIA.

STROKE AND THE SLEEP–WAKE CYCLE

Ischemic strokes most often occur in the morning hours, typically after awakening. Several factors may explain this pattern. Krantz et al. demonstrated that this phenomenon occurred independently of physical and mental activity changes.[48] Physiological mechanisms that may contribute to the early morning increase in ischemia include variations in blood pressure, heart rate, vascular tone, plasma catecholamine levels, and platelet and fibrinolytic activity.[48] In addition, the autonomic instability of rapid eye movement (REM) sleep may play a role, and REM sleep is most sustained in the early morning hours.[49]

While sleep timing affects stroke occurrence, incidence of stroke can also affect sleep–wake regulation. Sleep–wake disturbance occurs in up to half of stroke patients, and may manifest as hypersomnia, fatigue, or insomnia. While these complaints have been observed independently in stroke, they often occur together. Less frequently, parasomnias and sleep-related movement disorders are noted. New-onset restless legs syndrome (RLS) was noted in 12% of patients 1 month following a stroke,[50] however, the etiology of this is unclear.

Hypersomnia is defined as reduced sleep latency, excessive daytime sleepiness (EDS), or increased sleep quantity.[15] Poststroke hypersomnia is observed most frequently in paramedian thalamic and mesencephalic stroke, large hemispheric stroke, and bilateral tegmental pontine

TABLE 6.1   The Epworth Sleepiness Scale (ESS)

How likely are you to doze off or fall asleep in the following situations, in contrast to just feeling tired? This refers to your usual way of life in recent times. Even if you have not done some of these things recently try to work out how they would have affected you. Use the following scale to choose the most appropriate number for each situation:

0 = would never doze
1 = slight chance of dozing
2 = moderate chance of dozing
3 = high chance of dozing

Situation	Value
Sitting and reading	
Watching TV	
Sitting, inactive in a public place (e.g., a theater or meeting)	
As a passenger in a car for an hour without a break	
Lying down to rest in the afternoon when circumstances permit	
Sitting and talking to someone	
Sitting quietly after a lunch without alcohol	
In a car, while stopped for a few minutes in traffic	

stroke. Strokes of the caudate, striatum, medial medulla, caudal pons, and cerebral hemispheres can also be associated with hypersomnia.[51] In severe cases, hypersomnia can develop into extreme apathy. Akinetic mutism is defined by profound apathy, and lack of spontaneous movement and motivation.[52] This presentation has been associated with bilateral thalamic infarcts. Patients with bilateral thalamic strokes may present with "pseudo-hypersomnia," manifesting with symptoms more akin to apathy than sleepiness.[53]

Objective determination of EDS typically involves the use of patient surveys, similar to those used to define pain objectively. The most commonly used scale is the Epworth Sleepiness Scale (ESS). The self-administered questionnaire consists of eight scenarios, each of which the respondent assigns a likelihood of dozing/falling asleep "in recent times." The eight scenarios are common sedentary activities (Table 6.1). For each scenario, a value of 0–3 is assigned, with 0 being unlikely and 3 being very likely to doze. The total of the eight values is the Epworth score.[54] Values less than 10 are considered to be in the normal range, while values of 10 or greater are indicative of EDS.[55]

Stroke patients may also experience fatigue. Fatigue is defined as a sensation of physical tiredness or lack of energy without increased sleepiness.[15] The distinction between sleepiness and fatigue is analogous to the comparison of a sleep-deprived resident physician and marathon finisher. Both might express feeling "tired," but the former would be far more likely to report an elevated ESS than the latter. Fatigue symptoms are sometimes reported in brainstem strokes.[56]

Rarely, stroke patients can complain of acute insomnia, defined as difficulty initiating or maintaining sleep, early awakenings, and/or insufficient sleep quality.[57] Cases of neurogenic insomnia have been reported with pontine strokes.[58] Recent investigation has demonstrated diminished health-related quality of life in stroke survivors with chronic insomnia.[59]

STROKE, PARASOMNIAS, AND SLEEP-RELATED MOVEMENT DISORDERS

The most recent edition of the International Classification of Sleep Disorders defines parasomnias as "undesirable physical events or experiences that occur during entry into sleep, within sleep, or during arousal from sleep." REM sleep behavior disorder (RBD) is a REM parasomnia that has been has been observed in patients after a pontine stroke, especially if the stroke involves the pontine tegmentum,[60,61] the area containing the cholinergic neurons responsible for normal REM atonia.

RLS and periodic limb movements of sleep (PLMS) have been observed following strokes, mainly of subcortical structures (pons, thalamus, basal ganglia) and the corona radiata.[50] As stated earlier, there is no explanation for why this occurs following stroke. RLS and PLMS have been reported in patients without prior history or family history of the phenomena. Primary factors in the pathophysiology of RLS and PLMS include brain iron deficiency, impairment of central nervous dopamine regulation, and genetic predisposition. Iron plays a key role in brain dopamine regulation and synaptic density.[57] First line therapy includes dopamine agonist therapy and alpha-2-delta ligands, such as gabapentin.[62]

DIAGNOSIS

Suspicion of SDB is based on clinical history. Classic signs and symptoms include snoring, excessive daytime sleepiness, pauses in breathing during sleep, and nocturnal awakenings. There are simple screening questionnaires that may be used such as the Berlin, STOP-BANG, and Nordic Sleep questionnaires (Tables 6.2 and 6.3). In stroke patients suffering from cognitive impairment, questionnaires should be completed by their bed partner.

However, many patients with SDB do not report daytime sleepiness, and therefore it is imperative that the clinician considers subtle clues suggestive of SDB. Frequent headaches that occur out of sleep or upon awaking should prompt the clinician to consider SDB. Nocturnal bruxism has also been associated with SDB, with bruxing serving as a defense mechanism to restore airway patency by thrusting the mandible forward. Temporomandibular joint (TMJ) pain and headaches can develop when bruxism becomes tonic and sustained for prolonged periods during sleep. Researchers have demonstrated that nocturnal bruxism symptoms are often relieved with the use of CPAP therapy, presumably because mandibular protrusion and clenching is no longer necessary to restore airway patency once the occlusive process is eliminated with positive airway pressure.[65,66]

Frequent episodes of nocturia may also suggest SDB. This is thought to be the result of increased urine production due to intrathoracic pressure swings that stimulate secretion of ANP and BNP. This may also manifest as sleep-related enuresis. Nasal congestion developing over the course of sleep can be a byproduct of vibratory airflow, leading to irritation and mucous production. Likewise, airway narrowing is associated with oral breathing, which can result in complaints of dry mouth. Occasionally, sleep fragmentation from SDB is associated with parasomnias such as sleep talking, sleepwalking, night terrors, sleep paralysis, and other sleep-related behaviors.[57]

Diagnosis of SDB requires diagnostic testing, either in-laboratory or at home. For cases involving classic symptoms without significant cardiovascular morbidity or sleep-related

TABLE 6.2   Berlin Questionnaire[63]

Category 1 (1–5)	Category 2 (6–8)

1. Do you snore?
  a. Yes (+1)
  b. No
  c. I don't know
	6. How often do you feel tired or fatigued after your sleep?
	  a. Nearly every day (+1)
	  b. 3–4 times a week (+1)
	  c. 1–2 times a week
	  d. 1–2 times a month
	  e. Never or nearly never

2. (If you snore:) Your snoring is:
  a. Slightly louder than breathing
  b. As loud as talking
  c. Louder than talking (+1)
  d. Very loud—can be heard in adjacent rooms (+1)
	7. During your waking time, do you feel tired, fatigued, or not up to par?
	  a. Nearly every day (+1)
	  b. 3–4 times a week (+1)
	  c. 1–2 times a week
	  d. 1–2 times a month
	  e. Never or nearly never

3. How often do you snore
  a. Nearly every day (+1)
  b. 3–4 times a week (+1)
  c. 1–2 times a week
  d. 1–2 times a month
  e. Never or nearly never
	8. Have you ever nodded off or fallen asleep while driving a vehicle?
	  a. Yes (+1)
	  b. No

4. Has your snoring ever bothered other people?
  a. Yes (+1)
  b. No
  c. I don't know
	9. (If your answer to #8 was yes:) How often does this occur?
	  a. Nearly every day
	  b. 3–4 times a week
	  c. 1–2 times a week
	  d. 1–2 times a month
	  e. Never or nearly never

5. Has anyone noticed that you quit breathing during your sleep?
  a. Nearly every day (+2)
  b. 3–4 times a week (+2)
  c. 1–2 times a week
  d. 1–2 times a month
  e. Never or nearly never
	Category 3 (#10)
	10. Do you have high blood pressure
	  a. Yes
	  b. No
	  c. I don't know

Scoring:
Category 1 (questions 1–5) is considered positive if the total score is 2 or greater.
Category 2 (questions 6–8) is considered positive if the total score is 2 or greater.
Category 3 (question 10) is considered positive if the answer is "yes."
The answer to question 9 should be noted separately.
High risk of OSA: 2 or more categories that are positive
Low risk of OSA: 0–1 categories that are positive

movement disorders, home sleep testing (HST) is adequate for diagnosis. HSTs are capable of detecting changes in airflow and oxygen desaturations, and they can estimate sleep efficiency using actigraphy sensors. In patients with cardiovascular, neuromuscular or pulmonary co-morbidities, sleep-related movement disorders, or where a HST is not diagnostic but pretest clinical suspicion remains high, an in-laboratory polysomnogram (PSG) is required.[67]

TABLE 6.3   Stop Bang Questionnaire[64]

Snoring: do you snore loudly (louder than talking or loud enough to be heard through closed doors? (YES or NO)	BMI: is your BMI more than 35 kg/m²? (YES or NO)
Tired: do you often feel tired, fatigued, or sleepy during the daytime? (YES or NO)	Age: are you over 50 years old? (YES or NO)
Observed: has anyone observed you stop breathing during your sleep? (YES or NO)	Neck circumference: is your neck circumference greater than 40 cm (15.5 in)? (YES or NO)
Blood Pressure: do you have or are you being treated for high blood pressure? (YES or NO)	Gender: are you male? (YES or NO)

High risk of OSA: answering yes to three or more items
Low risk of OSA: answering yes to less than three items

TREATMENT

Treatment options for OSA are determined based on the severity of the disorder. In mild to moderate OSA (AHI < 30), first-line treatments include positive airway pressure (PAP), oral appliance therapy (OAT), and various surgical interventions. Alternatives to PAP therapy have low success rates in severe OSA, and thus these are recommended as second-line options in these cases.[68–70]

PAP works to eliminate SDB by compressing room air and delivering it through a mask interface to the patient. The PAP device thus creates an increased pressure environment in the upper airway, keeping the relaxed musculature from collapsing during respiration. A common misperception about PAP therapy is that it "blows air" or is an oxygen-delivering device. Rather, the device delivers room air at a preset increased pressure. PAP therapy is considered the reference standard treatment for obstructive breathing and can be employed in all degrees of severity.[15]

However, not all patients can tolerate PAP therapy. While this is true for many nonstroke patients, stroke survivors may face additional challenges, including cognitive impairment, confusion, and dominant arm weakness.[71] Presence of a caregiver overnight may be among the most important factors for successful initiation of PAP therapy. For those who cannot tolerate or who are unlikely to be able to manage PAP therapy due to disability from stroke, alternative therapy can be considered.

OAT can be prescribed as a first-line treatment for mild-to-moderate OSA, as well as for severe OSA in patients who have failed PAP therapy. There are more than 100 OAT models currently available, and most act as mandibular advancement splints (MAS). A custom-made appliance is constructed, using the maxillary portion as the tension point to hold the mandible in a protruded position. This advancement typically is between 6 and 10 mm, and this helps to prevent the retroposition of the mandible while increasing the diameter of the upper airway.[72] A related concept is oral pressure therapy (OPT), which applies negative pressure to the tongue and soft tissue, pulling the tongue and soft palate away from the posterior nasopharynx. However, there is no data assessing how this device performs in stroke patients with impaired tongue and upper airway muscular function.

Surgical procedures may also be considered in certain patients. Nasal resistance can be improved by nasal septoplasty and inferior turbinate reduction. Adenotonsillectomy is a first-line treatment for pediatric cases,[73] but with variable results and risk of recurrence within 12–24 months.[74] Depending on the clinical scenario, this may also be considered in adults. Uvulopalatopharyngoplasty (UPPP) involves tonsillar resection, suturing of the tonsillar pillars, and partial or complete resection of the uvula, but this procedure is not recommended in patients with neurological impairment.

Maxillomandibular advancement (MMA) increases the size of the upper airway by advancing the maxillary and mandibular bones, but neurological impairment of the upper airway has been considered a contraindication to such surgery. Hypoglossal nerve stimulation (HGNS) can be considered in certain cases of PAP failure. A stimulator is implanted subcutaneously on the right breast, and it is connected to an intrathoracic pressure sensor and a wire implanted in the neck. The stimulator sends periodic electrical discharges to the hypoglossal nerve, maintaining genioglossal muscle tone to impede airway collapse.[70] However, depending on the degree and type of neurologic impairment, HGNS may not be appropriate in certain stroke patients. A thorough evaluation of the patient's neurologic condition and close collaboration between the neurologist, sleep specialist, and surgeon is advised.

While each of these treatment options has proven efficacy in treating OSA, only PAP therapy has a robust poststroke outcome data. Martinez-Garcia et al. conducted an observational study of 223 patients consecutively admitted for stroke. Patients with moderate-to-severe OSA (AHI > 20) who could not tolerate CPAP had an increased incidence of new ischemic strokes ($n = 68$, hazard ratio 2.87, $p = 0.03$), compared with patients with moderate-to-severe OSA who tolerated CPAP ($n = 28$), patients with mild disease (AHI < 20, $n = 36$), and patients without OSA (AHI <10, $n = 31$).[75]

Bravata et al. demonstrated a greater reduction in the NIH Stroke Scale (NIHSS) at 30 days in patients compliant with CPAP (NIHSS −3.0) when compared to control patients (NIHSS −1.0, $p = 0.03$).[76] Similarly, another study showed that early intervention with CPAP used for more than 6.5 h per night showed greater improvement in the NIHSS within the first 8 days than the controls (2.30 vs. 1.40, $p = 0.022$).[77] Parra et al. studied neurological improvement in patients with stroke and moderate-to-severe OSA (AHI > 20). The percentage of patients with neurological improvement (measured by Rankin scale) 1 month following stroke was significantly higher with CPAP use (90.9% vs. 56.3%, $p < 0.01$).[78]

As mentioned earlier, use of nasal PAP may be a challenge in subjects with hemiparesis, particularly of the dominant arm, and also persistent nocturnal confusional events (i.e., "sundowning"). Stroke patients also may present with a combination of obstructive and central sleep-disordered breathing events, and CPAP may induce the occurrence of additional central events (i.e., treatment-emergent central apneas). In these cases, other types of nasal PAP should be considered, such as Bi-Level PAP with a backup respiratory rate or adaptoservo ventilation (ASV). If ASV is considered as a treatment mode, then cardiac function must be assessed with performance of at least of an echocardiogram to evaluate the ejection fraction (EF). ASV is contraindicated in subjects with EF below 45%.[79]

CONCLUSIONS

Stroke is a major cause of morbidity and mortality. SDB imparts a greater risk of stroke, and stroke imparts a greater risk of SDB. Stroke is also associated with an increased incidence of other sleep disorders, including insomnia, circadian dysrhythmia, parasomnias, and sleep-related movement disorders. Since as many as 80% of all patients with OSA are neither diagnosed nor treated, it remains a major public health risk.

While preventative therapy is ideal (i.e., treating OSA prior to the development of stroke), the literature clearly indicates that poststroke treatment of OSA with CPAP is critical to reducing the recurrence of stroke. Since incidence of OSA is quite high in patients with TIA and stroke, we recommend screening for SDB in all patients with stroke or TIA. Readily available screening tools include the Epworth Sleepiness Scale, as well as the STOP-BANG, Berlin, and Nordic Sleep questionnaires. In patients in whom SDB is suspected, home or in-laboratory diagnostic testing should be performed.

Treatment options include PAP, oral appliance therapy, oral pressure therapy, hypoglossal nerve stimulator implantation, and surgical manipulation of the soft tissue of the throat and craniofacial bones. However, at this time only PAP therapy has proven efficacy in reducing stroke recurrence in patients with OSA. Therefore, PAP should be considered first-line therapy for stroke patients with OSA, while the other interventions should be reserved for patients unable to acclimate to this treatment.

References

1. Mohsenin V. Obstructive sleep apnea: a new preventive and therapeutic target for stroke a new kid on the block. *Am J Med*. 2015;128:811–816.
2. Mozaffarian D, Benjamin EJ, Go AS, et al. Heart disease and stroke statistics—2015 update: a report from the American Heart Association. *Circulation*. 2015;131:e29–322.
3. Gonzaga C, Bertolami A, Bertolami M, Amodeo C, Calhoun D. Obstructive sleep apnea, hypertension and cardiovascular diseases. *J Hum Hypertens*. 2015;29:705–712.
4. Somers VK, White DP, Amin R, Abraham WT, Costa F, Culebras A, et al. Sleep apnea and cardiovascular disease: an American Heart Association/American College of Cardiology Foundation Scientific Statement from the American Heart Association Council for High Blood Pressure Research Professional Education Committee, Council on Clinical Cardiology, Stroke Council, and Council On Cardiovascular Nursing. In collaboration with the National Heart, Lung, and Blood Institute National Center on Sleep Disorders Research (National Institutes of Health). *Circulation*. 2008;118:1080–1111.
5. Floras JS. Sleep apnea and cardiovascular risk. *J Cardiol*. 2014;63:3–8.
6. Young T, Finn L, Peppard PE, Szklo-Coxe M, Austin D, Nieto FJ, et al. Sleep disordered breathing and mortality: eighteen-year follow-up of the Wisconsin sleep cohort. *Sleep*. 2008;31:1071–1078.
7. Tufik S, Santos-Silva R, Taddei JA, Bittencourt LR. Obstructive sleep apnea syndrome in the Sao Paulo Epidemiologic Sleep Study. *Sleep Med*. 2010;11:441–446.
8. Wessendorf T, Teschler H, Wang Y-M, Konietzko N, Thilmann A. Sleep-disordered breathing among patients with first-ever stroke. *J Neurol*. 2000;247:41–47.
9. Turkington P, Bamford J, Wanklyn P, Elliott M. Prevalence and predictors of upper airway obstruction in the first 24 hours after acute stroke. *Stroke*. 2002;33:2037–2042.
10. Harbison J, Ford G, James OF, Gibson G. Sleep-disordered breathing following acute stroke. *Q J Med*. 2002;95: 741–747.
11. Colten H, Abboud F, Block G, Boat T, Litt I, Mignot E, Miller R, Nieto J, Pack A, Parker K, Potolicchio S, Redline S, Reynolds C, Saper C. *Sleep Disorders and Sleep Deprivation: An Unmet Public Health Problem*. Washington, DC: National Academy of Sciences; 2006.
12. Yaggi H, Concato J, Kernan WN, et al. Obstructive sleep apnea as a risk factor for stroke and death. *N Engl J Med*. 2005;359:2034–2041.
13. Arzt M, Young T, Finn L, Skatrud JB, Bradley C. Association of sleep-disordered breathing and the occurrence of stroke. *Am J Resp Crit Care Med.*. 2005;172:1447–1451.
14. Redline S, Yenokyan G, Gottlieb DJ, Shahar E, O'Connor GT, et al. Obstructive sleep apnea-hypopnea and incident stroke: The sleep heart health study. *Am J Respir Crit Care Med*. 2010;182(2):269–277.

15. M.H. Kryger, T. Roth, and W.C. Dement. *Principles and Practices of Sleep Medicine.*5th ed. W.B. Saunders Co., 2011.

16. Johnson KG, Johnson DC. Frequency of sleep apnea in stroke and TIA patients: a metaanalysis. *J Clin Sleep Med.* 2010;15:131–137.

17. Bassetti C, Milanova M, Gugger M. Sleep disordered breathing and acute stroke: diagnosis, risk factors, treatment, and long-term outcome. *Stroke.* 2006;37:967–972.

18. Remmers JE, deGroot WJ, Sauerland EK, et al. Pathogenesis of upper airway occlusion during sleep. *J Appl Physiol.* 1978;44(6):931–938.

19. Vandenbussche NL, Overeem S, van Dijk JP, Simons PJ, Pevernagie DA. Assessment of respiratory effort during sleep: Esophageal pressure versus noninvasive monitoring techniques. *Sleep Med Rev.* 2014;24C:28–36.

20. Shiomi T, Guilleminault C, Maekawa M, Nakamura A, Yamada K. Flow velocity paradoxus and pulsus paradoxus in obstructive sleep apnea syndrome. *Chest.* 1993;103(5):1629–1631.

21. Hu K, Tu ZS, Lü SQ, Li QQ, Chen XQ. Urodynamic changes in patients with obstructive sleep apnea-hypopnea syndrome and nocturnal polyuria. *Zhonghua Jie He He Hu Xi Za Zhi.* 2011;34(3):182–186.

22. Narkiewicz K, Somers VK. The sympathetic nervous system and obstructive sleep apnea: implications for hypertension. *J Hypertens.* 1997;15(12 Pt 2):1613–1619.

23. Imadojemu VA, Gleeson K, Gray KS, Sinoway LI, Leuenberger UA. Obstructive apnea during sleep is associated with peripheral vasoconstriction. *Am J Respir Crit Care Med V 165.* 2002;(1):61–66.

24. Somers VK, Dyken ME, Clary MP, et al. Sympathetic neural mechanisms in obstructive sleep apnea. *J Clin Invest.* 1995;96(4):1897–1904.

25. Dempsey JA, Veasey SC, Morgan BJ, O'Donnell CP. Pathophysiology of sleep apnea. *Physiol Rev.* 2010;90(1):47–112.

26. Fraser CL. Obstructive sleep apnea and optic neuropathy: is there a link?. *Curr Neurol Neurosci Rep.* 2014;14(8):465.

27. Lattimore JD, Celermajer DS, Wilcox I. Obstructive sleep apnea and cardiovascular disease. *J Am Cardiol.* 2003;41:1429–1437.

28. Drager LF, Togeiro SM, Polotsky VY. Lorenzi-Filho. Obstructive sleep apnea: a cardiometabolic risk in obesity and the metabolic syndrome. *J Am Coll Cardiol.* 2013;62:569–576.

29. Peppard PE, Young T, Palta M, Skatrud J. Prospective study of the association between sleep-disordered breathing and hypertension. *N Engl J Med.* 2000;342:1378–1384.

30. Lavergne F, Morin L, Armistead J, Benjafield A, Richards G, Woehrle H. Atrial fibrillation and sleep-disordered breathing. *J Thorac Dis.* 2015;7(12):E575–E584.

31. Shamsuzzaman AS, Gersh BJ, Somers VK. Obstructive sleep apnea. *JAMA.* 2003;290(14):1906–1914.

32. Gami AS, Pressmann G, Caples SM, et al. Association of atrial fibrillation and obstructive sleep apnea. *Circulation.* 2004;110(4):364–367.

33. Valham F, Mooe T, Rabben T, et al. Increased risk of stroke in patients with coronary artery disease and sleep apnea: a 10-year follow-up. *Circulation.* 2008;118(9):955–960.

34. Guilleminault C, Connolly SJ, Winkle RA. Cardiac arrhythmia and conduction disturbances during sleep in 400 patients with sleep apnea syndrome. *Am J Cardiol.* 1983;52:490–494.

35. Almendros I, Acerbi I, Puig F, Montserrat JM, Navajas D, Farre R. Upper airway inflammation triggered by vibration in a rat model of snoring. *Sleep.* 2007;30:225–227.

36. Puig F, Rico F, Almendros I, Montserrat JM, Navajas D, Farre R. Vibration enhances interleukin-8 release in a cell model of snoring-induced airway inflammation. *Sleep.* 2005;28:1312–1316.

37. Amatoury J, Howitt L, Wheatley JR, Avolio AP, Amis TC. Snoring-related energy transmission to the carotid artery in rabbits. *J Appl Physiol.* 2006;100:1547–1553.

38. Kim J, Pack A, Maislin G, Lee SK, Kim SH, Shin C. Prospective observation on the association of snoring with subclinical changes in carotid atherosclerosis over four years. *Sleep Med.* 2014;15(7):769–775.

39. Askenasy JJM, Goldhammer I. Sleep apnea as a feature of bulbar stroke. *Stroke.* 1988;19(5):637–639.

40. Chaudhary BA, Elguindi A, Kinf DW. Obstructive sleep apnea after lateral medullary syndrome. *South Med J.* 1982;75(1):65–67.

41. Morrell MJ, Heywood P, Moosawi SH, et al. Unilateral focal lesions in the rostral medulla influence chemosensitivity and breathing measured during wakefulness, sleep, and exercise. *J Neurol Neurosurg Psychiatry.* 1999;67(5):637–645.

42. Brown HW, Plum F. The neurologic basis of Cheyne-Stokes respiration. *Am J Med.* 1961;30(6):849–869.

43. Iranzo A, Santamaria J, Berenguer J, et al. Prevalence and clinical importance of sleep apnea in the first night after cerebral infarction. *Neurology.* 2002;58(6):911–916.

44. Dyken ME, Somers VK, Yamada T, et al. Investigating the relationship between stroke and obstructive sleep apnea. *Stroke.* 1996;27(3):401–407.

45. Yan-Fang S, Yu-Ping W. Sleep-disordered breathing: impact on functional outcome of ischemic stroke patients. *Sleep Med*. 2009;10(7):717–719.

46. Turkington PM, Allgar V, Bamford J, et al. Effect of upper airway obstruction in acute stroke on functional outcome at 6 months. *Thorax*. 2004;59(5):367–371.

47. Aaronson JA, van Bennekom CA, Hofman WF, van Bezeij T, van den Aardweg JG, Groet E, Kylstra WA, Schmand B. Obstructive sleep apnea is related to impaired cognitive and functional status after stroke. *Sleep*. 2015;38(9):1431–1437.

48. Krantz DS, Kop WJ, Gabbay FH, et al. Circadian variation of ambulatory myocardial infarction. Triggering by daily activities and evidence for an endogenous circadian component. *Circulation*. 1996;93:1364–1371.

49. Verrier RL, Muller JE, Hobson JA. Sleep, dreams, and sudden death: the case for sleep as an autonomic stress test for the heart. *Cardiovasc Res*. 1996;31:181–211.

50. Lee SJ, Kim JS, Song IU, et al. Poststroke restless legs syndrome and lesion location: anatomical considerations. *Mov Disord*. 2009;24(1):77–84.

51. Bassetti CL, Valko P. Poststroke hypersomnia. *Sleep Med Clinics*. 2006;1(1):139–155.

52. Cairns H, Oldfield RC, Pennybacker JB, Whitteridge D. Akinetic mutism with an epidermoid cyst of the third ventricle. *Brain*. 1941;64(4):273–290.

53. Guilleminault C, Quera-Salva MA, Goldberg MP. Pseudo-hypersomnia and pre-sleep behaviour with bilateral paramedian thalamic lesions. *Brain*. 1993;116(6):1549–1563.

54. Johns MW. A new method for measuring daytime sleepiness: the Epworth Sleepiness Scale. *Sleep*. 1991;14:540–555.

55. Johns MW, Hocking B. Daytime sleepiness and sleep habits of Australian workers. *Sleep*. 1997;20:844–949.

56. Staub F, Bogousslavksy J. Fatigue after stroke: a major but neglected issue. *Cerebrovasc Dis*. 2001;12(2):75–81.

57. American Academy of Sleep Medicine*International Classification of Sleep Disorders*. 3rd ed. Darien, IL: American Academy of Sleep Medicine; 2014.

58. Freemon FR, Salinas-Garcia RF, Ward JW. Sleep patterns in a patient with brainstem infarction involving the raphe nucleus. *Electroencephalogr Clin Neurophysiol*. 1974;36:657–660.

59. Tang WK, Grace Lau C, Mok V, Ungvari GS, Wong KS. Insomnia and health-related quality of life in stroke. *Top Stroke Rehabil*. 2015;22(3):201–207.

60. Culebras A, Moore JT. Magnetic resonance findings in REM sleep behavior disorder. *Neurology*. 1989;39(11):1519–1523.

61. Kimura K, Tachibana N, Kohyama J, et al. A discrete pontine ischemic lesion could cause REM sleep behavior disorder. *Neurology*. 2000;55(6):894–895.

62. Aurora RN, Kristo DA, Bista SR, Rowley JA, Zak RS, Casey KR, Lamm CI, Tracy SL, Rosenberg RS. The treatment of restless legs syndrome and periodic limb movement disorder in adults—an update for 2012: practice parameters with an evidence-based systematic review and meta-analyses. *Sleep*. 2012;35:1039–1062.

63. Netzer NC, Stoohs RA, Netzer CM, Clark K, Strohl KP. Using the Berlin Questionnaire to identify patients at risk for the sleep apnea syndrome. *Ann Intern Med*. 1999 Oct 5;131(7):485–491.

64. Chung F, Yegneswaran B, Liao P, Chung S, Vairavanathan S, Islam S, Khajehdehi A, Shapiro CM. STOP Questionnaire: a tool to screen patients for obstructive sleep apnea. *Anesthesiology*. 2008;108:812–821.

65. Oksenberg A, Arons E. Sleep bruxism related to obstructive sleep apnea: the effect of continuous positive airway pressure. *Sleep Med*. 2002;3(6):513–515.

66. Simmons, J. Prehn, R. Nocturnal bruxism as a protective mechanism against obstructive breathing during sleep. Abstract: 2009 APSS conference.

67. Collop NA, Anderson WM, Boehlecke B, Claman D, Goldberg R, Gottlieb DJ, Hudgel D, Sateia M, Schwab R. Clinical guidelines for the use of unattended portable monitors in the diagnosis of obstructive sleep apnea in adult patients. *J Clin Sleep Med*. 2007;3:737–747.

68. Kushida CA, MD, Littner MR, Hirshkowitz M, Morgenthaler TI, Alessi CA, Bailey D, Boehlecke B, Brown TM, Coleman J, Friedman L, Kapen S, Kapur VK, Kramer M, Lee-Chiong T, Owens J, Pancer JP, Swick TJ, Wise MS. Practice parameters for the use of continuous and bilevel positive airway pressure devices to treat adult patients with sleep-related breathing disorders. *Sleep*. 2006;29:375–380.

69. Kushida CA, Morgenthaler TI, Littner MR, Alessi CA, Bailey D, Coleman J, Friedman L, Hirshkowitz M, Kapen S, Kramer M, Lee-Chiong T, Owens J, Pancer JP. Practice parameters for the treatment of snoring and obstructive sleep apnea with oral appliances: an update for 2005. *Sleep*. 2006;29:240–243.

70. Aurora NR, Casey KR, Kristo D, Auerbach S, Bista SR, Chowdhuri S, Karippot A, Lamm C, Ramar K, Zak R, Morgenthaler TI. Practice parameters for the surgical modifications of the upper airway for obstructive sleep apnea in adults. *Sleep*. 2010;33:1408–1413.

71. Palombini L, Guilleminault C. Stroke and treatment with nasal CPAP. *Eur J Neurol*. 2006;13:198–200.
72. Cartwright R, Ferguson K, Rogers R, Schmidt-Nowara W. Oral appliances for snoring and obstructive sleep apnea: a review. *Sleep*. 2005;29:244–262.
73. Marcus C, et al. Diagnosis and management of childhood obstructive sleep apnea syndrome. *Pediatrics*. 2012;130(3):714–755.
74. Huang YS, Guilleminault C, Lee LA, Lin CH, Hwang FM. Treatment outcomes of adenotonsillectomy for children with obstructive sleep apnea: a prospective longitudinal study. *Sleep*. 2014;37:71–76.
75. Martínez-García M, Campos-Rodríguez F, Soler-Cataluña J, Catalán-Serra P, Román-Sánchez P, Montserrat J. Increased incidence of nonfatal cardiovascular events in stroke patients with sleep apnoea: effect of CPAP treatment. *Eur Respir J*. 2012;39:906–912.
76. Bravata D, Concato J, Fried T, Ranjbar N, Sadarangani T, McClain V, Struve F, Zygmunt L, Knight H, Lo A, Richerson G, Gorman M, Williams L, Brass L, Agostini J, Mohsenin V, Roux F, Yaggi H. Continuous positive airway pressure: evaluation of a novel therapy for patients with acute ischemic stroke. *Sleep*. 2011;34:1271–1277.
77. Minnerup J, Ritter M, Wersching H, Kemmling A, Okegwo A, Schmidt A, Schilling M, Ringelstein E, Schäbitz W, Young P, Dziewas R. Continuous positive airway pressure ventilation for acute ischemic stroke: a randomized feasibility study. *Stroke*. 2012;43:1137–1139.
78. Parra O, Sánchez-Armengol A, Bonnin M, Arboix A, Campos-Rodríguez F, Pérez-Ronchel J, Durán-Cantolla J, de la Torre G, González Marcos J, de la Peña M, Carmen Jiménez M, Masa F, Casado I, Luz Alonso M, Macarrón J. Early treatment of obstructive apnoea and stroke outcome: a randomised controlled trial. *Eur Respir J*. 2011;37:1128–1136.
79. Cowie M, Woehrle H, Wegscheider K, Angermann A, d'Ortho P, Erdmann E, Levy P, Simonds A, Somers V, Zannad F, Teschler H. Adaptive servo-ventilation for central sleep apnea in systolic heart failure. *N Engl J Med*. 2015;373:1095–1105.

Sleep and Epilepsy

B. Razavi, R.S. Fisher

Department of Neurology and Neurological Sciences, Palo Alto, CA, USA

INTRODUCTION

An epileptic seizure is defined as a transient occurrence of signs and/or symptoms due to abnormal excessive or synchronous neuronal activity in the brain.[1] Epilepsy is a chronic neurological condition characterized by recurrent epileptic seizures.[2] Seizures can be focal in onset (with or without secondary generalization) or they can be primarily generalized (i.e., idiopathic).[3,4] Focal onset seizures may either be associated with impaired awareness (complex partial seizures) or without impaired awareness (simple partial seizures). Epilepsy is the fourth most common neurological disorder in the United States, with approximately 150,000 new cases each year.[5] One out of every 26 individuals will develop epilepsy at some point in their lifetime. Thus, epilepsy poses a significant burden on patients, their families, and the health care system.

The purpose of this chapter is to review the intricate relationship between sleep and epilepsy, with a special focus on the impact of sleep on seizure timing and frequency. Understanding this relationship is an important step toward successful evaluation and management of patients with epilepsy.

Sleep and Neurologic Disease. http://dx.doi.org/10.1016/B978-0-12-804074-4.00007-8

CIRCADIAN PATTERNS

Seizures

Timing of seizures often relates to the sleep cycle. This was recognized in the 1800s by Gowers, who reported that approximately 45% of seizures occurred during the day, 22% were nocturnal, and 33% showed a mixed circadian pattern.[6] These findings were later corroborated by others, including Langdon-Down in 1929.[7] Over time, seizures are more likely to occur during both day and night in a given patient.[7,8] More recent studies have reported that approximately 43% of focal onset seizures[9] and 45% of tonic–clonic seizures[10] start during sleep.

Circadian patterns for seizures can also depend on the epileptogenic lobe of origin. For example, most frontal lobe seizures tend to occur during sleep[11,12] (Fig. 7.1A). In a study of 90 seizures from 26 patients undergoing video scalp electroencephalogram (EEG) recording, Pavlova found that 50% of temporal lobe seizures clustered between 3 p.m. and 7 p.m., which was not likely to be due to chance.[13] In contrast, 47% of nontemporal lobe seizures occurred

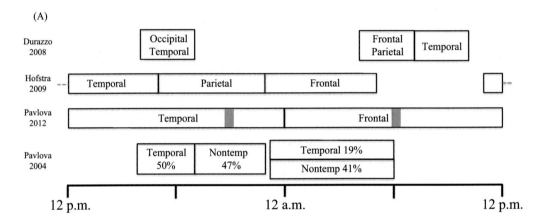

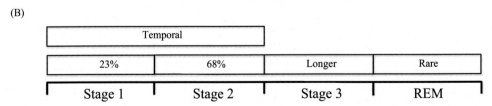

FIGURE 7.1 (A) Circadian patterns for seizures based on several representative studies. In general, frontal lobe seizures tend to be nocturnal and temporal lobe seizures diurnal. Percentages for the Pavlova et al.[13] study from 2004 indicate proportion of seizures from that brain region occurring during the delimited time window. *Shaded areas* for the Pavlova et al.[14] study from 2012 indicate when most seizures for that brain region clustered. (B) Key characteristics of seizures in relation to different stages of sleep. For example, the great majority of seizures occur during stages 1 and 2 of sleep, much more so during stage 2.

between 7 p.m. and 11 p.m. Only 19% of temporal lobe seizures occurred during sleep, compared to 41% of those originating from nontemporal regions. Similar circadian patterns were observed in the outpatient setting using ambulatory EEG measurements.[14] Frontal lobe seizures were most frequent between 12 a.m. and 12 p.m., whereas temporal lobe seizures had the highest incidence between 12 p.m. and 12 a.m. Frontal lobe seizures typically clustered around 6:30 a.m., and temporal lobe seizures around 8:50 p.m. Invasive intracranial EEG recordings have also documented a tendency for parietal seizures to be nocturnal, while temporal and occipital seizures are more likely to occur in the afternoon.[15,16]

Seizures during sleep may be more likely to propagate. One-third of complex partial seizures during sleep progress to secondary generalization, compared to only one-fifth during wakefulness.[9,11] Complex partial seizures that arise from the frontal lobe are an exception, and approximately one-fifth of these seizures secondarily generalize during both sleep and wakefulness. In contrast, 45% of complex partial seizures originating from the temporal lobe secondarily generalize during sleep, compared to 19% during wakefulness, though interpretation of comparative data must account for ascertainment bias due to subtle nocturnal seizures.

Certain sleep stages are more susceptible to seizures, especially temporal lobe seizures, which are more prevalent during stages 1 and 2 sleep, rare during slow wave sleep, and rare or absent during REM[9,11,17,18] (Fig. 7.1B). Furthermore, seizures are 3× more likely to begin during stage 2 (68%) than stage 1 (23%) of sleep.[9,18] Seizures arising from slow wave sleep are significantly longer than those occurring during other stages of sleep or wakefulness.[11] Clinical findings that seizures are rare or absent during REM sleep are complemented by similar findings in animals studies. Increasing the duration of REM sleep raises the threshold for seizures (i.e., electrical stimulation at higher current is needed to provoke seizures).[19] This effect extends to other stages of the sleep–wake cycle, suggesting that REM sleep may have long-term protective effects against cortical hyperexcitability and seizures.

The great majority of seizures originating from the mesial temporal lobes (and starting during stage 2 of sleep) are followed by arousals, though in some cases a seizure can occur after an arousal.[17] Time to arousal is shorter for left temporal lobe seizures.[20] The seizure's lobe of origin can also affect sleep characteristics. In a study of patients undergoing inpatient video EEG telemetry, the sleep efficiency index was lower in temporal lobe (0.84) epilepsy patients than it was in those with frontal lobe seizures (0.94),[12] however there were no significant differences in sleep architecture between the temporal and frontal lobe epilepsy patients.

Certain epilepsy syndromes also follow strong circadian patterns. Seizures in juvenile myoclonic epilepsy typically occur early in the morning or shortly after awakening.[21] Seizures related to autosomal dominant nocturnal frontal lobe epilepsy[22–24] and electrical status epilepticus in sleep[25,26] occur only during sleep. Seizures associated with benign partial epilepsy with centrotemporal spikes[27,28] and Panayiotopoulos syndrome[29] usually emerge from sleep.

Interictal Discharges

Epileptiform interictal discharges also follow a circadian pattern, and are more frequent during sleep[30,31] (Fig. 7.2). An older study by Gibbs reported that 82% of spikes occurred in sleep and only 36% during wakefulness.[32] In another study, 57% of patients had more interictal discharges during sleep.[33] About one third of patients with refractory temporal lobe epilepsy may have epileptiform discharges only in sleep.[30] Interictal discharges are most

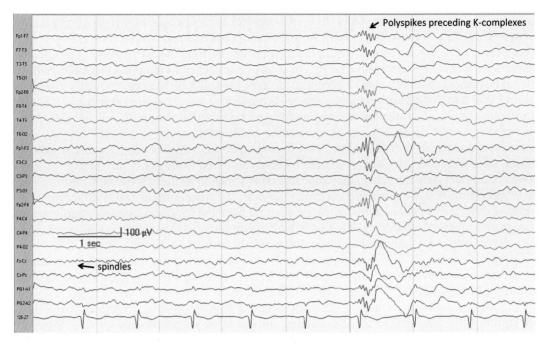

FIGURE 7.2 **An example of sleep activating epileptiform discharges.** In this case, polyspikes occur in conjunction with K-complexes, which is not unusual. EEG traces are displayed in bipolar (double banana) montage.

prevalent during slow wave sleep, but less likely to occur in REM sleep.[34] This is in contrast to an older study by White et al. that reported more prominent interictal abnormalities during lighter stages of sleep.[35] The scarcity of abnormal interictal activity during REM sleep is also seen in childhood epileptic syndromes, such as those of West,[36,37] Landau-Kleffner,[38] and Lennox-Gastaut.[39] Although interictal discharges are less prevalent in REM sleep, when they are present they carry more localizing value for the seizure focus.[40]

SLEEP DEPRIVATION

Seizures

It has been demonstrated in animal models that sleep deprivation can increase susceptibility to seizures.[41,42] This observation is nothing new. The importance of adequate sleep on seizure control was recognized centuries ago by Hippocrates,[43] and multiple investigations have consistently substantiated this observation. In an observational study of more than 3000 patients starting in 1946, Janz reported that approximately 60% of seizures occurring upon awakening were precipitated by sleep deprivation.[10] Untimely awakening was identified as a trigger in close to 40% of cases (other less prevalent precipitating factors included menses, alcohol, over-exertion, high temperatures, infection, and emotional shock). Sleep deprivation was an important precipitating factor for generalized convulsions,[44] as has been demonstrated in military populations.[45,46] Sleep deprivation may not necessarily be the precipitating factor for seizures in every study population, and its role may depend on multiple factors including epilepsy type and seizure lobe of origin.[47] In some settings, stress is ranked as a more important precipitating factor. Sleep deprivation is particularly deleterious for older patients, patients with idiopathic partial or generalized epilepsy, and patients with temporal lobe seizures. Increasing the duration of sleep deprivation increases the likelihood of seizure occurrence, with the greatest risk occurring in those sleep deprived more than 37 h.[46] In addition, seizure risk is highest within the first 48 h of sleep deprivation.[48] This suggests that the impact of sleep deprivation on seizures has a temporally cumulative effect. In a systematic study of sleep duration and seizure frequency in temporal lobe epilepsy, sleep deprivation was associated with an approximately 6 times increase in the probability of seizures.[49] Interestingly, oversleeping increased the probability of seizures three-fold in these patients. These findings imply that any disturbance in sleep duration can precipitate seizures, though sleep restriction is the most deleterious.

Sleep deprivation is often used as a provocative maneuver for increasing the likelihood of seizures so that video-EEG monitoring can be performed in a reasonable time frame. However, as mentioned earlier, sleep deprivation often occurs in conjunction with other factors, such as psychological or emotional stress, physical stress, exhaustion or fatigue, heat, dehydration, menses, and alcohol use. This arguably raises the possibility that sleep deprivation alone may not be sufficient to trigger seizures, and multiple simultaneous factors must be present over time. Alternatively, sleep deprivation itself may be a secondary result of these other factors (or stressors) and not always the precipitant.

This question was partially addressed by a study of 84 patients with medically refractory complex partial seizures with secondary generalization undergoing inpatient video-EEG monitoring.[50] In this study, a subgroup of patients did not sleep from 10 p.m. to 6 a.m. every other day. This subgroup and the other remaining patients were awake from 6 a.m. to 10 p.m. Overall study duration was comparable between the two groups, as was the reduction in antiseizure medications. The mean number of complex partial seizures per day was similar for the sleep deprived (0.52) and normal sleep (0.53) groups. Paradoxically, the rate of secondarily generalized seizures was slightly higher (0.21) in the normal sleep group as opposed to the sleep deprived group (0.14), though this difference was not statistically significant.

This data has two implications. One, sleep deprivation alone may not be sufficient to increase seizure frequency. Inpatient video-EEG telemetry units may be relatively low stress environments. Conclusions based on the study's data should be limited since sleep deprivation was not chronic or cumulative and only 24 h at a time. Two, using sleep deprivation may have questionable utility for provoking seizures, especially given that this maneuver may promote adverse mood changes in patients.[51] In turn, this may negatively impact patients' quality of the experience, potentially prompting them to shorten their inpatient stay before video-EEG monitoring could be completed.

Interictal Discharges

Sleep deprivation has an activating effect on interictal discharges, similar to its effect on seizures.[52–55] An older study showed that 34% of patients with a clinical diagnosis of epilepsy and normal (or borderline) nonsleep deprived EEGs had clear epileptiform activity after sleep deprivation.[56] This is similar for men and women. About one-third of patients who demonstrated activation after sleep deprivation were not on antiseizure medication. In contrast, about one-half of the patients who did not have activation were taking antiseizure medications. This suggests that antiseizure medications may reduce the likelihood of capturing epileptiform discharges in the setting of sleep deprivation. Sleep deprivation also increased the amount of EEG abnormalities in 56% of patients with abnormal baseline EEGs. The probability of capturing epileptiform abnormalities does not seem to be influenced by the clinical seizure types or etiologies.[57] The increased likelihood of epileptiform discharges due to sleep deprivation is thought to be independent of activation related to sleep itself.[58] In fact, there is evidence to suggest that sleep deprivation is more effective for activating epileptiform discharges than sleep itself.[59]

The impact of sleep deprivation on seizures and epileptiform discharges has also prompted some interesting work on cortical excitability. In humans, much of this effort has utilized transcranial magnetic stimulation (TMS). This technology has facilitated noninvasive quantification of cortical excitability, often based on activity of the motor system.[60,61] In normal individuals, TMS results in decreased cortical inhibition as well as decreased facilitation (i.e., decreased excitation) in the setting of sleep deprivation.[62] A change in the inhibition-facilitation balance may have an overall activating (net excitation) effect. In epilepsy patients, sleep deprivation increases cortical excitability in newly diagnosed, but untreated, idiopathic generalized epilepsy.[63] The decrease in cortical inhibition and increase in facilitation (hence, increased excitability) is also associated with increased paroxysmal activity and sleep deprivation.[64]

Patients with focal onset seizures—newly diagnosed but untreated—typically demonstrate increased cortical excitability on the same side of the seizure origin.[63] However, the increase in excitability is more pronounced in patients with idiopathic generalized epilepsy. There is also a concurrent increase in the prevalence of interictal discharges with sleep deprivation, suggesting that interictal discharges are themselves a marker of cortical excitability. Patterns of cortical excitability may be less consistent in the absence of sleep deprivation. About one third of patients with drug-naïve focal onset seizures demonstrate impaired cortical inhibition, which also seems related to more severe interictal abnormalities.[65] The remaining two-thirds have cortical inhibitions comparable to normal controls.

SLEEP RELATED DISORDERS

Epilepsy is greatly influenced by sleep- disorders. Obstructive sleep apnea (OSA) has been studied extensively in epilepsy patients, and is of particular interest for two reasons. One, the arousals and fragmentation of sleep resulting from apneas, hypopneas, and hypoxia prevent restorative sleep. This effectively leads to chronic sleep deprivation, which by itself may increase the likelihood of seizure occurrence. Two, it is reasonable to postulate that the hypoxia associated with hypopneas or apneas may be a trigger for seizures. Thus, it is incumbent on the neurologist to screen every patient with epilepsy for OSA.

In the general adult population, OSA has an estimated prevalence of 3–7%.[66] However, estimates as high as 24% in men and 9% in women are reported[67] (Table 7.1). Prevalence can be as high as 11% in children, depending on the study.[68] In addition, OSA is much more common in patients with epilepsy than it is in the general population. One-third of those with medically refractory seizures are diagnosed with OSA,[69,70] and 13% with severe OSA.[69] Higher estimates have also been reported.[71,72] However, it is important to note that selection bias based on

TABLE 7.1   Prevalence of Sleep Apnea

Author	Year	Adults	Children	General	Epilepsy	Prevalence
Young	1993	✓		✓		24% male, 9% female
Lumeng	2008		✓	✓		≤11%
Beran	1999	✓	✓		✓	42%
Malow	2000	✓			✓	33%
Manni	2003	✓			✓	15.4% male, 5.4% female
Foldvary-Schaefer	2012	✓			✓	30%
Jain	2012		✓		✓	≤45.8%
Vendrame	2014	✓			✓	75%

pretest suspicion of OSA may artificially elevate the prevalence of OSA in these populations. Prevalence rates as low as 10% have been reported in some populations with epilepsy, with the great majority of cases still being mild to moderate in severity.[73] OSA is more common in men than women with epilepsy, as it is in the general population.[73] In children with epilepsy, the prevalence of OSA based on polysomnography (PSG) can be as high as 65%.[74,75] In addition, central sleep apnea (CSA) and complex sleep apnea (CompSA) have also been studied in epilepsy patients. One retrospective study found CSA in 3.7% and CompSA in 7.9% of patients based on PSG.[71] In that study, OSA was very common, with a prevalence of 75%. Focal seizures were more common in patients with CSA. Another study demonstrated that patients with temporal lobe epilepsy are at a higher risk of sleep apnea than those with nontemporal lobe epilepsy, based on a screening questionnaire.[76]

OSA has also been implicated as an exacerbating factor for seizures. Higher seizure frequency is associated with a higher likelihood of OSA,[77] though this does not necessarily suggest a casual relationship. A common underlying process may be driving seizure frequency but also OSA risk. Some investigators have attempted to determine whether apneas directly trigger seizures. In one cross-sectional study, Chihorek compared the prevalence of OSA in 2 groups of patients with epilepsy.[78] Group 1 included patients with seizure onset at age 50 or older or seizure onset before age 50 with increased seizure frequency at or after age 50, defined by a 20% or greater increase in seizure frequency over the preceding 3 months. Group 2 included patients with seizure onset before age 50 with stable or improved seizure frequency at or after age 50. All patients underwent PSG for evaluation of OSA. Patients in group 1 had a greater apnea–hypopnea index (AHI, average in this group of 23.2). In contrast, patients in the younger group had an AHI of 3.1, which is considered normal. This led the authors to conclude that OSA may contribute to worsening seizure control. However, the late-life-onset seizure group did have 4 times as many men as women, which is a potential confounding variable.

There is also evidence to support the converse relationship, that epilepsy increases the risk of OSA. One case study demonstrated resolution of OSA after a left premotor frontal resection, which could not be attributed to other factors.[79] The resection was associated with marked reduction in spikes, suggesting that interictal discharges may impair upper airway control. Seizures are also directly implicated as a source of central apneas.[80–82] The correlation between seizures and airway control has been suggested as a potential factor in sudden unexpected death in epilepsy.[83,84]

Epilepsy patients with OSA are more likely to have seizures during sleep,[69] suggesting a more direct link between respiratory disturbance and seizures. In fact, a direct temporal relationship between OSA and seizures was reported in a patient with predominantly nocturnal seizures,[85] in whom 6 out of 12 seizures occurred during wakefulness immediately or shortly after an OSA-related arousal. Another seizure occurred during an OSA event without any associated arousals or K-complexes. The apneas were not necessarily associated with hypoxia, suggesting that sleep–wake transitions prompted by an OSA event may be a trigger and not the hypoxia itself. At baseline, the patient also had left temporal epileptiform discharges. However, some instances of OSA and related arousals were associated with generalized spike-wave discharges that seemed to begin on the left before spreading to the right. This implies that OSA events may promote rapid bilateral synchrony of focal epileptiform discharges. Seizures (and the generalized epileptiform discharges) greatly improved with initiation of continuous positive airway pressure (CPAP) therapy in this patient. Another case report also demonstrated a stereotyped temporal relation between seizures and CSA,[86] in which seizures followed the apnea events but preceded arousal. This patient's seizures also improved with positive airway pressure therapy. One case demonstrated severe oxygen desaturation associated with CSA as the trigger for seizures, also with resolution of seizures after initiation of CPAP and weaning off all antiseizure medications.[87]

The benefits of treating OSA to improve seizure control have been demonstrated consistently across multiple studies in larger populations. The response is often quite remarkable, regardless of changes to antiepileptic medications.[88–90] In a retrospective study of 41 patients, Vendrame compared seizure frequency in patients who were compliant with CPAP therapy ($n = 28$) to those who were not ($n = 13$; <21 days per month, and <4 h per day of CPAP use).[91] Participants were followed for at least 6 months. In the compliant group, seizure frequency decreased on average from 1.8 to 1.0 per month, whereas in the noncompliant group seizure frequency decreased from 2.1 to 1.8. Furthermore, 57% of the compliant group became seizure-free, compared to 23% of the noncompliant group. Overall, the odds of improved seizure control can increase 10-fold with CPAP.[92] Even with medically refractory epilepsies, CPAP therapy can improve seizure control in up to 100% of cases, and sometimes can lead to seizure freedom.[77] A randomized pilot trial of 35 subjects found at least a 50% reduction in seizures in 28% of participants in the therapeutic group, as opposed to 15% in the Sham group.[93] However, this difference was not statistically significant, which may be attributed to the small sample size and exclusion of subjects with severe OSA. Treatment of OSA in children also results in marked reduction in seizure frequency.[94]

CPAP therapy reduces rates of abnormal interictal discharges.[95] Spike rate reductions during sleep (around 78%) can be as great as that of wakefulness (around 39%) with CPAP use.[96] AHI also correlates with spike rates during sleep (but not wakefulness); that is, more severe OSA is associated with more prevalent spikes. Sleep continuity and the amount of oxygen desaturation do not seem to correlate with spike rates, though this finding may be confounded by the small patient population.

The markedly reduced seizure frequency and interictal discharges with CPAP therapy support the notion that sleep fragmentation caused by OSA may be contributing to epileptogenicity. This finding, along with the relatively high prevalence OSA in epilepsy patients, strongly warrants screening this patient population for OSA, and treating when appropriate.

CONCLUSIONS

Epilepsy is a relatively common neurological disorder, and knowledge of its relationship to sleep is an important step toward better understanding, evaluation, and treatment of seizures. Seizures generally follow strong circadian patterns. Nocturnal seizures are most likely to occur during stage 2 of sleep, and REM sleep tends to have a protective effect against seizures. Sleep deprivation is a major trigger for seizures. Interictal epileptiform discharges demonstrate circadian patterns similar to those of seizures, and discharges are activated by sleep deprivation. Sleep apnea is very common in epilepsy patients and its treatment can significantly improve seizure control.

FUTURE DIRECTIONS

Despite a substantial body of literature about sleep and epilepsy, many questions remain unanswered. What is the impact of sleep and sleep related disorders on seizures during pregnancy? Does the recurrent hypoxia due to sleep apnea over many years lead to cortical injury, in turn resulting in epilepsy? Does treatment of other sleep fragmenting disorders, such as periodic leg movements, also improve seizure control? Does treatment of sleep apnea reduce risk of SUDEP? Addressing these topics would certainly lead to new and exciting answers but they would also lead to many new questions.

References

1. Fisher RS, van Emde Boas W, Blume W, et al. Epileptic seizures and epilepsy: definitions proposed by the International League Against Epilepsy (ILAE) and the International Bureau for Epilepsy (IBE). *Epilepsia.* 2005;46(4):470–472.
2. Fisher RS, Acevedo C, Arzimanoglou A, et al. ILAE official report: a practical clinical definition of epilepsy. *Epilepsia.* 2014;55(4):475–482.
3. Berg AT, Berkovic SF, Brodie MJ, et al. Revised terminology and concepts for organization of seizures and epilepsies: report of the ILAE Commission on Classification and Terminology, 2005–2009. *Epilepsia.* 2010;51(4):676–685.
4. Shorvon SD. The etiologic classification of epilepsy. *Epilepsia.* 2011;52(6):1052–1057.
5. England MJ, Liverman CT, Schultz AM, Strawbridge LM. Epilepsy across the spectrum: promoting health and understanding. A summary of the Institute of Medicine report. *Epilepsy Behav.* 2012;25(2):266–276.
6. Gowers W. Course of epilepsy. In: Gowers W, ed. *Epilepsy and Other Chronic Convulsive Diseases: Their Causes, Symptoms and Treatment.* New York: William Wood; 1885:157–164.
7. Langdon-Down M, Russell Brain W. Time of day in relation to convulsion in epilepsy. *The Lancet.* 1929;213(5516):1029–1032.
8. Gibberd FB, Bateson MC. Sleep epilepsy: its pattern and prognosis. *Br Med J.* 1974;2(5916):403–405.
9. Herman ST, Walczak TS, Bazil CW. Distribution of partial seizures during the sleep–wake cycle: differences by seizure onset site. *Neurology.* 2001;56(11):1453–1459.
10. Janz D. The grand mal epilepsies and the sleeping-waking cycle. *Epilepsia.* 1962;3:69–109.
11. Bazil CW, Walczak TS. Effects of sleep and sleep stage on epileptic and nonepileptic seizures. *Epilepsia.* 1997;38(1):56–62.
12. Crespel A, Baldy-Moulinier M, Coubes P. The relationship between sleep and epilepsy in frontal and temporal lobe epilepsies: practical and physiopathologic considerations. *Epilepsia.* 1998;39(2):150–157.
13. Pavlova MK, Shea SA, Bromfield EB. Day/night patterns of focal seizures. *Epilepsy Behav.* 2004;5(1):44–49.
14. Pavlova MK, Lee JW, Yilmaz F, Dworetzky BA. Diurnal pattern of seizures outside the hospital: is there a time of circadian vulnerability?. *Neurology.* 2012;78(19):1488–1492.
15. Durazzo TS, Spencer SS, Duckrow RB, Novotny EJ, Spencer DD, Zaveri HP. Temporal distributions of seizure occurrence from various epileptogenic regions. *Neurology.* 2008;70(15):1265–1271.

16. Hofstra WA, Spetgens WP, Leijten FS, et al. Diurnal rhythms in seizures detected by intracranial electrocortico-graphic monitoring: an observational study. *Epilepsy Behav*. 2009;14(4):617–621.

17. Malow A, Bowes RJ, Ross D. Relationship of temporal lobe seizures to sleep and arousal: a combined scalp-intracranial electrode study. *Sleep*. 2000;23(2):231–234.

18. Minecan D, Natarajan A, Marzec M, Malow B. Relationship of epileptic seizures to sleep stage and sleep depth. *Sleep*. 2002;25(8):899–904.

19. Kumar P, Raju TR. Seizure susceptibility decreases with enhancement of rapid eye movement sleep. *Brain Res*. 2001;922(2):299–304.

20. Gumusyayla S, Erdal A, Tezer FI, Saygi S. The temporal relation between seizure onset and arousal-awakening in temporal lobe seizures. *Seizure-Eur J Epilep*. 2016;39:24–27.

21. Asconapé J, Penry JK. Some clinical and EEG aspects of benign juvenile myoclonic epilepsy. *Epilepsia*. 1984;25(1):108–114.

22. Scheffer IE, Bhatia KP, Lopes-Cendes I, et al. Autosomal dominant nocturnal frontal lobe epilepsy. A distinctive clinical disorder. *Brain*. 1995;118(Pt 1):61–73.

23. Steinlein OK, Mulley JC, Propping P, et al. A missense mutation in the neuronal nicotinic acetylcholine receptor alpha 4 subunit is associated with autosomal dominant nocturnal frontal lobe epilepsy. *Nat Genet*. 1995;11(2):201–203.

24. Zucconi M, Oldani A, Smirne S, Ferini-Strambi L. The macrostructure and microstructure of sleep in patients with autosomal dominant nocturnal frontal lobe epilepsy. *J Clin Neurophysiol*. 2000;17(1):77–86.

25. Nickels K, Wirrell E. Electrical status epilepticus in sleep. *Semin Pediatr Neurol*. 2008;15(2):50–60.

26. Tassinari CA, Rubboli G, Volpi L, Billard C, Bureau M. Electrical status epilepticus during slow sleep (ESES or CSWS) including acquired epileptic aphasia (Landau–Kleffner syndrome). In: Roger J, Bureau M, Dravet C, Genton P, Tassinari CA, Wolf P, eds. *Epileptic Syndromes in Infancy, Childhood and Adolescence*. London: John Libbey; 2005:295–314.

27. Lerman P. Benign partial epilepsy with centro-temporal spikes. In: Roger J, Dravet C, Bureau M, Dreifuss FE, Wolf P, eds. *Epileptic Syndromes in Infancy, Childhood and Adolescence*. London: John Libbey; 1985:150–158.

28. Bernardina BD, Beghini G. Rolandic spikes in children with and without epilepsy. (20 subjects polygraphically studied during sleep). *Epilepsia*. 1976;17(2):161–167.

29. Covanis A. Panayiotopoulos syndrome: a benign childhood autonomic epilepsy frequently imitating encephalitis, syncope, migraine, sleep disorder, or gastroenteritis. *Pediatrics*. 2006;118(4):e1237–1243.

30. Niedermeyer E, Rocca U. The diagnostic significance of sleep electroencephalograms in temporal lobe epilepsy. A comparison of scalp and depth tracings. *Eur Neurol*. 1972;7(1):119–129.

31. Silverman D. Sleep as a general activation procedure in electroencephalography. *Electroencephalogr Clin Neurophysiol*. 1956;8(2):317–324.

32. Gibbs EL, Gibbs FA. Diagnostic and localizing value of electroencephalographic studies in sleep. *J Nerv Ment Dis*. 1947;26:336–376.

33. Gloor P, Tsai C, Haddad F. An assessment of the value of sleep-electroencephalography for the diagnosis of temporal lobe epilepsy. *Electroencephalogr Clin Neurophysiol*. 1958;10(4):633–648.

34. Malow BA, Lin X, Kushwaha R, Aldrich MS. Interictal spiking increases with sleep depth in temporal lobe epilepsy. *Epilepsia*. 1998;39(12):1309–1316.

35. White P, Dyken M, Grant P, Jackson L. Electroencephalographic abnormalities during sleep as related to the temporal distribution of seizures. *Epilepsia*. 1962;3(2):167–174.

36. Hrachovy RA, Frost Jr JD, Kellaway P. Hypsarrhythmia: variations on the theme. *Epilepsia*. 1984;25(3):317–325.

37. Watanabe K, Negoro T, Aso K, Matsumoto A. Reappraisal of interictal electroencephalograms in infantile spasms. *Epilepsia*. 1993;34(4):679–685.

38. Pearl PL, Carrazana EJ, Holmes GL. The Landau-Kleffner syndrome. *Epilepsy Curr*. 2001;1(2):39–45.

39. Beaumanoir A, Blume W. The Lennox-Gastaut syndrome. In: Roger J, Bureau M, Dravet C, Genton P, Tassinari CA, Wolf P, eds. *Epileptic Syndromes in Infancy, Childhood and Adolescence*. London: John Libbey; 2005:125–148.

40. Sammaritano M, Gigli GL, Gotman J. Interictal spiking during wakefulness and sleep and the localization of foci in temporal lobe epilepsy. *Neurology*. 1991;41(2 (Pt 1)):290–297.

41. Shouse MN. Sleep deprivation increases susceptibility to kindled and penicillin seizure events during all waking and sleep states in cats. *Sleep*. 1988;11(2):162–171.

42. McDermott CM, LaHoste GJ, Chen C, Musto A, Bazan NG, Magee JC. Sleep deprivation causes behavioral, synaptic, and membrane excitability alterations in hippocampal neurons. *J Neurosci*. 2003;23(29):9687–9695.

43. Lloyd G. Hippocrates, the sacred disease, aphorisms, and prognosis. In: Lloyd GER, ed. *Hippocratic Writings*. Boston, MA: Penguin; 1983:170–251.

44. Friis ML, Lund M. Stress convulsions. *Arch Neurol*. 1974;31(3):155–159.

45. Bennett DR. Sleep deprivation and major motor convulsions. *Neurology*. 1963;13:953–958.

46. Gunderson CH, Dunne PB, Feyer TL. Sleep deprivation seizures. *Neurology*. 1973;23(7):678–686.

47. Frucht MM, Quigg M, Schwaner C, Fountain NB. Distribution of seizure precipitants among epilepsy syndromes. *Epilepsia*. 2000;41(12):1534–1539.

48. Rodin E. Sleep deprivation and epileptological implications. In: Degen R, Niedermeyer E, eds. *Epilepsy, Sleep and Sleep Deprivation*. Amsterdam: Elsevier; 1984:293–300.

49. Rajna P, Veres J. Correlations between night sleep duration and seizure frequency in temporal lobe epilepsy. *Epilepsia*. 1993;34(3):574–579.

50. Malow BA, Passaro E, Milling C, Minecan DN, Levy K. Sleep deprivation does not affect seizure frequency during inpatient video-EEG monitoring. *Neurology*. 2002;59(9):1371–1374.

51. Bristol K, Natarajan A, Lin X, Malow B. Effects of long-term video-electroencephalographic monitoring on mood in epilepsy patients. *Epilepsy Behav*. 2001;2(5):433–440.

52. Veldhuizen R, Binnie CD, Beintema DJ. The effect of sleep deprivation on the EEG in epilepsy. *Electroencephalogr Clin Neurophysiol*. 1983;55(5):505–512.

53. Degen R. A study of the diagnostic value of waking and sleep EEGs after sleep deprivation in epileptic patients on anticonvulsive therapy. *Electroencephalogr Clin Neurophysiol*. 1980;49(5–6):577–584.

54. Ellingson RJ, Wilken K, Bennett DR. Efficacy of sleep deprivation as an activation procedure in epilepsy patients. *J Clin Neurophysiol*. 1984;1(1):83–101.

55. Halasz P, Filakovszky J, Vargha A, Bagdy G. Effect of sleep deprivation on spike-wave discharges in idiopathic generalised epilepsy: a 4 × 24 h continuous long term EEG monitoring study. *Epilepsy Res*. 2002;51(1–2):123–132.

56. Mattson RH, Pratt KL, Calverley JR. Electroencephalograms of epileptics following sleep deprivation. *Arch Neurol*. 1965;13(3):310–315.

57. Pratt KL, Mattson RH, Weikers NJ, Williams R. EEG activation of epileptics following sleep deprivation: a prospective study of 114 cases. *Electroencephalogr Clin Neurophysiol*. 1968;24(1):11–15.

58. Fountain NB, Kim JS, Lee SI. Sleep deprivation activates epileptiform discharges independent of the activating effects of sleep. *J Clin Neurophysiol*. 1998;15(1):69–75.

59. Roupakiotis SC, Gatzonis SD, Triantafyllou N, et al. The usefulness of sleep and sleep deprivation as activating methods in electroencephalographic recording: contribution to a long-standing discussion. *Seizure*. 2000;9(8):580–584.

60. Badawy RA, Loetscher T, Macdonell RA, Brodtmann A. Cortical excitability and neurology: insights into the pathophysiology. *Funct Neurol*. 2012;27(3):131–145.

61. Cantello R, Civardi C, Cavalli A, et al. Cortical excitability in cryptogenic localization-related epilepsy: interictal transcranial magnetic stimulation studies. *Epilepsia*. 2000;41(6):694–704.

62. Civardi C, Boccagni C, Vicentini R, et al. Cortical excitability and sleep deprivation: a transcranial magnetic stimulation study. *J Neurol Neurosurg Psychiatry*. 2001;71(6):809–812.

63. Badawy RA, Curatolo JM, Newton M, Berkovic SF, Macdonell RA. Sleep deprivation increases cortical excitability in epilepsy: syndrome-specific effects. *Neurology*. 2006;67(6):1018–1022.

64. Manganotti P, Bongiovanni LG, Fuggetta G, Zanette G, Fiaschi A. Effects of sleep deprivation on cortical excitability in patients affected by juvenile myoclonic epilepsy: a combined transcranial magnetic stimulation and EEG study. *J Neurol Neurosurg Psychiatry*. 2006;77(1):56–60.

65. Varrasi C, Civardi C, Boccagni C, et al. Cortical excitability in drug-naive patients with partial epilepsy: a cross-sectional study. *Neurology*. 2004;63(11):2051–2055.

66. Punjabi NM. The epidemiology of adult obstructive sleep apnea. *Proc Am Thorac Soc*. 2008;5(2):136–143.

67. Young T, Palta M, Dempsey J, Skatrud J, Weber S, Badr S. The occurrence of sleep-disordered breathing among middle-aged adults. *N Engl J Med*. 1993;328(17):1230–1235.

68. Lumeng JC, Chervin RD. Epidemiology of pediatric obstructive sleep apnea. *Proc Am Thorac Soc*. 2008;5(2):242–252.

69. Malow BA, Levy K, Maturen K, Bowes R. Obstructive sleep apnea is common in medically refractory epilepsy patients. *Neurology*. 2000;55(7):1002–1007.

70. Foldvary-Schaefer N, Andrews ND, Pornsriniyom D, Moul DE, Sun Z, Bena J. Sleep apnea and epilepsy: who's at risk?. *Epilepsy Behav*. 2012;25(3):363–367.

71. Vendrame M, Jackson S, Syed S, Kothare SV, Auerbach SH. Central sleep apnea and complex sleep apnea in patients with epilepsy. *Sleep Breath.* 2014;18(1):119–124.
72. Beran RG, Plunkett MJ, Holland GJ. Interface of epilepsy and sleep disorders. *Seizure.* 1999;8(2):97–102.
73. Manni R, Terzaghi M, Arbasino C, Sartori I, Galimberti CA, Tartara A. Obstructive sleep apnea in a clinical series of adult epilepsy patients: frequency and features of the comorbidity. *Epilepsia.* 2003;44(6):836–840.
74. Gogou M, Haidopoulou K, Eboriadou M, Pavlou E. Sleep apneas and epilepsy comorbidity in childhood: a systematic review of the literature. *Sleep Breath.* 2015;19(2):421–432.
75. Jain SV, Simakajornboon S, Shapiro SM, Morton LD, Leszczyszyn DJ, Simakajornboon N. Obstructive sleep apnea in children with epilepsy: prospective pilot trial. *Acta Neurol Scand.* 2012;125(1):e3–6.
76. Yildiz FG, Tezer FI, Saygi S. Temporal lobe epilepsy is a predisposing factor for sleep apnea: a questionnaire study in video-EEG monitoring unit. *Epilepsy Behav..* 2015;48:1–3.
77. Li P, Ghadersohi S, Jafari B, Teter B, Sazgar M. Characteristics of refractory vs. medically controlled epilepsy patients with obstructive sleep apnea and their response to CPAP treatment. *Seizure-Eur J Epilep.* 2012;21(9):717–721.
78. Chihorek AM, Abou-Khalil B, Malow BA. Obstructive sleep apnea is associated with seizure occurrence in older adults with epilepsy. *Neurology.* 2007;69(19):1823–1827.
79. Foldvary-Schaefer N, Stephenson L, Bingaman W. Resolution of obstructive sleep apnea with epilepsy surgery? Expanding the relationship between sleep and epilepsy. *Epilepsia.* 2008;49(8):1457–1459.
80. Andrade EO, Arain A, Malow BA. Partial epilepsy presenting as apneic seizures without posturing. *Pediatr Neurol.* 2006;35(5):359–362.
81. Lee HW, Hong SB, Tae WS, Seo DW, Kim SE. Partial seizures manifesting as apnea only in an adult. *Epilepsia.* 1999;40(12):1828–1831.
82. Coulter DL. Partial seizures with apnea and bradycardia. *Arch Neurol.* 1984;41(2):173–174.
83. Schuele SU, Afshari M, Afshari ZS, et al. Ictal central apnea as a predictor for sudden unexpected death in epilepsy. *Epilepsy Behav.* 2011;22(2):401–403.
84. So EL, Sam MC, Lagerlund TL. Postictal central apnea as a cause of SUDEP: evidence from near-SUDEP incident. *Epilepsia.* 2000;41(11):1494–1497.
85. Nguyen-Michel V-H, Pallanca O, Navarro V, Dupont S, Baulac M, Adam C. How are epileptic events linked to obstructive sleep apneas in epilepsy?. *Seizure-Eur J Epilep.* 2015;24:121–123.
86. Khachatryan SG, Prosperetti C, Rossinelli A, et al. Sleep-onset central apneas as triggers of severe nocturnal seizures. *Sleep Med.* 2015;16(8):1017–1019.
87. Miliauskas S, Liesiene V, Zemaitis M, Sakalauskas R. Late-onset nocturnal intractable seizure during sleep: what is the origin?. *Medicina (Kaunas).* 2010;46(2):120–124.
88. Devinsky O, Ehrenberg B, Barthlen GM, Abramson HS, Luciano D. Epilepsy and sleep apnea syndrome. *Neurology.* 1994;44(11):2060–2064.
89. Vaughn BV, D'Cruz OF, Beach R, Messenheimer JA. Improvement of epileptic seizure control with treatment of obstructive sleep apnoea. *Seizure.* 1996;5(1):73–78.
90. Malow BA, Fromes GA, Aldrich MS. Usefulness of polysomnography in epilepsy patients. *Neurology.* 1997;48(5):1389–1394.
91. Vendrame M, Auerbach S, Loddenkemper T, Kothare S, Montouris G. Effect of continuous positive airway pressure treatment on seizure control in patients with obstructive sleep apnea and epilepsy. *Epilepsia.* 2011;52(11):e168–e171.
92. Pornsriniyom D, Kim H, Bena J, Andrews ND, Moul D, Foldvary-Schaefer N. Effect of positive airway pressure therapy on seizure control in patients with epilepsy and obstructive sleep apnea. *Epilepsy Behav.* 2014;37:270–275.
93. Malow BA, Foldvary-Schaefer N, Vaughn BV, et al. Treating obstructive sleep apnea in adults with epilepsy: a randomized pilot trial. *Neurology.* 2008;71(8):572–577.
94. Malow BA, Weatherwax KJ, Chervin RD, et al. Identification and treatment of obstructive sleep apnea in adults and children with epilepsy: a prospective pilot study. *Sleep Med.* 2003;4(6):509–515.
95. Oliveira AJ, Zamagni M, Dolso P, Bassetti MA, Gigli GL. Respiratory disorders during sleep in patients with epilepsy: effect of ventilatory therapy on EEG interictal epileptiform discharges. *Clin Neurophysiol.* 2000;111(suppl 2):S141–S145.
96. Pornsriniyom D, Shinlapawittayatorn K, Fong J, Andrews ND, Foldvary-Schaefer N. Continuous positive airway pressure therapy for obstructive sleep apnea reduces interictal epileptiform discharges in adults with epilepsy. *Epilepsy Behav.* 2014;37:171–174.

8

Central Nervous System Hypersomnias

J. Cheung, C.M. Ruoff, E. Mignot

The Stanford Center for Sleep Sciences and Medicine, Redwood City, CA, USA

O U T L I N E

CLASSIFICATION OF CNS HYPERSOMNIA DISORDERS

The International Classification of Sleep Disorders,[1] currently in its third edition (ICSD-3), distinguishes eight subtypes of central nervous system (CNS) hypersomnia disorders (Table 8.1). The nosology of narcolepsy has also been revised, subdividing the disorder into type 1 and type 2 narcolepsy, replacing narcolepsy with and without cataplexy, respectively.

Sleep and Neurologic Disease. http://dx.doi.org/10.1016/B978-0-12-804074-4.00008-X

TABLE 8.1   ICSD 3 Classification of CNS Hypersomnias

Narcolepsy type 1

Narcolepsy type 2

Idiopathic hypersomnia

Kleine–Levin syndrome

Hypersomnia due to a medical disorder

Hypersomnia due to a medication or substance

Hypersomnia associated with a psychiatric disorder

Insufficient sleep syndrome

This reflects a change of focus from diagnosis based on symptoms to diagnosis based on pathophysiology, in this case hypocretin (orexin) deficiency status. This change was predicated on the notion that almost all patients with cataplexy have hypocretin deficiency. In addition, "narcolepsy with cataplexy" is improper because some patients with hypocretin deficiency do not have cataplexy or have yet to develop cataplexy. Even though this revised classification focuses on pathophysiology, little evidence exists in distinguishing narcolepsy type 2 (Na-2) from idiopathic hypersomnia (IH).

HISTORICAL PERSPECTIVES AND PATHOPHYSIOLOGY

Among CNS hypersomnias, our understanding of the pathophysiology is mostly limited to Na-1. The first descriptions of narcolepsy with cataplexy were made by Westphal in 1877 and Gelineau in 1880.[2,3] It was at this time that narcolepsy was first recognized as a unique syndrome distinct from epilepsy or other neurological disorders. In the late 1950s and early 1960s, a number of studies established the association between narcolepsy and REM sleep abnormalities.[4,5] It was discovered that individuals with narcolepsy entered into REM sleep rapidly after sleep onset, in contrast with normal individuals who typically entered their first REM cycle 90 min after sleep onset. In 1973, the identification of a narcoleptic dog was first reported.[6,7] This led to the breeding and studying of narcoleptic dog colonies with Labradors and Dobermans. Dement and Carskadon developed the MSLT, which continues to be the standard diagnostic test for narcolepsy.[8] In 1983, Honda and coworkers identified the association of narcolepsy with the human leukocyte antigen locus (HLA-DR2) in a Japanese cohort.[9] HLA genes encode for the major histocompatibility complexes found on the surface of antigen-presenting cells, immune cells that present antigens to other immune cells such as T cells. Mignot and coworkers later identified that HLA-DQB1*06:02 was a better marker than DR2 for narcolepsy across all ethnic groups, notably African Americans.[10,11]

In 1998, a hypothalamic specific neuropeptide called hypocretin/orexin was simultaneously discovered by DeLecea et al. and Sakurai et al.[12,13] Its exact function was unknown until 1999 when a mutation in the hypocretin (orexin) receptor 2 gene was identified by Lin, Mignot and coworkers in canine narcolepsy,[14] connecting hypocretin with sleep and narcolepsy. In 2000, Nishino et al. found undetectable levels of CSF hypocretin in human patients with cataplexy, indicating a lack of hypocretin rather than a receptor defect in human cases.[15,16] Neuropathological work on human brain tissues followed and identified a loss of hypocretinergic cells in the posterior hypothalamus in human Na-1 (Fig. 8.1).[17,18] Although hypocretin deficiency has been identified as the cause of Na-1, the pathophysiologic mechanisms underlying sleepiness, cataplexy, and other REM-related disturbances are complicated and incompletely understood.

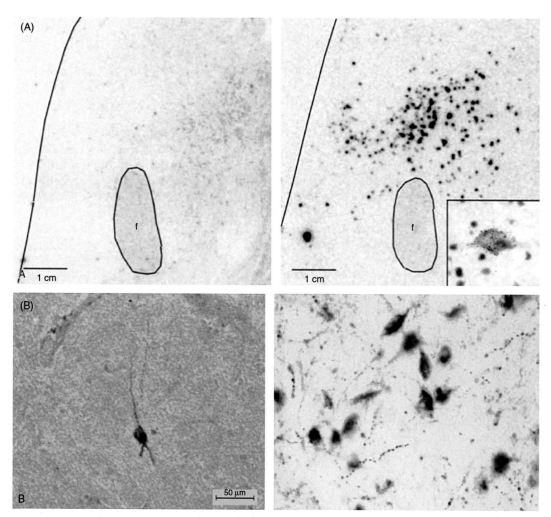

FIGURE 8.1 **Hypocretin peptides and neurons in the lateral hypothalamus of narcoleptics versus controls.** (A) Significant reduction of hypocretin mRNA expression in the lateral hypothalamus in a narcoleptic (left) versus a control brain (right). (B) Significant reduction of hypocretin-stained peptides in hypocretin cells (in the lateral hypothalamus) in a narcoleptic (left) versus a control brain (right). Narcoleptics have an 85–95% reduction in the number of hypocretin neurons. *Source: (A) Kryger M et al. Atlas of Clinical Sleep Medicine, originally modified from Peyron C, Faraco J, Rogers W, et al. A mutation in a case of early onset narcolepsy and a generalized absence of hypocretin peptides in human narcoleptic brains. Nat Med. 2000;6:991–997. (B) Modified from Thannickal TC, Moore RY, Nienjuis R, et al. Reduced number of hypocretin neurons in human narcolepsy. Neuron. 2000;27(3):469–474.*

Hypocretin is a neuropeptide found in the posterolateral hypothalamus that regulates arousal, wakefulness, and appetite. The brain contains only 70,000 of these neurons on average, relatively few in terms of cell populations. Hypocretin is secreted as two homologous peptides, hypocretin-1 and 2, also known as orexin-A and B.[12,13] Hypocretin neurons send axons to the entire cerebral cortex, including the brainstem and basal forebrain, with intense input to the tuberomammillary nucleus and the locus coeruleus.[19] Hypocretin neurons are most active during wakefulness and silent during REM sleep. They also help maintain normal muscle tone by exciting monoaminergic neurons, motor neurons, and neurons in the ventrolateral periaqueductal gray matter/lateral pontine tegmentum (vlPAG/LPT). In Na-1, the loss of the hypocretin neurons together with strong, positive emotions can trigger cataplexy. One possibility may be that positive emotions activate neurons in the amygdala that subsequently excite the sublaterodorsal tegmental nucleus (SLD) and inhibit the vlPAG/LPT. The SLD then excites neurons in the medial medulla and spinal cord that strongly hyperpolarize downstream motor neurons, resulting in cataplexy.[20]

The extremely high association between narcolepsy type 1 and the HLA-DQB1*06:02 genotype, along with other recent findings, support the hypothesis of an autoimmune basis for the disease, with hypocretin cells as the logical target. As seen in many other autoimmune diseases, infections such as group A Streptococcus and influenza A have been identified as possible triggers for Na-1. Titers of the antistreptolysin antibodies, a marker for recent Streptococcus infection, have been found to be elevated in early onset narcolepsy cases.[21] In addition, a sudden increase in the number of new narcolepsy cases was found in 2010, soon after the occurrence of the H1N1 pandemic. In China, there was a 3–5× increase in the number of children newly diagnosed with narcolepsy when compared to prior years.[22] These cases appeared 4–6 months after the peak of H1N1 infections. In parallel with this, in both Finland and Sweden, a 10× increase in the cases of childhood-onset narcolepsy was reported a few months following vaccination with Pandemrix, a particular pH1N1 vaccine containing a potent adjuvant called AS03.[23,24] Remarkably, in the positive cases who were HLA-typed, they were found to all test positive for the narcolepsy risk allele HLA-DQB1*06:02. Several other European studies later confirmed that this particular vaccine had similar effects.

In monozygotic twins, there is a 25–32% concordance for narcolepsy, and first-degree relatives of patients with narcolepsy-cataplexy carry a 1% risk of developing the disorder, compared to 0.03% in the general population.[25] To date, the HLA-DQB1*06:02 allele is the most specific genetic marker for Na-1 and is found in ~98% of cases with CSF hypocretin-1 deficiency, compared to 12–38% in the general population, depending on ethnicity. In addition to HLA-DQB1*06:02, polymorphisms in the T-cell receptor (TCR) α- and β- locus, P2RY11 purinergic receptors, Cathepsin-H, and OX40L which are all involved in immune processes, increase the susceptibility for Na-1.[26–28] Altogether, it is thought that perhaps a peptide unique to hypocretin cells is mistakenly recognized by the immune system in the context of DQB1*06:02, which may lead to autoimmune destruction of hypocretin neurons.

Secondary cases, however, referred by the ICSD-3 as narcolepsy due to a medical condition, may result from direct insults to the hypocretin system rather than an autoimmune mechanism. This is supported by cases of hypothalamic tumors seen in association with narcolepsy.[29] In other cases, narcolepsy may be the result of complex genetic pathology, including disorders such as myotonic dystrophy, Prader–Willi syndrome, Niemann-Pick disease, and autosomal dominant cerebellar ataxia, deafness and narcolepsy (ADCADN).[30]

In contrast to Na-1, the pathophysiologies of Na-2 and IH are unknown at the present time. There is little evidence for differentiating Na-2 (with normal CSF hypocretin-1) and IH at the clinical level. In patients with IH, reduced CSF histamine levels have been reported, which led to the suggestion that CSF histamine is a biomarker of hypersomnia. But in a recent study using a new, validated assay, Dauvilliers et al. did not find any differences in levels of CSF histamine comparing various hypersomnia disorders including IH and neurological controls.[31] Similarly, other neurochemical studies measuring monoamine metabolites in the CSF have been inconclusive. In 2012, Rye et al. studied CSF samples of 32 hypersomnolent patients and found a gain of function in the GABA-A system in vitro.[32] They also showed that this effect can be reversed in some cases by flumazenil, a GABA-A receptor antagonist, and improve symptoms of sleepiness in patients. The authors reasoned that the presence of a positive allosteric modulator of GABA-A receptors in hypersomnolent patients may be at play, although the molecular culprit has yet to be identified. A genome-wide association study of Japanese subjects with "essential hypersomnia," equivalent to Na-2 and IH, identified risk alleles in three gene loci: NCKAP5, SPRED1, and CRAT.[33] The role of these genes remains to be determined.

Even though the pathophysiologic mechanism remains unknown, a number of neuroimaging studies have provided some interesting insights. Typically, patients with KLS have normal brain morphology on MRI and head CT.[34] However, in functional imaging studies, consistent abnormalities have been reported. Studies utilizing single photon emission CT scanning during patients' symptomatic periods have demonstrated hypoperfusion in the thalamus, hypothalamus, temporal lobes, orbitofrontal and parasagittal frontal lobes, basal ganglia and occipital regions—thalamic hypoperfusion being most commonly found.[35,36] When patients are asymptomatic, the thalamic hypoperfusion appears to resolve but persistent hypoperfusion in the mesial temporal lobe, frontal lobe, and basal ganglia has been found. It is conceivable that abnormalities in these brain regions correlate with the symptomatology seen in KLS. For instance, sensation of derealization and memory deficits may result from dysfunction in the temporal lobes, while behavioral symptoms of apathy, hyperphagia, and hypersexuality may result from abnormalities in the orbitofrontal and anterior parasagittal regions. Nevertheless, these correlations have not been confirmed.

Kleine–Levin Syndrome

Kleine–Levin syndrome (KLS) is a recurrent hypersomnia that is characterized by recurrent (relapsing-remitting) episodes of severe hypersomnolence separated by intervening periods of normal behavior, in association with cognitive, psychiatric, and behavioral disturbances (Table 8.2). KLS is a rare disorder, with an estimated prevalence of 1 to 5 per million individuals.[34] Most patients are teenagers at disease onset, although rare cases with onset as young as 4 years of age have been reported. KLS is uncommon in those older than age 35, and is 4× more likely to occur in males. The typical duration of an episode is, on average, 10 days (range from 2.5 to 80 days), with rare episodes lasting multiple weeks to months. The first episode is often triggered by an infection or associated with a fever. Further episodes recur every 1–12 months (median recurrence interval of 3 months).

During symptomatic episodes, patients complain of sleepiness despite sleeping up to 16–21 h per day. Patients get up only to eat and void, and spend the majority of the day and night

TABLE 8.2   Diagnostic Criteria for Kleine–Levin Syndrome (ICSD-3)

Criteria A to E must be met:

    A. The patient experiences at least two recurrent episodes of excessive sleepiness and sleep duration, each persisting for 2 days to 5 weeks.

    B. Episodes recur usually more than once a year and at least once every 18 months.

    C. The patient has normal alertness, cognitive function, behavior, and mood between episodes.

    D. The patient must demonstrate at least one of the following during episodes:

        1. Cognitive dysfunction

        2. Altered perception

        3. Eating disorder (anorexia or hyperphagia)

        4. Disinhibited behavior (such as hypersexuality)

    E. The hypersomnolence and related symptoms are not better explained by another sleep disorder, other medical, neurologic, or psychiatric disorder (especially bipolar disorder), or use of drugs or medications.

in bed. They remain arousable but are irritable if prevented from sleeping. When awake, patients become apathic and report impairment in communication, concentration, and memory. Partial or total amnesia episodes is generally reported following resolution of the episodes. Patients often withdraw from social interaction and display transient symptoms of depression and anxiety, and as such that they can be misdiagnosed with psychiatric conditions. This is an area of current debate and some researchers believe that psychiatric disease, especially bipolar disorder, may be more prominent in these individuals.

The classic triad of hypersomnia, hyperphagia, and hypersexuality is only seen in a quarter of patients with KLS. The most specific findings are cognitive abnormalities, notably a sensation of depersonalization or derealization—an altered, dream-like state. Toward the end of an episode, patients may experience transient rebound insomnia. Once symptoms have completely resolved, patients may feel elated, and can sometimes border on hypomania. Patients are completely normal between episodes. KLS typically resolves after a median of 14 years from initial onset, and most adults are unaffected after age 35, although long term sequelae have been suggested by some studies, especially in patients who experience their first symptoms as adults.

Treatment

The combination of behavioral and pharmacologic treatments provides the best therapeutic option to control EDS in CNS hypersomnias. A key behavioral treatment is the maintenance of a regular, adequate sleep schedule. For patients with narcolepsy, institution of scheduled naps can be very effective in ameliorating EDS. This strategy is, however, less successful in patients with IH who typically experience unrefreshing naps. Educating the patient and family about the disorder helps promote treatment compliance and a supportive network. Patients may find helpful resources through organized patient advocacy support groups. Importantly, clinicians should provide counseling to hypersomnolent patients regarding the risk and avoidance of drowsy driving. A number of pharmacologic treatments are available to treat EDS, cataplexy, disrupted nocturnal sleep, and other REM-related phenomena. These medications include stimulants, antidepressants, and sodium oxybate (Table 8.3). As many of these medications have a significant side effect profile and a potential for abuse, clinicians

TABLE 8.3   Pharmacologic Treatments for CNS Hypersomnias

Medication	Pharmacologic properties	Typical dosing regimen	Side effects/remarks
STIMULANTS			
Modafinil	Mode of action not completely clear, likely involves selective DA reuptake inhibition, and has minimal addiction potential Less effective than amphetamine-based stimulants	Modafinil, 100–400 mg in the morning or as divided doses (Armodafinil, 50–250 mg in the morning)	Headaches are common but minimized by starting at lower doses and slow up-titration May decrease efficacy of hormonal oral contraceptives Possible allergic effects (rash) Armodafinil is twice as potent at steady state
Amphetamines	Bind to dopamine and norepinephrine transporter, preventing their reuptake. Substitute for monoamines via vesicular monoamine transporter, disrupt storage and increase concentrations of dopamine, norepinephrine, and serotonin (DA > NE >> 5-HT)	Dextroamphetamine (Dexedrine) Dextroamphetamine/ amphetamine (Adderall) Methamphetamine (Desoxyn) 5–60 mg daily or as divided doses in all of the above	Monitor blood pressure and heart rate Preference given to longer lasting formulations Can be more effective than methylphenidate in some patients Methamphetamine has higher brain penetration and potency than amphetamine
Methylphenidate	Blocks monoamine uptake Increases release of norepinephrine and dopamine Short half-life Potential for addiction, especially for immediate release preparation More effective than modafinil	10–60 mg daily or as divided doses	Various formulas can have different interindividual effects Relatively low cost Immediate release (5–10 mg) can be helpful on an as-needed basis Monitor blood pressure and heart rate
ANTIDEPRESSANTS			
Fluoxetine (SSRI)	Selective serotonin reuptake inhibitor Active metabolite norfluoxetine has more adrenergic effects	20–60 mg	Long half-life, more stable effect on cataplexy Useful with concomitant anxiety disorder Abrupt discontinuation of antidepressants may lead to rebound cataplexy
Venlafaxine (SNRI)	Selective serotonin-norepinephrine reuptake inhibitor Short half-life; extended release formulation preferred Mild stimulant effect	37.5–150 mg, max 300 mg/day Once daily dose with extended release formulation	Gastrointestinal upset (main reason for discontinuation) Potential withdrawal side effects on abrupt cessation

(Continued)

TABLE 8.3   Pharmacologic Treatments for CNS Hypersomnias (*cont.*)

Medication	Pharmacologic properties	Typical dosing regimen	Side effects/remarks
Atomoxetine (NRI)	Selective norepinephrine reuptake inhibitor Mild stimulant effect	10–60 mg, usually as divided doses, maximum 80 mg/day	Side effects include nausea, reduced appetite and urinary retention Short half-life Monitor blood pressure and heart rate
OTHER			
Sodium oxybate	Exact mechanism unknown γ-hydroxybutyric acid (GHB) is the active ingredient GHB is a gamma-aminobutyric acid (GABA-B) receptor agonist	4.5–9 g (at a minimum divided into bi-nightly doses)	Side effects include nausea, weight loss, parasomnia, enuresis Treats all aspects of narcolepsy, including disrupted nocturnal sleep, EDS, cataplexy Use with caution in presence of hypoventilation or significant sleep apnea Immediate effects on disrupted nocturnal sleep Only FDA-indicated treatment for cataplexy
Clarithromycin	An antibiotic, likely acts as a negative allosteric modulator of GABA-A receptor Exact mechanism unclear	Typical dose at 500 mg taken with breakfast and 500 mg taken with lunch (start with a 2-week trial period). Dose may be increased to as much as 1000 mg twice daily depending on therapeutic response	Side effects include gastrointestinal distress, altered taste perception Cases of neurotoxicity including mania, psychosis, and nonconvulsive status epilepticus had been reported Metabolized via cytochrome P450 3A subfamily

need to be judicious in choosing a pharmacologic regimen, particularly if the etiology is unclear or multifactorial in nature like in Na-2 and IH.

The nonamphetamine, wake-promoting agent modafinil is often considered a first line standard therapy for EDS in narcolepsy. It is sometimes more effective taken twice daily as compared to its advertised once daily dosing. Armodafinil is the R-enantiomer of modafinil and is twice as potent as modafinil at steady state. It is typically dosed once daily in the morning. Both modafinil and armodafinil are also Food and Drug Administration (FDA) approved for the treatment of residual EDS in OSA and shift work disorders. Currently, they are not FDA approved for IH, although they are often prescribed off-label in these patients. One possible side effect of modafinil is the possibility of an allergic reaction, often manifesting as a rash. Clinicians and patients should also be aware that modafinil and armodafinil can decrease the efficacy of oral contraceptives due to their effect on the cytochrome P450 system, Alternative contraceptive methods should be proposed to patients who wish to remain on low dose hormonal therapy. Although the mode of action of modafinil is controversial, it is likely to act through inhibition of the dopamine transporter (DAT). The compound is of particular interest as its abuse potential is low.

Amphetamines and amphetamine-like CNS stimulants are alternative effective pharmacologic treatments of EDS in narcolepsy, and have been in use since 1935. These include dextroamphetamine/amphetamine, methylphenidate, and methamphetamine. Amphetamines' main mechanism of action is increased dopamine release, although weaker effects on the release of norepinephrine and serotonin are also observed. Increased release occurs because amphetamine inhibits vesicular storage of monoamine, resulting in a reverse efflux of dopamine via DAT into the synaptic cleft. Other compounds such as methylphenidate increase dopamine transmission by inhibiting the reuptake of monoamines, thereby increasing the availability of already released dopamine. As a class, these compounds have the greatest effects on EDS; however, they also have a high potential for abuse and tolerance. This is especially notable for short acting formulations. For these reasons, amphetamine-like stimulants should be prescribed at the minimum effective doses, and with caution in patients without a clear etiology.

Sodium oxybate (γ-hydroxybutyrate) is now considered a standard therapy for narcolepsy. One advantage of the compound is that it treats multiple symptoms, including EDS, cataplexy, and disrupted nocturnal sleep. Sodium oxybate is a sedative anesthetic compound known to increase slow-wave sleep and, to a lesser extent, REM sleep. The mode of action of sodium oxybate in narcolepsy is unclear. Most studies suggest that its sedative-hypnotic effect is mediated through agonism at the GABA-B receptor.[37] Sodium oxybate is FDA approved for the treatment of cataplexy in Na-1, as well as for the treatment of EDS in Na-1 and Na-2. It is taken twice nightly in divided dose because it has a short half-life (~30 min). Sodium oxybate has been found to be effective in reducing cataplexy and sleep attacks, thereby improving patients' overall daytime function.[38] Interestingly, it may take weeks of treatment and dose adjustments to achieve a full therapeutic effect. Because of its potential for abuse and possible adverse effects with heavy sedation and respiratory depression, it is dispensed through a central pharmacy in the United States.

Cataplexy can also be treated with antidepressants (although not FDA approved), which have REM-suppressing properties. Tricyclic antidepressants including imipramine, protriptyline, and clomipramine have been used since the 1970s and have established immediate effects on cataplexy. Unfortunately, these compounds also have significant anticholinergic side effects leading to dry mouth, tachycardia, sexual dysfunction, and difficulty with urination. SSRIs/SNRIs are also useful in the management of cataplexy and have fewer side effects than tricyclics. Among this class, venlafaxine, a serotonin-norepinephrine reuptake inhibitor (also a weak inhibitor of dopamine reuptake), is very effective and is therefore commonly used. The extended release formulation is preferred due to the drug's short half-life. Atomoxetine, a specific noradrenergic reuptake inhibitor (NRI) is also used in the treatment of cataplexy and EDS, particularly in children. Although the mode of action of these medications in suppressing cataplexy is not known, it is speculated that increased noradrenergic and serotonergic signaling may lead to a suppression of REM sleep. In addition, tricyclics, SSRIs, and venlafaxine may be helpful in treating sleep paralysis, and hypnagogic and hypnopompic hallucinations. Clinicians and patients should be aware that although these compounds have immediate therapeutic effects, abrupt discontinuation of a chronic antidepressant therapy can trigger severe and dramatic rebound cataplexy.

There are no currently FDA approved drugs for IH and, therefore, treatment is usually off-label and adapted from experience with narcolepsy patients. The use of stimulants such as modafinil are commonly prescribed. Antidepressant therapy may be useful if there is suggestion of associated anxiety or depression. Sodium oxybate, a compound primarily used for cataplexy, was also recently found to help some patients with IH.[39] In another study, Trotti et al. showed that patients taking clarithromycin—a negative allosteric modulator of the GABA-A receptor—at a dose of 500 mg bid (taken in morning and lunch time) had significant improvement in subjective sleepiness.[40]

At the time of this writing, several other new compounds are currently undergoing clinical trials in the United States for narcolepsy and IH. Pitolisant, a histamine H3 receptor inverse agonist which activates histamine release in the brain, is in Phase III trial for treatment of EDS in Na-1 and Na-2. Another compound, JZP-110 ([R]-2-amino-3-phenylpropylcarbamate hydrochloride; formerly known as ADX-N05) a phenylalanine derivative that enhances dopaminergic and noradrenergic neurotransmission, is currently undergoing Phase III trial for treatment of EDS in narcolepsy and OSA. Flumazenil, a benzodiazepine antagonist, is being studied in patients with IH. BTD-001 (pentylenetetrazole), a GABA-A receptor antagonist, is undergoing Phase II trial for treatment of IH and Na-2.

The results of drug trials in KLS patients have been inconsistent and disappointing. Unless episodes are particularly severe and frequent, the best treatment is supportive, educating patients and parents about the disorder. Patients should be permitted to rest during the symptomatic periods; school, and work-related activities should be adjusted as well. The use of stimulant medications such as modafinil, methylphenidate, and amphetamines is rarely beneficial and may unmask cognitive and behavioral symptoms. Lithium has been proposed as a treatment option, especially in cases when episodes of recurrent hypersomnia are severe and frequent. However, in one metaanalysis it was found to be helpful in only 20–40% of cases.[41,42] When lithium is used, serum levels should be between 0.8 and 1.2 mmol/L, and ongoing consultation with a psychiatrist is advised. The effects of other mood stabilizers, such as valproic acid or carbamazepine are less well documented. Antipsychotics such as risperidone have been used when prolonged psychotic symptoms are present.

CLINICAL FEATURES

Excessive Daytime Sleepiness

The prevalence of excessive daytime sleepiness (EDS) in the general population ranges from 4% to 28%, depending on the definition used.[43] EDS, a primary feature of all CNS hypersomnia disorders, is defined as an "inability to stay awake and alert during the major waking episode (typically daytime), resulting in periods of irrepressible need for sleep or unintended lapses into drowsiness or sleep."[1]

Sleepiness typically occurs during monotonous, sedentary activity or after a heavy meal, but can also occur during regular activities such as talking and eating. Duration and frequency of sleepiness episodes can be variable in CNS hypersomnias. Although sleepiness is a core symptom of narcolepsy and IH, it can also be found in association with many other conditions such as insufficient sleep, obstructive sleep apnea, and circadian rhythm disorders.

It may also be a side effect of medication use, or it may be secondary to medical or neurological disorders. During episodes of sleepiness or "sleep attacks," an individual may experience "microsleep"—a brief, unintended episode of sleep lasting only seconds—which can lead to domestic, occupational and vehicular accidents. Sleep attacks occur more often in narcolepsy than in IH. Naps are generally reported to be refreshing in patients with narcolepsy, while this is more variable in IH, explaining why scheduled naps are an effective behavioral management strategy in narcolepsy. Patients with IH may instead report prolonged episodes of unrefreshing sleep or "sleep drunkenness" and often wake up in a confused state. Severe sleepiness may also result in "automatic behaviors," carrying out routine activities in a semi-automatic manner, lasting from a period of seconds to 30 min, for which patients are amnestic after the episode.

Cataplexy

Cataplexy occurs in 65–75% of individuals with narcolepsy,[44,45] although cases without cataplexy are being recognized with increasing frequency. In its typical presentation, cataplexy is defined as a sudden and transient loss of skeletal muscle tone triggered by strong emotions (Fig. 8.2). It is the core symptom of narcolepsy type 1 (Na-1, with hypocretin deficiency) and is almost always absent in narcolepsy type 2 (Na-2, without hypocretin deficiency). Cataplexy

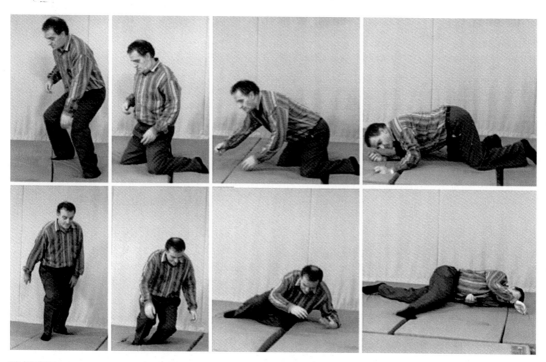

FIGURE 8.2 **A cataplectic episode in an adult demonstrating buckling of the knees and falling to the floor.** *Source: Courtesy of Overeem S, Mignot E, van Dijk JB, Lammers GJ. Narcolepsy: clinical features, new pathophysiologic insights, and future perspectives.* J Clin Neurophysiol. 2001;18(2):78–105.

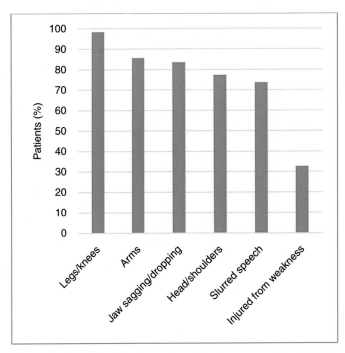

FIGURE 8.3   **Muscle groups affected in typical cataplexy. The most commonly affected muscle groups involve the legs/knees.** *Source: Modified from Anic-Labat S, Guilleminault C, Kraemer HC, et al. Validation of a cataplexy questionnaire in 983 sleep disordered patients. Sleep. 1999;22(1):77–87.*

is most specifically triggered by positive emotions, such as laughter and joking, but can rarely occur in the context of anger or fright. A typical cataplexy episode lasts from a few seconds to a minute, and very rarely more than a minute. It most often results in buckling of the knees, head dropping, sagging of the jaw, slurred speech or weakness in the arms (Fig. 8.3).[46] In the vast majority of attacks, cataplexy affects muscle groups bilaterally, although patients sometimes report one side of the body to be more involved than the other. Falls and injury are rare as patients typically have time to find support or to sit down while the attack is escalating. If time permits and reflexes can be assessed (which is rare because the duration is very brief), they are diminished or absent. Respiration is uncompromised, although choking may be reported if the head drops forward. Awareness is preserved throughout the episode. Recovery is complete and immediate. Partial attacks, for example, a brief grimace, slurred speech or facial drooping, can be quite subtle and only recognizable by an experienced observer.

Given that patients rarely exhibit cataplexy during examination, the diagnosis is usually made by history. A good initial question may be: "Does anything unusual happen when you tell a joke, hear something funny or laugh?" It is also helpful to ask patients to describe their first and last episodes, where and in what kinds of situation it occurred, trying to evoke a real story rather than a textbook clinical description. In Na-1, cataplexy typically occurs within a year of the onset of excessive sleepiness,[47] although cases where cataplexy has developed more than 20 years after sleepiness have been described. Interestingly, the frequency of cataplexy often decreases with age.

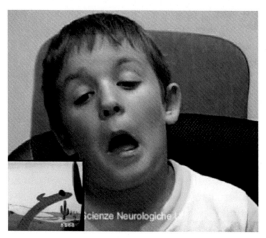

FIGURE 8.4 **"Cataplectic facies" in a child watching a cartoon.** *Source: Courtesy from Serra L, Montagna P, Mignot E, Lugaresi E, Plazzi G. Cataplexy features in childhood narcolepsy. Mov Disord. 2008;23(6):858–865.*

In children with narcolepsy, cataplexy can have a very atypical presentation when the onset of the disease is recent. The symptom may occur without any clear emotional trigger, although association with laughter and other typical emotions often develop in the following 6 months. In these cases, cataplexy often presents as facial hypotonia with droopy eyelids, mouth opening, and a protruded tongue, a symptom referred as "cataplectic facies" (Fig. 8.4). These patients may also experience "status cataplecticus," a severe form of rebound cataplexy characterized by a string of cataplectic attacks lasting several hours per day, confining the patient to the bedroom. Status cataplecticus can also occur as a result of the abrupt withdrawal of an anticataplectic medication, typically an antidepressant. Although clear-cut cataplexy is a pathognomonic sign of Na-1, cataplexy or cataplexy-like episodes can occur in other disorders, such as Niemann-Pick disease type C, Prader–Willie and Norrie's disease in children and DNA methyltransferase 1 (DNMT1) mutations in adults.[30,48] It can also be seen in association with anti-Ma2 paraneoplastic syndrome, a rare limbic encephalitis associated with seminomas.[49]

Sleep Paralysis

Sleep paralysis occur frequently in narcoleptic patients, either when falling asleep or upon awakening. Because many normal individuals also experience sleep paralysis, it is not a useful symptom diagnostically, especially if it occurs sporadically. In one study, 8% of the general population and 28% of students reported having isolated sleep paralysis.[50] The occurrence of isolated sleep paralysis may be precipitated by stress, sleep deprivation, and an irregular sleep-wake schedule. It occurs when there is an incomplete, abrupt transition between wakefulness and REM sleep. When this happens, the individual finds himself paralyzed, unable to move the limbs, speak, or open the eyes, despite being awake and able to recall the event later. The experience may last for several minutes and can be very frightening. Breathing may appear more difficult than usual because intercostal muscles are paralyzed and the chest feels heavier. In some cases dream-like hallucinations can occur, making the experience even more terrifying. Sleep paralysis can be so distressing to some individuals that they may begin to fear going to sleep. After multiple episodes, however, patients generally learn that the episodes are benign, rarely lasting longer than a few minutes and always end spontaneously. Sleep paralysis is more frequently reported in narcolepsy than in IH. It is, however, more rarely reported in prepubertal children with Na-1 in comparison to later ages.

Hypnagogic and Hypnopompic Hallucinations

Visual hallucinations either at sleep onset (hypnagogic) or upon awakening (hypnopompic) are another typical symptom of narcolepsy. The hallucinations are dream-like and may be unpleasant, pleasant or neutral. A repetitive theme may emerge and in these cases they can be very distressing, especially if they have a nightmarish quality. At times, these experiences can be so realistic that the patient may act upon them after awakening, believing them to be real. For example, patients with narcolepsy have been known to call the police convinced that intruders were in their home, only to find out it was just a hallucination or a dream afterward. Auditory hallucinations are also common and can range from simple sounds to an elaborate tune. These may then be difficult to distinguish from auditory hallucinations reported in schizophrenia, a possible comorbidity of narcolepsy, although in this case hallucinations are more likely part of a more complex network of delusions and the patient cannot be easily convinced these are just dreams. Hypnagogic and hypnopompic hallucinations can be thought of as dissociative states between sleep (dreaming) and wake, where the distinct boundary is unclear. Because hypnagogic hallucinations are at times difficult to differentiate from sleep onset mentation, or nightmares, they are not a very useful diagnostic sign, and they can also be found in IH.

Disrupted Nocturnal Sleep

Even though daytime symptoms such as EDS and cataplexy are the most common chief complaints in narcolepsy, up to 80% of narcolepsy patients experience disruptions in nocturnal sleep.[51] Therefore, clinicians should inquire about nocturnal symptoms as well as daytime symptoms. Patients often report more problems maintaining sleep than initiating sleep, with frequent nocturnal awakenings. Vivid dreaming, REM behavior disorder, and periodic limb movements may also disrupt patient's sleep. It is a misconception that patients with narcolepsy sleep more than usual. In fact, despite having sleep attacks during the day, patients with Na-1 may even sleep less than normal individuals over the course of a 24 h day. Their major problem is staying awake for long periods of time without napping. Individuals with IH, on the contrary, typically report sleeping soundly and efficiently throughout the night.

DIAGNOSTIC PROCEDURES

When evaluating patients with EDS, the Epworth Sleepiness Scale (ESS) is one of the most widely used, validated, subjective self-administered questionnaires.[8] When using this scale, patients rate their usual chances of dozing off or falling asleep in eight different daily situations on a 4-point scale (0–3). The total ESS score is the sum of the eight item-scores and can range between 0 and 24, with a higher score reflecting a higher level of daytime sleepiness. It provides a subjective measure of an individual's usual level of daytime sleepiness and sleep propensity in daily life. The ESS has not been validated in children, and the Pediatric Daytime

Sleepiness Scale is more appropriate for use in this population.[52] In evaluating EDS, clinicians must differentiate sleepiness from fatigue and exhaustion, as patients often confuse the two. Fatigue presents as a lack of physical energy or body "tiredness." The use of the Fatigue Severity Scale together with the ESS is often helpful in distinguishing EDS from fatigue.[53]

Several tests have been designed to objectively evaluate sleepiness. The multiple sleep latency test (MSLT) was first developed by Carskadon and Dement in 1977.[54] It was designed to measure physiologic sleep tendencies in the absence of external alerting factors. The MSLT consists of a series of four to five napping tests conducted during the day and performed at 2 h intervals. To obtain a clinically valid MSLT, the test must be conducted under specific conditions.[55] First, it is recommended that patients keep a regular sleep schedule, allowing for adequate sleep in the days leading up to the test. To ensure this, patients are often asked to complete a sleep diary or to wear a wrist actigraphy for 2 weeks prior to the MSLT. Second, patients are asked to withhold taking medications such as sedatives, stimulants, and those that affect the propensity to enter REM sleep, particularly tricyclic antidepressants and SSRIs/SNRIs, for 1–2 weeks prior to testing. Interrupting therapy too close to the test may also affect the results by creating rebound sleepiness or REM sleep. A drug screen should be performed on the morning of the test. Third, a polysomnography (PSG) study must be performed on the night preceding the MSLT test. The PSG serves to evaluate for alternative and coexisting causes of chronic daytime sleepiness, such as undiagnosed obstructive sleep apnea. It is also used to keep track of the patients' total sleep time; ideally at least 360 min of sleep must be observed to exclude inadequate sleep as a confounder.[56] Finally, the PSG is useful as patients with narcolepsy may have sleep onset REM periods (SOREMPs)—REM sleep which occurs within the first 15 min of sleep onset.

During PSG and MSLT testing, the patient is monitored in a comfortable, soundproof, and dark bedroom. During the MSLT, patients are asked to stay awake between each nap. The initial nap opportunity begins 1.5–3 h after termination of nocturnal PSG, which ideally occurs at the patient's usual wake up time. Prior to each nap, the patient is instructed to lie quietly and attempt to fall asleep. The MSLT records the latency for each 4–5 min nap opportunity (time between lights out and sleep onset). If no sleep is observed for 20 min, then the nap ends. If the patient falls asleep, then the nap continues for another 15 min to evaluate for the possibility of a SOREMP. At the end of the test, the mean sleep latency (MSL) is calculated for all naps. A MSL of 10–20 min is generally seen in healthy, rested subjects, while a MSL ≤ 8 min indicates sleepiness. This cut off is used for the diagnosis of both narcolepsy and IH, although ~22% of the general population meets this criteria, and thus is not a very specific finding.[57] More importantly, in narcolepsy, patients generally exhibit multiple SOREMPs during the test, with at least two instances during nocturnal sleep onset and daytime naps considered diagnostic for the condition (Fig. 8.5). Whereas a MSL ≤ 8 min is not a very specific finding, MSL ≤ 8 min and the observation of ≥ 2 SOREMPs is only found in approximately 2–4% of the population.

The PSG-MSLT has several limitations. While it has been validated in the context of Na-1, its use in diagnosis of Na-2 and IH have mostly been based on consensus and by extension. Clinicians should be mindful of the fact that SOREMPs are common in shift workers and can occur in other disorders that increase pressure for REM sleep, such as in insufficient sleep, untreated sleep apnea, or delayed sleep phase syndrome. It must also be performed free of any neuroactive substances, which is often difficult for patients with psychiatric or pain disorders. Further, whereas a positive MSLT in the context of Na-1 is reliable, repeatability in the context of Na-2 or IH is extremely poor.[58,59] Because there are no clear clinical differentiating

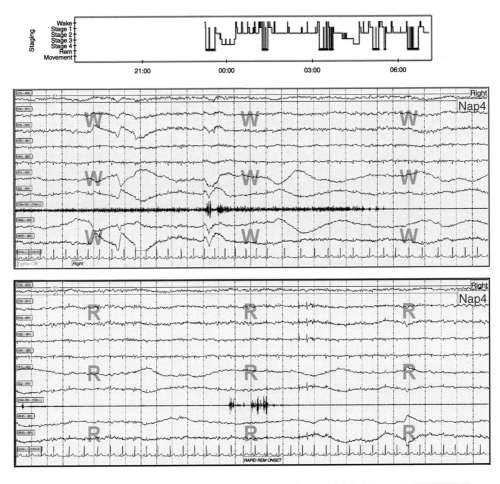

	Nap 1	Nap 2	Nap 3	Nap 4	Nap 5	Means
Start time	09:17:53	11:21:23	01:18:53	03:16:53	05:21:23	
End time	09:33:23	11:36:23	01:34:23	03:32:23	05:36:53	
Recording length	15.5	15.0	15.5	15.5	15.5	15.4
Total sleep time	15.0	15.0	15.0	15.0	15.0	15.0
Onset to sleep	0.5	0.0	0.5	0.5	0.5	0.4
Onset to REM sleep	0.5	1.0	3.5	0.0	1.0	1.2

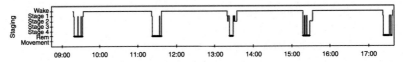

FIGURE 8.5 **MSLT and PSG result from a narcolepsy type 1 patient.** (Top) A PSG performed the night prior to the MSLT with a SOREMP (REM sleep occurs in the first 15 min of sleep)—a common finding seen in patients with narcolepsy type 1. (Middle panels) Continuous two 30 s epochs from a MSLT recording of a narcoleptic patient showing an abrupt transition from wake to REM sleep. (Bottom) A summary report of a MSLT from a patient with a mean sleep latency of 0.4 min across five naps and five out of five recorded SOREMPs.

features between IH and Na-2,[60] the two disorders are likely best considered as a spectrum and they should be treated similarly.

Another polysomnographic test which can be used in the evaluation of EDS is the maintenance of wakefulness test (MWT).[61] As the name implies, the MWT assesses one's ability to maintain wakefulness. It was developed based on the assumption that the volitional ability to remain awake provides important information regarding one's ability to do so. The MWT is conducted during the day and a variety of protocols have been used. The recommended protocol consists of four 40 min trials performed at 2 h intervals, with the first trial beginning at 1.5–3 h after the patient's usual wake up time. Patients are instructed to maintain wakefulness while sitting comfortably in bed in a dark room. In normal controls, the mean sleep latency (to the first epoch of sleep) in a MWT was found to be at 30.4 ± 11.2 min, and a MSL ≤ 8.0 min is considered abnormal.[56] The MWT can be a useful tool in pharmacologic trials in evaluating response to a treatment for EDS, and in evaluating the risk of falling asleep associated with specific jobs or activities.

A seldom-used test to document sleepiness is the continuous 24 or 36 h PSG. This test aims to obtain information about the frequency, timing and duration of daytime sleep episodes, as well as documenting nighttime sleep disruptions. This test is typically performed using ambulatory equipment. Although the test is excellent in distinguishing Na-1 from other pathology and gives detailed information on the nature of each patient's hypersomnia, it is difficult to perform and thus rarely used. The long polysomnographic recording may capture an episode of cataplexy in the evaluation of a narcolepsy by showing the absence of chin and muscle twitches in the awake patient. Moreover, a 24 h PSG can also be used to help evaluate patients who are suspected to have IH or Kleine–Levin syndrome (KLS) while they are experiencing an episode of hypersomnia.

INDIVIDUAL PATHOLOGIES

Narcolepsy Type 1

Narcolepsy type 1 (Na-1) typically presents with a pentad of symptoms: EDS, cataplexy, sleep paralysis, hypnagogic and hypnopompic hallucinations, and disrupted nocturnal sleep. The diagnosis of Na-1 is most often apparent from the clinical history alone, notably the presence of cataplexy, which should be typical, for example, triggered by usual emotions (most often positive emotions such as laughter or joking). Patients with Na-1 also experience daily episodes of an irrepressible need to sleep or lapses into sleep, leading to naps, which are generally refreshing. Until recently (current update of the ICSD-3), Na-1 was known as narcolepsy with cataplexy but, because most of these cases are caused by a deficiency of hypothalamic hypocretin (orexin) signaling,[26] it now represents cases with documented biological abnormality, even in the absence of cataplexy (Table 8.4). Given that CSF hypocretin-1 measurements are not systematically performed, a diagnosis of Na-1 can be made on the basis of both cataplexy and a positive MSLT test—these patients would be positive in most cases if the test was performed.[15] It is only in cases without cataplexy that low CSF hypocretin-1 is mandated to diagnose Na-1.

Given that large scale CSF hypocretin-1 evaluation is not possible, only the prevalence of narcolepsy with cataplexy is known. It is estimated to occur in about 0.03% (1 in 3000) of individuals in the United States, Europe, and Korea.[62] The prevalence of narcolepsy may be higher in Japan (0.16%), and lowest in Israel (0.0002%).[63,64] There is a slight male predominance, and

TABLE 8.4 ICSD 3 Diagnostic Criteria for Narcolepsy Type 1, Type 2, and Idiopathic Hypersomnia[1]

Narcolepsy type 1 (criteria A and B must be met)	Narcolepsy type 2 (criteria A to E must be met)	Idiopathic hypersomnia (criteria A to F must be met)
A. Daily periods of irrepressible need to sleep or daytime lapses into sleep occurring for at least 3 months	A. Daily periods of irrepressible need to sleep or daytime lapses into sleep occurring for at least 3 months	A. Daily periods of irrepressible need to sleep or daytime lapses into sleep occurring for at least 3 months[c]
B. The presence of one or both of the following: 1. Cataplexy and a mean sleep latency (MSL) of ≤8 min and two or more sleep onset REM periods (SOREMPs) on an MSLT. A SOREMP on the preceding nocturnal polysomnogram may replace one of the SOREMPs on the MSLT. 2. CSF hypocretin-1 concentration, measured by immunoreactivity, is either ≤110 pg/mL or <1/3 of mean values obtained in normal subjects with the same standardized assay	B. A MSL of ≤8 min and two or more SOREMPs on a MSLT. A SOREMP on the preceding nocturnal polysomnogram may replace one of the SOREMPs on the MSLT	B. Cataplexy is absent
	C. Cataplexy is absent[a]	C. An MSLT shows fewer than two SOREMPS on MSLT and nocturnal PSG combined
	D. Either CSF hypocretin-1 concentration has not been measured or CSF hypocretin-1 concentration measured by immunoreactivity is either >110 pg/mL or >1/3 of mean values obtained in normal subjects with the same standardized assay[b]	D. The presence of at least one of the following: 1. MSLT shows a MSL of ≤8 min 2. Total 24 h sleep time is ≥660 min (typically 12–14 h) on 24 h polysomnographic monitoring (performed after correction of chronic sleep deprivation), or by wrist actigraphy in association with a sleep log (averaged over at least seven days with unrestricted sleep)
	E. The hypersomnolence and/or MSLT findings are not better explained by other causes such as insufficient sleep, obstructive sleep apnea, delayed sleep phase disorder, or the effect of medication or substances or their withdrawal	E. Insufficient sleep syndrome is ruled out
		F. The hypersomnolence and/or MSLT findings are not better explained by another sleep disorder, other medical or psychiatric disorder, or use of drugs or medications

[a]If cataplexy develops later, then it should be reclassified as narcolepsy type 1.

[b]If the CSF Hcrt-1 concentration is tested later and found to be either ≤110 pg/mL or <1/3 of mean values obtained in normal subjects, then it should be reclassified as narcolepsy type 1.

[c]Sleep drunkenness and/or long (>1 h) unrefreshing naps are additional supportive clinical features.

the age of onset varies from early childhood up to 50 years of age with a bimodal distribution, including a large peak between 15 and 25 years of age and a second, smaller peak between 35 and 45 years of age.[65] Narcolepsy symptoms have been reported in patients as young as 2 years of age, and in one case of a hypocretin gene mutation, symptoms were identified at 6 months of age.[66] It is also not uncommon to see a long gap, often more than 10 years, between the emergence of symptoms and the correct diagnosis, especially when cataplexy is initially absent. Increased awareness of narcolepsy among health-care professionals and the general public has helped to shorten that gap in recent decades. Epidemiological studies have shown that obesity is a common symptom in Na-1, and an unexplained increase in body weight is often seen at disease onset, especially when symptoms are acute.[67] Increased frequency of several other sleep abnormalities has also been described in narcolepsy, including REM sleep behavior disorder (RBD), periodic limb movements of sleep, and sleep disordered breathing. In addition, there is an increased prevalence of depressive symptoms and anxiety disorders in patients with narcolepsy, with about 20% of patients experiencing panic attacks or social phobias.[68]

Ethnic differences in symptom manifestation are also seen. A recent article of a large cohort of subjects found that significant racial group differences exist around the age of symptom onset, including those of cataplexy, sleepiness, and hypnagogic hallucinations.[69] In subjects with low CSF hypocretin-1, African Americans (28.3%) were 3.5 fold more likely to be without cataplexy when compared to Caucasians (8.1%). African Americans also appear to have a younger age of onset of sleepiness and higher subjective sleepiness scores when compared to other racial groups. These findings suggest that narcolepsy may present differently in African Americans, or alternatively there could be differences in referral patterns across ethnic groups.

Narcolepsy Type 2

Narcolepsy type 2 (Na-2) is characterized by EDS and $\geq$ 2 SOREMPs on the PSG-MSLT. Patients with Na-2 typically have elevated ESS scores. In a Japanese series, the mean ESS score in 62 patients with Na-2 was 14.9 ± 3.5, similar to the mean ESS (14.6 ± 3.7) of 52 patients with Na-1.[70] By definition, cataplexy is absent in Na-2, although some atypical sensations of weakness triggered by unusual emotions such as stress or anger may be reported. In contrast to Na-1, Na-2 is a more challenging diagnosis. This uncertainty arises from the nonspecific nature of the symptoms, the poor reliability of the MSLT, and the lack of a clear pathophysiology. Another problem is that some patients develop cataplexy many years after initial presentation. In these cases, should CSF hypocretin-1 have been measured, it would likely have been low and these patients would have been characterized as Na-1.[71] For these reasons, clinicians must obtain a thorough history and evaluation in identifying alternative, confounding diagnoses such as insufficient sleep syndrome, shift work disorder, circadian phase delay, and obstructive sleep apnea. Na-2 is estimated to occur in approximately 0.03% of the population, although reliable data are lacking.[45,72] In the Wisconsin Sleep Cohort, a population based sample of ~1500 individuals, many of whom have been subjected to repeat sleep studies and MSLT testing, only three subjects were found to have consistent symptoms and repeat positive MSLTs that were not confounded by shift work or chronic sleep restriction, suggesting a prevalence of approximately 0.16–0.3%.[73]

Little is known about the natural history of Na-2. As mentioned previously, reliability of the MSLT in this population is poor when hypocretin deficiency is not present,[59] therefore the disorder should not be considered a life-long condition. One study of 171 patients with Na-2 followed for several years found that of those with intermediate levels of CSF hypocretin-1 (between 110 and 200 pg/mL), 18% went on to develop cataplexy, but among those with normal hypocretin levels, cataplexy occurred in only one subject.[71] This suggests that in some individuals with hypocretin deficiency the symptoms may develop gradually, further highlighting the diagnostic challenge of Na-2.

Idiopathic Hypersomnia

In the past, IH was considered to be a rather rare disorder characterized by hypersomnolence with long and unrefreshing naps, prolonged nocturnal sleep time, high sleep efficiency, absence of cataplexy, and great difficulty waking from sleep. More recently, an increasing number of subjects have been diagnosed with isolated daytime sleepiness documented by the MSLT, and a new category was created in the ICSD2: idiopathic hypersomnia (IH) without long sleep time. In the ICSD-3, IH is no longer separated between cases with and without long sleep time. Rather, the MSLT must either document a MSL of ≤ 8 min and less than 2 SOREMPs, or a long sleep time (24 h sleep time of ≥ 660 min) must be documented using a 24 h PSG or wrist actigraphy. In all cases, confounding disorders which might cause daytime sleepiness should be considered and excluded, particularly insufficient sleep syndrome. Historically, two forms of IH were described.[74] The monosymptomatic form was characterized by excessive daytime sleep of one to several hours duration, while the polysymptomatic form was characterized by excessive daytime sleep, prolonged nocturnal sleep and great difficulty upon awakening in the morning.

The EDS experienced by IH patients is typically not as irresistible as in narcolepsy (Table 8.5 for a comparison of clinical features seen in CNS hypersomnias). Disease onset commonly occurs between 10 and 30 years of age. A familial history in association with IH has been noted, although rigorous studies are still lacking. IH can be a disabling condition and is usually chronic, although spontaneous improvement in EDS may be observed in up to one quarter of patients. Patients often report having an extreme form of sleep inertia with tremendous difficulty waking, irritability, automatic behavior, and confusion commonly known as "sleep drunkenness." The presence of sleep drunkenness has been reported in 36–52% of patients with IH.[75,76] Patients often take long naps, longer than 60 min, and report waking unrefreshed. Patients with IH typically have a high sleep efficiency on PSG studies with ≥ 90%. CSF hypocretin-1 levels are normal in IH patients. In the absence of systematic studies, the exact prevalence of IH remains to be determined.

Other CNS Hypersomnias

The remaining categories of CNS hypersomnias represent a collection of conditions clinicians must consider and exclude when making a diagnosis of narcolepsy or IH. Hypersomnia due to a medical disorder would be an appropriate diagnosis when patients' symptoms are attributable to a coexisting medical or neurological disorder. For example, patients with sleep related breathing disorders may present with EDS. In this case, a diagnosis of hypersomnia due to a medical disorder should be made. Nevertheless, in some patients with OSA, residual EDS may exist despite adequate sleep and optimal treatment of their sleep apnea. It would be

TABLE 8.5 Common Features of Narcolepsy, Idiopathic Hypersomnia and Kleine–Levin Syndrome

	Excessive daytime sleepiness	Cataplexy	Sleep paralysis[a]	Sleep hallucination[a]	Fragmented nocturnal sleep	Neurocognitive changes	Pathophysiology
Narcolepsy type 1	Present	Present (except in rare cases)	Present in 67%	Present in 74%	Decreased sleep efficiency compared to other hypersomnias	Absent	98% with hypocretin deficiency and HLA-DQB1*06:02
Narcolepsy type 2	Present	Absent	Present in 49%	Present in 57%	May be common	Absent	Unknown, heterogeneous 40% with HLA-DQB1*06:02
Idiopathic hypersomnia	Present	Absent	Present in 24%	Present in 32%	Not typical	Sleep drunkenness associated with automatic behavior and confusion	Unknown, likely heterogeneous
KLS	Present (recurrent episodes)	Absent	Uncommon	Present in 50%	Not typical	Present during periods of hypersomnia	Unknown, abnormalities in thalamus and hypothalamus have been found in imaging studies

[a]Data obtained from the Stanford Center for Narcolepsy Research database (unpublished).

advisable to fully treat patients for at least 3 months, monitor the efficacy of treatment with compliance data download, and reassess at this time. In some cases, a positive airway pressure titration PSG should be performed to ensure adequate treatment. One should consider an alternative hypersomnia diagnosis only if the hypersomnolence persists after adequate treatment. Hypersomnolence has been described in association with a large number of other neurological and medical conditions, such as head trauma, stroke, brain tumors, encephalitis, myotonic dystrophy, multiple sclerosis, systemic infection, metabolic encephalopathy and neurodegenerative diseases such as Parkinson's disease (PD) and Lewy body dementia (DLB).

In the case of PD and DLB, pathology in the nigrostriatal pathway, pedunculopontine tegmental nucleus, ventrolateral tegmental area, basal forebrain, and thalamic nuclei can affect the regulation of sleep, wakefulness and circadian rhythm, resulting in a myriad of symptoms. In one study, nocturnal sleep fragmentation was found to occur about three times more frequently in patients with PD than in healthy controls (38.9% vs. 12%).[77] Another study found that PD patients had significantly less total sleep time and reduced sleep efficiency as measured by PSG.[78] Moreover, the CNS pathophysiology in PD may also affect the circadian system resulting in sleep-wake dysrhythmia. Sleepiness is common, and SOREMPs may be seen if an MSLT is performed. The sleep disturbance in PD has been demonstrated to correlate with disease severity.[79] However, the cases of hypersomnolence and sedation that are due to the side effects of dopaminergic agents used for treatment of motor symptoms should be better classified as hypersomnia due to a medication or substance. For a more thorough review of sleep disorders associated with these conditions, the reader is directed to Chapter 5.

In cases of post traumatic brain injury (TBI), symptoms of hypersomnolence are common, with one metaanalysis reporting a frequency of 30%.[80] In some cases, injuries to hypothalamic hypocretin neurons, midbrain or the pons may impact wake-promoting systems. Furthermore, it has been found that there is an increased prevalence of sleep disordered breathing in patients with TBI, at a frequency of 23–25%.[81] The exact mechanism underlying the relationship between TBI and OSA remains unclear.

Patients with symptoms of hypersomnia attributable to sedating medications, alcohol, or drugs of abuse should be diagnosed with hypersomnia due to a medication or substance. This diagnosis also includes symptoms associated with withdrawal from stimulant medications such as amphetamines. Common prescription medications associated with hypersomnolence include benzodiazepines, opioids, barbiturates, anticonvulsants, antipsychotics, anticholinergics, antihistamines, and hypnotics.

Patients who suffer from a psychiatric disorder may also present with symptoms of hypersomnia. In this scenario, a diagnosis of hypersomnia associated with a psychiatric disorder would be appropriate. Psychiatric conditions most commonly associated with hypersomnolence include bipolar II disorder and atypical depression.[82]

Insufficient sleep syndrome is a behaviorally induced syndrome characterized by chronic sleep deprivation and EDS. It occurs when a person persistently fails to obtain the adequate amount of sleep required to maintain wakefulness, resulting in chronic sleep deprivation. A thorough review of the individual's sleep pattern should demonstrate a significant disparity between the need for sleep and the amount actually acquired. In addition, there is often a lack of insight and the patient may not realize that they are not obtaining adequate sleep. Patients who may markedly restrict their sleep time during the workweek and significantly extend sleep on the weekends or during holidays may be at risk of this disorder. A therapeutic trial of sleep extension can reverse the symptoms and, therefore, should be recommended as an initial step. Finally, clinicians must be judicious in excluding the presence of insufficient sleep before considering other hypersomnia diagnoses.

CONCLUSIONS

CNS hypersomnias are a collection of disorders characterized by excessive sleepiness. Major advances in the past decade provided a better understanding of the pathophysiology of Na-1, likely mediated by hypocretin deficiency. Medications such as sodium oxybate can be very effective in treating patients with Na-1. On the other hand, a better understanding of the pathophysiology of Na-2, IH, and KLS is much needed. Research in these hypersomnia disorders will not only benefit patients but may also lead to a better understanding of fundamental sleep-regulatory mechanisms.

References

1. American Academy of Sleep Medicine*International Classification of Sleep Disorders*. 3rd ed. Darien, IL: American Academy of Sleep Medicine; 2014.
2. Westphal C. Eigenthumliche mit Einschlafen verbundene Anfalle. *Arch Psychiat*. 1877;7:631–635.
3. Gelineau J. De la narcolepsie. *Gaz des Hop*. 1880;53:626–628.
4. Yoss RE, Daly DD. Criteria for the diagnosis of the narcoleptic syndrome. *Proc Staff Meet Mayo Clin*. 1957;32(12): 320–328.
5. Rechtschaffen A, Wolpert EA, Dement WC, Mitchell SAFC. Noctural sleep of narcoleptics. *Electroencephalogr Clin Neurophysiol*. 1963;15:599–609.
6. Knecht CD, Oliver JE, Redding R, Selcer R, Johnson G. Narcolepsy in a dog and a cat. *J Am Vet Med Assoc*. 1973;162(12):1052–1053.
7. Mitler MM, Boysen BG, Campbell L, Dement WC. Narcolepsy-cataplexy in a female dog. *Exp Neurol*. 1974;45(2):332–340.
8. Johns MW. A new method for measuring daytime sleepiness: the epworth sleepiness scale. *Sleep*. 1991;14(6): 540–545.
9. Juji T, Satake M, Honda Y, Doi Y. HLA antigens in Japanese patients with narcolepsy. All the patients were DR2 positive. *Tissue Antigens*. 1984;24(5):316–319.
10. Matsuki K, Grumet FC, Lin X, Gelb M, Guilleminault C, Dement WC, Mignot E. DQ (rather than DR) gene marks susceptibility to narcolepsy. *Lancet*. 1992;339(8800):1052.
11. Mignot E, Lin X, Arrigoni J, Macaubas C, Olive F, Hallmayer J, Underhill P, Guilleminault C, Dement WC, Grumet FC. DQB1*0602 and DQA1*0102 (DQ1) are better markers than DR2 for narcolepsy in Caucasian and black Americans. *Sleep*. 1994;17(8 suppl):S60–S67.
12. de Lecea L, Kilduff TS, Peyron C, Gao X, Foye PE, Danielson PE, Fukuhara C, Battenberg EL, Gautvik VT, Bartlett FS, Frankel WN, van den Pol AN, Bloom FE, Gautvik KM, Sutcliffe JG. The hypocretins: hypothalamus-specific peptides with neuroexcitatory activity. *Proc Natl Acad Sci USA*. 1998;95(1):322–327.
13. Sakurai T, Amemiya A, Ishii M, Matsuzaki I, Chemelli RM, Tanaka H, Williams SC, Richardson JA, Kozlowski GP, Wilson S, Arch JR, Buckingham RE, Haynes AC, Carr SA, Annan RS, McNulty DE, Liu WS, Terrett JA, Elshourbagy NA, Bergsma DJ, Yanagisawa M. Orexins and orexin receptors: a family of hypothalamic neuropeptides and G protein-coupled receptors that regulate feeding behavior. *Cell*. 1998;92(4):573–585.
14. Lin L, Faraco J, Li R, Kadotani H, Rogers W, Lin X, Qiu X, de Jong PJ, Nishino S, Mignot E. The sleep disorder canine narcolepsy is caused by a mutation in the hypocretin (orexin) receptor 2 gene. *Cell*. 1999;98(3):365–376.
15. Mignot E, Lammers GJ, Ripley B, Okun M, Nevsimalova S, Overeem S, Vankova J, Black J, Harsh J, Bassetti C, Schrader H, Nishino S. The role of cerebrospinal fluid hypocretin measurement in the diagnosis of narcolepsy and other hypersomnias. *Arch Neurol*. 2002;59(10):1553–1562.

16. Nishino S, Ripley B, Overeem S, Lammers GJ, Mignot E. Hypocretin (orexin) deficiency in human narcolepsy. *Lancet*. 2000;355(9197):39–40.

17. Peyron C, Faraco J, Rogers W, Ripley B, Overeem S, Charnay Y, Nevsimalova S, Aldrich M, Reynolds D, Albin R, Li R, Hungs M, Pedrazzoli M, Padigaru M, Kucherlapati M, Fan J, Maki R, Lammers GJ, Bouras C, Kucherlapati R, Nishino S, Mignot E. A mutation in a case of early onset narcolepsy and a generalized absence of hypocretin peptides in human narcoleptic brains. *Nat Med*. 2000;6(9):991–997.

18. Thannickal TC, Moore RY, Nienhuis R, Ramanathan L, Gulyani S, Aldrich M, Cornford M, Siegel JM. Reduced number of hypocretin neurons in human narcolepsy. *Neuron*. 2000;27(3):469–474.

19. Saper CB, Fuller PM, Pedersen NP, Lu J, Scammell TE. Sleep state switching. *Neuron*. 2010;68(6):1023–1042.

20. Burgess CR, Scammell TE, Narcolepsy:. Neural mechanisms of sleepiness and cataplexy. *J Neurosci*. 2012;32(36):12305–12311.

21. Aran A, Lin L, Nevsimalova S, Plazzi G, Hong SC, Weiner K, Zeitzer J, Mignot E. Elevated anti-streptococcal antibodies in patients with recent narcolepsy onset. *Sleep*. 2009;32(8):979–983.

22. Han F, Lin L, Warby SC, Faraco J, Li J, Dong SX, An P, Zhao L, Wang LH, Li QY, Yan H, Gao ZC, Yuan Y, Strohl KP, Mignot E. Narcolepsy onset is seasonal and increased following the 2009 H1N1 pandemic in China. *Ann Neurol*. 2011;70(3):410–417.

23. Nohynek H, Jokinen J, Partinen M, Vaarala O, Kirjavainen T, Sundman J, Himanen S-L, Hublin C, Julkunen I, Olsén P, Saarenpää-Heikkilä O, Kilpi T. AS03 adjuvanted AH1N1 vaccine associated with an abrupt increase in the incidence of childhood narcolepsy in Finland. *PLoS One*. 2012;7(3):e33536.

24. Persson I, Granath F, Askling J, Ludvigsson JF, Olsson T, Feltelius N. Risks of neurological and immune-related diseases, including narcolepsy, after vaccination with Pandemrix: a population- and registry-based cohort study with over 2 years of follow-up. *J Intern Med*. 2014;275(2):172–190.

25. Mignot E. Genetic and familial aspects of narcolepsy. *Neurology*. 1998;50(2 suppl 1):S16–S22.

26. Mignot EJM. History of narcolepsy at Stanford University. *Immunol Res*. 2014;58:315–339.

27. Hallmayer J, Faraco J, Lin L, Hesselson S, Winkelmann J, Kawashima M, Mayer G, Plazzi G, Nevsimalova S, Bourgin P, Hong S-C, Hong SS-C, Honda Y, Honda M, Högl B, Longstreth WT, Montplaisir J, Kemlink D, Einen M, Chen J, Musone SL, Akana M, Miyagawa T, Duan J, Desautels A, Erhardt C, Hesla PE, Poli F, Frauscher B, Jeong J-H, Lee S-P, Ton TGN, Kvale M, Kolesar L, Dobrovolná M, Nepom GT, Salomon D, Wichmann H-E, Rouleau GA, Gieger C, Levinson DF, Gejman PV, Meitinger T, Young T, Peppard P, Tokunaga K, Kwok P-Y, Risch N, Mignot E. Narcolepsy is strongly associated with the T-cell receptor alpha locus. *Nat Genet*. 2009;41(6):708–711.

28. Kornum BR, Kawashima M, Faraco J, Lin L, Rico TJ, Hesselson S, Axtell RC, Kuipers H, Weiner K, Hamacher A, Kassack MU, Han F, Knudsen S, Li J, Dong X, Winkelmann J, Plazzi G, Nevsimalova S, Hong S-C, Honda Y, Honda M, Högl B, Ton TGN, Montplaisir J, Bourgin P, Kemlink D, Huang Y-S, Warby S, Einen M, Eshragh JL, Miyagawa T, Desautels A, Ruppert E, Hesla PE, Poli F, Pizza F, Frauscher B, Jeong J-H, Lee S-P, Strohl KP, Longstreth WT, Kvale M, Dobrovolna M, Ohayon MM, Nepom GT, Wichmann H-E, Rouleau GA, Gieger C, Levinson DF, Gejman PV, Meitinger T, Peppard P, Young T, Jennum P, Steinman L, Tokunaga K, Kwok P-Y, Risch N, Hallmayer J, Mignot E. Common variants in P2RY11 are associated with narcolepsy. *Nat Genet*. 2011;43(1):66–71.

29. Aldrich MS, Naylor MW. Narcolepsy associated with lesions of the diencephalon. *Neurology*. 1989;39(11):1505–1508.

30. Winkelmann J, Lin L, Schormair B, Kornum BR, Faraco J, Plazzi G, Melberg A, Cornelio F, Urban AE, Pizza F, Poli F, Grubert F, Wieland T, Graf E, Hallmayer J, Strom TM, Mignot E. Mutations in DNMT1 cause autosomal dominant cerebellar ataxia, deafness and narcolepsy. *Hum Mol Genet*. 2012;21(10):2205–2210.

31. Dauvilliers Y, Delallée N, Jaussent I, Scholz S, Bayard S, Croyal M, Schwartz J-C, Robert P. Normal cerebrospinal fluid histamine and tele-methylhistamine levels in hypersomnia conditions. *Sleep*. 2012;35(10):5–8.

32. Rye DB, Bliwise DL, Parker K, Trotti LM, Saini P, Fairley J, Freeman A, Garcia PS, Owens MJ, Ritchie JC, Jenkins A. Modulation of vigilance in the primary hypersomnias by endogenous enhancement of GABAA receptors. *Sci Transl Med*. 2012;4:161ra151.

33. Khor S-S, Miyagawa T, Toyoda H, Yamasaki M, Kawamura Y, Tanii H, Okazaki Y, Sasaki T, Lin L, Faraco J, Rico T, Honda Y, Honda M, Mignot E, Tokunaga K. Genome-wide association study of HLA-DQB1*06:02 negative essential hypersomnia. *PeerJ*. 2013;1:e66.

34. Miglis MG, Guilleminault C. Kleine-Levin syndrome: a review. *Nat Sci Sleep*. 2014;6:19–26.

35. Huang Y-S, Guilleminault C, Kao P-F, Liu F-Y. SPECT findings in the Kleine-Levin syndrome. *Sleep*. 2005;28(8):955–960.

36. Haba-Rubio J, Prior JO, Guedj E, Tafti M, Heinzer R, Rossetti AO. Kleine-Levin syndrome: functional imaging correlates of hypersomnia and behavioral symptoms. *Neurology*. 2012;79(18):1927–1929.

37. Snead OC. Evidence for a G protein-coupled gamma-hydroxybutyric acid receptor. *J Neurochem.* 2000;75(5):1986–1996.

38. Alshaikh MK, Tricco AC, Tashkandi M, Mamdani M, Straus SE, BaHammam AS. Sodium oxybate for narcolepsy with cataplexy: systematic review and meta-analysis. *J Clin Sleep Med.* 2012;8(4):451–458.

39. Leu-Semenescu S, Louis P, Arnulf I. Benefits and risk of sodium oxybate in idiopathic hypersomnia versus narcolepsy type 1: a chart review. *Sleep Med.* 2015;.

40. Trotti LM, Saini P, Bliwise DL, Freeman AA, Jenkins A, Rye DB. Clarithromycin in γ-aminobutyric acid-related hypersomnolence: a randomized, crossover trial. *Ann Neurol.* 2015;78(3):454–465.

41. Arnulf I, Zeitzer JM, File J, Farber N, Mignot E. Kleine-Levin syndrome: a systematic review of 186 cases in the literature. *Brain.* 2005;128(Pt 12):2763–2776.

42. Arnulf I, Lin L, Gadoth N, File J, Lecendreux M, Franco P, Zeitzer J, Lo B, Faraco JH, Mignot E. Kleine-Levin syndrome: a systematic study of 108 patients. *Ann Neurol.* 2008;63(4):482–493.

43. Ohayon MM, Dauvilliers Y, Reynolds CF. Operational definitions and algorithms for excessive sleepiness in the general population: implications for DSM-5 nosology. *Arch Gen Psychiatry.* 2012;69(1):71–79.

44. Mignot E, Hayduk R, Black J, Grumet FC, Guilleminault C. HLA DQB1*0602 is associated with cataplexy in 509 narcoleptic patients. *Sleep.* 1997;20(11):1012–1020.

45. Silber MH, Krahn LE, Olson EJ, Pankratz VS. The epidemiology of narcolepsy in Olmsted County, Minnesota: a population based study. *Sleep.* 2002;25(2):197–202.

46. Anic-Labat S, Guilleminault C, Kraemer HC, Meehan J, Arrigoni J, Mignot E. Validation of a cataplexy questionnaire in 983 sleep disorders patients. *Sleep.* 1999;22(1):77–87.

47. Okun ML, Lin L, Pelin Z, Hong S, Mignot E. Clinical aspects of narcolepsy-cataplexy across ethnic groups. *Sleep.* 2002;25(1):27–35.

48. Khan Z, Trotti LM. Central disorders of hypersomnolence. *Chest J.* 2015;148(1):262.

49. Bourgin P, Zeitzer JM, Mignot E. CSF hypocretin-1 assessment in sleep and neurological disorders. *Lancet Neurol.* 2008;7(7):649–662.

50. Sharpless BA, Barber JP. Lifetime prevalence rates of sleep paralysis: a systematic review. *Sleep Med Rev.* 2011;15(5):311–315.

51. Roth T, Dauvilliers Y, Mignot E, Montplaisir J, Paul J, Swick T, Zee P. Disrupted nighttime sleep in narcolepsy. *J Clin Sleep Med.* 2013;9(9):955–965.

52. Drake C, Nickel C, Burduvali E, Roth T, Jefferson C, Pietro B. The pediatric daytime sleepiness scale (PDSS): sleep habits and school outcomes in middle-school children. *Sleep.* 2003;26(4):455–458.

53. Krupp LB, LaRocca NG, Muir-Nash J, Steinberg AD. The fatigue severity scale. Application to patients with multiple sclerosis and systemic lupus erythematosus. *Arch Neurol.* 1989;46(10):1121–1123.

54. Carskadon MA, Harvey K, Dement WC. Multiple sleep latency tests during the development of narcolepsy. *West J Med.* 1981;135(5):414–418.

55. Littner MR, Kushida C, Wise M, Davila DG, Morgenthaler T, Lee-Chiong T, Hirshkowitz M, Daniel LL, Bailey D, Berry RB, Kapen S, Kramer M. Practice parameters for clinical use of the multiple sleep latency test and the maintenance of wakefulness test. *Sleep.* 2005;28:113–121.

56. Littner MR, Kushida C, Wise M, Davila DG, Morgenthaler T, Lee-Chiong T, Hirshkowitz M, Daniel LL, Bailey D, Berry RB, Kapen S, Kramer M. Practice parameters for clinical use of the multiple sleep latency test and the maintenance of wakefulness test. *Sleep.* 2005;28(1):113–121.

57. Carskadon MA, Dement WC, Mitler MM, Roth T, Westbrook PR, Keenan S. Guidelines for the multiple sleep latency test (MSLT): a standard measure of sleepiness. *Sleep.* 1986;9(4):519–524.

58. Chervin RD, Aldrich MS. Sleep onset REM periods during multiple sleep latency tests in patients evaluated for sleep apnea. *Am J Respir Crit Care Med.* 2000;161(2 Pt 1):426–431.

59. Trotti LM, Staab BA, Rye DB. Test-retest reliability of the multiple sleep latency test in narcolepsy without cataplexy and idiopathic hypersomnia. *J Clin Sleep Med.* 2013;9(8):789–795.

60. Bassetti C, Aldrich MS. Idiopathic hypersomnia. A series of 42 patients. *Brain.* 1997;120(Pt 8):1423–1435.

61. Mitler MM, Gujavarty KS, Browman CP. Maintenance of wakefulness test: a polysomnographic technique for evaluating treatment efficacy in patients with excessive somnolence. *Electroencephalogr Clin Neurophysiol.* 1982;53(6):658–661.

62. Dement W, Carskadon MLR. The prevalence of narcolepsy II. *Sleep Res.* 1973;2:147.

63. Honda Y. Census of narcolepsy, cataplexy and sleep life among teenagers in Fujisawa city. *Sleep Res.* 1979;8:191.

64. Lavie P, Peled R. Narcolepsy is a rare disease in Israel. *Sleep.* 1987;10(6):608–609.

65. Dauvilliers Y, Montplaisir J, Molinari N, Carlander B, Ondze B, Besset A, Billiard M. Age at onset of narcolepsy in two large populations of patients in France and Quebec. *Neurology*. 2001;57(11):2029–2033.
66. Guilleminault C, Pelayo R. Narcolepsy in prepubertal children. *Ann Neurol*. 1998;43(1):135–142.
67. Kotagal S, Krahn LE, Slocumb N. A putative link between childhood narcolepsy and obesity. *Sleep Med*. 2004;5(2):147–150.
68. Fortuyn HAD, Lappenschaar MA, Furer JW, Hodiamont PP, Rijnders CAT, Renier WO, Buitelaar JK, Overeem S. Anxiety and mood disorders in narcolepsy: a case-control study. *Gen Hosp Psychiatry*. 2010;32(1):49–56.
69. Kawai M, O'Hara R, Einen M, Lin L, Mignot E. Narcolepsy in African Americans. *Sleep*. 2015;38(11):1673–1681.
70. Takei Y, Komada Y, Namba K, Sasai T, Nakamura M, Sugiura T, Hayashida K, Inoue Y. Differences in findings of nocturnal polysomnography and multiple sleep latency test between narcolepsy and idiopathic hypersomnia. *Clin Neurophysiol*. 2012;123(1):137–141.
71. Andlauer O, Moore H, Hong S-C, Dauvilliers Y, Kanbayashi T, Nishino S, Han F, Silber MH, Rico T, Einen M, Kornum BR, Jennum P, Knudsen S, Nevsimalova S, Poli F, Plazzi G, Mignot E. Predictors of hypocretin (orexin) deficiency in narcolepsy without cataplexy. *Sleep*. 2012;.
72. Shin YK, Yoon IY, Han EK, No YM, Hong MC, Yun YD, Jung BK, Chung SH, Choi JB, Cyn JG, Lee YJ, Hong SC. Prevalence of narcolepsy-cataplexy in Korean adolescents. *Acta Neurol Scand*. 2008;117(4):273–278.
73. Goldbart A, Peppard P, Finn L, Ruoff CM, Barnet J, Young T, Mignot E. Narcolepsy and predictors of positive MSLTs in the Wisconsin Sleep Cohort. *Sleep*. 2014;37(6):1043–1051.
74. Roth B. Narcolepsy and hypersomnia: review and classification of 642 personally observed cases. *Schweizer Arch für Neurol Neurochir und Psychiatr*. 1976;119(1):31–41.
75. Anderson KN, Pilsworth S, Sharples LD, Smith IE, Shneerson JM. Idiopathic hypersomnia: a study of 77 cases. *Sleep*. 2007;30(10):1274–1281.
76. Vernet C, Arnulf I. Idiopathic hypersomnia with and without long sleep time: a controlled series of 75 patients. *Sleep*. 2009;32(6):753–759.
77. Adler CH, Thorpy MJ. Sleep issues in Parkinson's disease. *Neurology*. 2005;64(12 suppl 3):S12–S20.
78. Martinez-Ramirez D, De Jesus S, Walz R, Cervantes-Arriaga A, Peng-Chen Z, Okun MS, Alatriste-Booth V, Rodríguez-Violante M. A polysomnographic study of Parkinson's disease sleep architecture. *Parkinsons Dis*. 2015;2015:570375.
79. Roychowdhury S, Forsyth DR. Sleep disturbance in Parkinson disease. *J Clin Gerontol Geriatr*. 2012;3(2):53–61.
80. Masel BE, Scheibel RS, Kimbark T, Kuna ST. Excessive daytime sleepiness in adults with brain injuries. *Arch Phys Med Rehabil*. 2001;82(11):1526–1532.
81. Viola-Saltzman M, Watson NF. Traumatic brain injury and sleep disorders. *Neurol Clin*. 2012;30(4):1299–1312.
82. Kaplan KA, Harvey AG. Hypersomnia across mood disorders: a review and synthesis. *Sleep Med Rev*. 2009;13(4):275–285.

Sleep and Multiple Sclerosis

D.J. Kimbrough*, T.J. Braley**

*Harvard Medical School; Partners Multiple Sclerosis Center at Brigham & Women's
Hospital, Boston, MA, USA
**University of Michigan School of Medicine; University of Michigan Multiple Sclerosis
Center and Sleep Disorders Center, Ann Arbor, MI, USA

INTRODUCTION

Multiple sclerosis (MS) is a chronicautoimmune disease of the central nervous system (CNS) that causes myelin destruction and axonal damage. As the most common cause of non-traumatic neurological disability in young adults, MS affects approximately 250,000–400,000 patients in the United States, with worldwide median incidence and prevalence estimates of 4.2 per 100,000 and 0.9 per 1,000, respectively.[1] Women are affected at least twice as often as men, and individuals are typically diagnosed in the 2nd to 5th decades of life. The clinical history and a physical exam are still the most crucial components of the diagnosis, which can be aided with a specialized set of recently revised diagnostic criteria (McDonald), and supported by radiological and laboratory data, typically magnetic resonance imaging (MRI) and spinal fluid analysis.[2–4] Although MS has historically been described as a disease of the CNS white matter tracts, recent pathological and radiological research has unequivocally demonstrated

Sleep and Neurologic Disease. http://dx.doi.org/10.1016/B978-0-12-804074-4.00009-1

early and profound cortical, and deep gray matter involvement.[5-8] Cerebrospinal fluid studies (CSF) often reveal evidence of an active intrathecal humoral immune response.[9-11]

Among diagnostic subtypes, approximately 85% of MS cases present with relapsing-remitting multiple sclerosis (RRMS), characterized by acute episodes of focal neurological dysfunction, in which symptoms may vary depending on the affected neuroanatomical region. Recovery usually occurs within days to weeks of onset. Recovery from a relapse (also commonly called an exacerbation, flare, or attack) is either spontaneous or hastened by treatment with corticosteroids, although in both cases the natural history remains unaffected. In the majority of patients, if left untreated, this relapsing-remitting course eventually yields to an insidious secondary progressive phase (SPMS) during which previously acquired symptoms may gradually worsen without clinically obvious inflammatory attacks. Fixed disability may ensue, manifested by deficits in neurological functions such as impaired visual acuity, ambulatory disability, spasticity, paresthesias and/or dysethesia, ataxia, and cognitive dysfunction. To forestall exacerbations and prevent accrual of disability, long-term immunotherapies (also known as disease-modifying agents) are recommended for RRMS. Two additional diagnostic categories round out the classification of MS subtypes. Primary progressive MS (PPMS) begins with insidious neurological deterioration from the outset without obvious clinical relapses. The less commonly described progressive-relapsing MS (PRMS) essentially has a progressive course that is rarely punctuated by relapse activity. Treatment of these subtypes largely involves the management of attendant symptoms, although there is cautious optimism about development of new agents to eventually treat progressive forms of MS.[12-15]

In addition to focal physical deficits, MS is also associated with a number of other chronic symptoms that contribute to poor functional status and impaired quality of life. Occasionally, these issues may be the cardinal symptoms that bring the patient to medical attention, but more often they are chronic issues in previously diagnosed patients that may not be addressed adequately without proper recognition. Such symptoms include fatigue, depression, pain, and cognitive impairment, which may in part arise from MS, but in many cases may be exacerbated by other comorbid conditions, including sleep disorders.

FATIGUE AND THE ASSOCIATION OF SLEEP DISORDERS IN MS

Chronic fatigue is the most common complaint of persons with MS, affecting nearly 90% of patients,[16-19] and is a leading cause of diminished quality of life.[19] Although MS-related fatigue is often multifactorial, identification of treatable causes that may contribute to its severity is an essential element of care for all MS patients.

While postulated primary mechanisms underlying fatigue in MS include neuronal network dysfunction and cytokine dysregulation, several treatable comorbidities—including sleep disorders—are well-recognized as important secondary causes of fatigue in MS.[20-22] Recent studies suggest that insomnia, obstructive sleep apnea, and restless legs syndrome are independent predictors of fatigue in MS.[23-25] Early data also suggest that successful treatment of these conditions may improve fatigue, yet these conditions remain underrecognized in this population.[25-28]

Given that fatigue is a highly individualized symptom with overlapping descriptors and concomitant symptoms, accurate characterization and identification of the most likely exacerbating factors can be challenging.[24] Although common synonyms include tiredness, lack of energy, lassitude, decreased motivation, weariness, and asthenia, patients with sleep disorders may also use the term fatigue interchangeably with sleepiness, a term more readily recognized as an indication that problems with sleep may exist.[29] Consequently, an improved understanding of these subjective symptoms will provide the best chance of identifying all treatable causes of fatigue and related symptoms in patients with MS. One useful strategy is to start by asking the patient about his or her level of energy. Often patients will endorse terms such as "low energy," "fatigue," or "tiredness", but they should be asked to provide their own descriptions of what these terms mean to avoid any misconceptions. Given that patients with sleep disorders often prefer terms other than sleepiness to describe their symptoms, it is possible that the term "sleepiness" might not be endorsed by the patient, even with close-ended questioning.[24,29]

More specific queries regarding associated symptoms, as well as aggravating and alleviating factors, are also recommended. For example, if the patient states that their problem is worse during sedentary, monotonous activities rather than during extended physical or cognitive activity, or if the patient endorses a propensity to doze off during sedentary activities, excessive sleepiness is more likely than fatigue.

To quantify fatigue and sleepiness severity—which can be useful to assess the effects of treatment—brief self-report instruments are available. The fatigue severity scale (FSS) is a nine-item questionnaire that assesses the impact of fatigue on multiple outcomes. It is validated for use in MS, and is sensitive to change over time, making it a useful took to track the effects of interventions.[30] Other instruments that have been employed in efforts to quantify fatigue severity include the fatigue descriptive scale, modified fatigue impact scale, neurological fatigue index, and visual analog scale for fatigue.[30–34]

The Epworth sleepiness scale (ESS) is an eight-item questionnaire that measures subjective sleepiness, or likelihood of dozing in various situations. While not specifically validated in MS patients, it has been extensively utilized in a variety of outpatient settings and as a primary measure of sleepiness in MS research studies. Given its ability to distinguish hypersomnolence for the patient, it may serve as a useful adjunct to the FSS to aid in symptom assessment.[25,26,35,36]

Ultimately, as fatigue and hypersomnolence have the potential to coexist in MS and primary sleep disorders, it is incumbent on the provider to successfully differentiate these symptoms in order to eventually identify reversible underlying etiologies.

Sleep Disorders in MS

Approximately half of patients with MS report one or more sleep disturbances, often with delays in recognition and/or adequate treatment by physicians.[37–40] Pertinent sleep disorders discussed in the following sections include insomnia, nocturnal movement disorders, sleep-disordered breathing, rapid eye movement sleep behavior disorder (RBD), and narcolepsy.

Insomnia

Insomnia is characterized by difficulty initiating or maintaining sleep. Insomnia can exist as a symptom or as a disorder, in which case symptoms of insomnia must be associated with some form of distress about poor sleep, or lead to impairments in social, academic, or vocational functioning. The *International Classification of Sleep Disorders Diagnostic Manual-Third Edition (ICSD-3)*, defines chronic insomnia as "a persistent difficulty with sleep initiation, duration, consolidation, or quality that occurs despite the opportunity and circumstances for sleep, and results in some form of daytime impairment," which must be present for at least 3 months.[41] Symptoms of daytime impairment include fatigue, impaired concentration or memory, mood disturbances, excessive daytime sleepiness, behavioral problems, reduced motivation or energy, impaired social, family, academic, or occupational performance, susceptibility to errors, and concerns or dissatisfaction with sleep. The ICSD-3 also delineates specific subtypes of insomnia based on etiology, such as insomnia due to a medical condition and insomnia due to another mental disorder.[41] Some of the most common symptoms experienced by patients with MS, including nocturia from neurogenic bladder, pain syndromes, spasticity, and mood disorders (e.g., depression and anxiety) commonly contribute to sleep initiation or sleep maintenance insomnia (Table 9.1).[42]

Given the substantial overlap between insomnia-related symptoms of daytime impairment and chronic symptoms of MS, all MS patients who endorse symptoms of daytime impairment should be queried about their sleep. The evaluation should start with a thorough sleep history that simultaneously addresses the most common symptoms of MS. Patients should be asked about the frequency and severity of nocturnal pain, spasticity, urinary frequency, depression, and anxiety, preferably in conjunction with the sleep history. Patients should also be asked about associated daytime distress or impairment. As with non-MS patients, further inquiries should be directed toward external factors or habits that may interfere with sleep hygiene. The Insomnia Severity Index (ISI) is also a useful tool to screen for symptoms of insomnia that also measures the response to pharmacological or behavioral interventions.[43] As in non-MS patients, polysomnography (PSG) is not typically necessary to establish an insomnia diagnosis, unless there is suspicion that sleep-disordered breathing or sleep-related movement disorders are contributing factors.

A thorough medication assessment (including over-the-counter medications) is also important in the insomnia assessment, as selective serotonin reuptake inhibitors, stimulants/wake-promoting agents, and over-the-counter antihistamines are frequently used in patients with MS and may interfere with sleep. Corticosteroids, which are typically used for to treat MS exacerbations, may also contribute to insomnia.

Treatment should start with amelioration of any precipitating causes (Table 9.1), either prior to or in tandem with first-line therapies. Medications or substances that may contribute to insomnia should be minimized, if possible. Chronic symptoms such as nocturnal spasticity, neuropathic pain, or nocturia should be treated. If neuropathic pain or spasticity are contributing factors, effective medications that also have sedating properties (such as tricyclic antidepressants, alpha-2 delta ligands, or antispasmodics) may be reasonable first options. Dosages of wake promoting agents may also require adjustment, and alternative dosing schedules that allow for earlier administration should be discussed.

If comorbid symptoms or concomitant medications are not significant contributing factors, then psychological and behavioral therapies should be considered. Cognitive behavioral therapy for insomnia (CBT-I) is an ideal treatment modality for patients with MS, particularly in patients with comorbid depression. Depression affects approximately 50% of MS patients and shares a bidirectional relationship with insomnia.[44] Recent data show that rates of

TABLE 9.1   Factors That may Exacerbate Common Sleep Disorders in MS

Insomnia	Sleep-disordered breathing	Restless legs syndrome
Medications	Medications	Medications/Substances
- Corticosteroids	- Opioids	- 1st generation antihistamines
- Stimulants/Wake-promoting agents	- Benzodiazepines	- Dopamine antagonists
- SSRIs	- Antispasmodics	- Lithium
- Beta-interferons	Brainstem dysfunction	- Alcohol
Depression/Anxiety	Increased disability level	- Tobacco
Pain	Progressive MS subtype	- Caffeine
Spasticity	General risk factors	Increased disability level
Nocturia	- Age > 50 years	Cervical cord lesions
Other sleep disorders	- Obesity	Primary progressive MS subtype
- Sleep-disordered breathing	- Increased neck circumference	Iron deficiency
- Restless legs syndrome	- Crowded oropharyngeal outlet	Family history
Poor sleep hygiene	- Retrognathia/micrognathia	

insomnia are higher in depressed MS patients than nondepressed MS patients, and suggest that insomnia management should extend beyond the treatment of the underlying comorbid psychiatric disorder.[45]

Pharmacological therapies can be considered if more conservative strategies have been exhausted or are not fully effective. While no MS-specific contraindications exist to use such agents, selection should be made with consideration of other comorbid symptoms or diseases. Benzodiazepine receptor agonists (zolpidem, zolpidem extended release, zaleplon, eszopiclone), melatonin receptor agonists (ramelteon), and newer orexin receptor antagonists (contraindicated in patients with concomitant narcolepsy) are all potentially useful therapies for MS patients with insomnia. Although over-the-counter antihistamine agents are frequently used by patients for insomnia, use of these agents is discouraged due to their long half-lives and subsequent hangover effects.[46,47]

Sleep-Disordered Breathing: Obstructive Sleep Apnea and Central Sleep Apnea

Obstructive sleep apnea (OSA) is defined as a collapse of the upper airway that leads to a reduction in airflow during sleep, despite the patient's attempt to resume normal respiration.[48] The episodes, which by definition last for 10 s or longer, can be complete (apneas) or partial (hypopneas), despite continuous respiratory effort. Repeated arousals are required to reestablish sufficient muscular tone in the pharynx to reopen the upper airways and normalize oxygen levels. As a result, patients often have reduced sleep efficiency and quality, which is reflected by a decrease in slow-wave sleep and REM sleep.

Recent data suggest an increased prevalence of OSA among patients with MS. Two recently published studies that assessed the frequency of patients with MS already diagnosed with OSA demonstrated a prevalence ranging from 4% to 21%.[26,49] Furthermore, both studies also showed a strikingly high prevalence (38–56%) of patients who were at elevated risk for OSA based on a validated screening tool.[50] While the cause of this relationship still requires further investigation, underlying neuroanatomical and immunological features associated with MS may in part explain this elevated prevalence. Maintenance of upper airway patency during sleep requires an increase in pharyngeal tone, that is, primarily mediated by efferent motor output from cranial nerves X and XII to the palatal and genioglossus muscles, respectively. This process is largely influenced by afferent sensory input from pressure receptors in the upper airway, peripheral chemoreceptors, and brainstem respiratory generators.[51,52] Processes that disrupt these tightly regulated brainstem pathways have the potential to impair nocturnal respiration.[51,53] In a previous cross-sectional study, subjects with MS referred for overnight polysomnography (PSG), and particularly those patients with radiographic evidence of brainstem involvement, were found to have more severe OSA than control subjects also referred to a sleep laboratory but without MS.[24] Among the subjects with MS, progressive MS subtype and MS disease-modifying therapy also use predicted apnea severity. Disease-modifying therapy use in particular emerged as a strong predictor of reduced apnea severity, raising interesting possibilities about the role of local and/or systemic inflammation in OSA.

In contrast to OSA, central sleep apnea (CSA) results from complete or partial impairment of airflow due to a lack of respiratory effort.[54] This is typically due to impaired respiratory control at the level of the brainstem, namely the medullary reticular formation. Although CSA is much less common than OSA in MS patients, it is anatomically correlated with lesions affecting the pontine and medullary respiratory centers, and CSA severity may be worse in patients with MS.[24] Furthermore, central alveolar hypoventilation syndrome, eponymously described as "Ondine's curse," has been rarely reported to affect patients with MS. This condition is marked by normal respiration during wakefulness but hypoventilation and hypercapnea during non-REM sleep.[55,56] Although most often referenced in its congenital form, the acquired type is associated with lesions in pontine and medullary respiratory generators along with the solitary nucleus. Autopsy case reports of two patients with MS who died during sleep revealed the presence of plaques within the medullary reticular formation.[55]

Given the high prevalence of sleep-disordered breathing in MS, all patients with MS should be queried about common symptoms of OSA and CSA, including snoring, pauses in breathing witnessed by a bed partner, gasping or choking upon awakening, nonrestorative sleep, excessive daytime hypersomnolence or fatigue, cognitive disturbances, and night-time awakenings. Dysarthria or dysphagia—signs of brainstem dysfunction—may also suggest an increased risk of OSA or CSA. Obesity, large neck circumference, crowded oropharyngeal inlet, retrognathia, or micronathia are common physical exam findings associated with OSA and should, in conjunction with the above symptoms, prompt clinicians to consider a sleep clinic referral.

Currently, no MS-specific algorithm for OSA risk has been validated. However, considerable literature from the general population is likely to be relevant to screening among patients with MS. To facilitate a thorough screen of the most common symptoms and physical exam findings associated with OSA, several questionnaires are commonly used. The STOP-BANG questionnaire is an eight-item instrument useful in screening for OSA both in outpatient and in perioperative settings.[57] In a pilot study demonstrating its applicability to MS, nearly half of queried patients were identified as having high risk of OSA.[58]

To confirm the diagnosis of OSA or CSA, a full-night PSG is necessary to demonstrate the presence of obstructive respiratory events during sleep. Management strategies for sleep-disordered breathing in MS should take into account the patient's primary apnea subtype, apnea severity, comorbidities and behaviors, and other MS-specific symptoms or limitations. Conservative approaches include modification of behavioral factors such as weight loss, avoidance of alcohol, and limiting the use of sedating compounds. The gold standard treatment for OSA, positive airway pressure (PAP) therapy, is delivered by a mechanical device and mask to splint the upper airway open during sleep. Although the general benefits of treatment of OSA by continuous PAP are likely to be similarly beneficial among patients with MS, rigorous demonstration of effectiveness in patients with MS specifically has yet to be published. A previous observational study suggests that compliant PAP use ameliorates fatigue in patients with MS, but large-scale, controlled clinical trials that minimize selection bias are still needed.[27]

If PAP therapy is selected, existing neurological deficits and symptoms should be taken into consideration when selecting a mask interface. For example, if patients have significant dexterity issues or hemiparesis, masks that involve a complex strap apparatus should be avoided. Patients with a history of trigeminal neuralgia may benefit from masks that minimize facial contact. In select cases, oral appliances may also be considered for the treatment of OSA. These devices are considered by many as a useful conservative nonsurgical approach for the treatment of mild to moderate OSA, or snoring without OSA, due to their size and portability.[59] Although there are generally no MS-specific contraindications to oral appliances if otherwise clinically indicated, additional studies are necessary to determine their clinical utility and feasibility in MS. These devices may also be inappropriate for patients with trigeminal neuralgia who experience pain triggered by oral stimulation.

Treatment of CSA is similar to treatment approaches non-MS patients, and the current level of evidence suggests that treatment options be tailored to the underlying etiology. CNS depressant medications such as opiates or antispasmodics, which may also worsen CSA, should be minimized whenever possible.

Restless Legs Syndrome and Periodic Limb Movements of Sleep

The motor disorders of sleep, including restless legs syndrome (RLS, eponymously named Willis-Ekbom Disease) and periodic limb movement disorder (PLMD) may occur in primary (idiopathic) and secondary (symptomatic) forms, and in the latter case, are experienced more frequently by patients with MS than by the general population.[60-63] Prevalence rates of RLS among patients with MS have been reported in ranges from approximately 10–36%.[62,64-66] Cervical spinal cord damage has been associated with a higher incidence of RLS in MS.[67]

RLS is characterized by lower extremity (or, less often, upper extremity) discomfort that is alleviated by movement, worsened by inactivity, and occurs most prominently during rest periods and in the evening hours (circadian component). Periodic limb movements of sleep (PLMs) consist of rhythmic, stereotyped movements of the lower extremities during sleep. PLMs are frequently associated with RLS, but can also occur in the absence of RLS. PSG is required to diagnose PLMs, whereas RLS is a clinical diagnosis that does not require PSG. If frequent PLMs are associated with significant sleep disruption or impairments in daytime functioning, in the absence of RLS or another sleep, neurological, or medical disorder, a diagnosis of PLMD may be made. Approximately 80% of patients with RLS also have PLMs, yet only a limited proportion of individuals with PLMs have RLS.[68] Patients with MS and higher frequencies of periodic limb movements are more likely to have greater disability measured by the Expanded Disability Status Scale (EDSS), a composite score that is commonly used to quantify disability level in patients with MS.[69,70]

Although the cause of idiopathic RLS/PLMD is unknown, a genetic etiology has been suggested, and dysfunction of brain circuits that require the neurotransmitter dopamine have been implicated.[71-73] Impaired iron metabolism is also thought to contribute to the pathogenesis of RLS, as iron is a cofactor for a rate-limiting step in the synthesis of dopamine, and impaired iron metabolism has been corroborated by serum and CSF ferritin deficiencies noted in patients with idiopathic RLS.[74] Others have suggested a role for dysfunction of downstream dopaminergic projections to the spinal cord, namely diencephalospinal and reticulospinal pathways.[75] These dopaminergic pathways are responsible for the suppression of sensory inputs and motor excitability, and are susceptible to injury from spinal cord disorders. This hypothesis may explain the increased prevalence of secondary RLS in certain neurologic conditions, including spinal cord injury and MS, and is supported clinically by work from Manconi and coworkers, who have demonstrated associations between RLS and decreased myelin integrity in the cervical spinal cord (Fig. 9.1).[64,67] Additional clinical evidence from Manconi's group also demonstrates a link between RLS and primary progressive MS and increased levels of neurologic disability.[62]

RLS is a clinical diagnosis with four required criteria:[76]

1. An urge to move the legs (or less often, the arms) is accompanied or caused by uncomfortable sensations.
2. The urge to move worsens at times of rest or inactivity.
3. The aforementioned symptoms are partially or totally relieved by movement.
4. The aforementioned symptoms are exacerbated or are solely present at night.

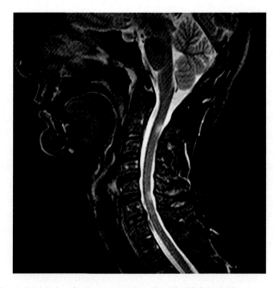

FIGURE 9.1   **MS-related cervical spine lesions associated with RLS in MS.**

Many descriptors have been used to define the uncomfortable sensation experienced by RLS patients, including creeping, crawling, itching, burning, tightening, or tingling. Some patients will have trouble characterizing the symptoms beyond a description of discomfort or an urge to move, while others will describe this sensation as painful.[76,77] Several other symptoms that may be particularly common in MS, such as cramping, clonus, spasticity, or in particular, neuropathic pain, can be difficult to discern from RLS and often require detailed, direct questioning.[78] Neuropathic pain may be described as more noticeable to patients at night, when they are not distracted, which may suggest a circadian predilection, thereby mimicking RLS. In this case, endorsement of relief with movement, even if the relief is only temporary while the movement continues, provides support for RLS. Persistent pain that is not ameliorated by movement suggests neuropathic pain. Spasticity or clonus may also become more noticeable to the patient at night, and during times of fatigue later in the day. In this case, queries to assess the nature of the leg movements are paramount. Symptoms of leg tightness relieved by voluntary movement suggest RLS, whereas involuntary spasms, even if a circadian component is endorsed, suggests spasticity. In this regard, the Restless Legs Syndrome Diagnostic Index (RLS-DI) may be a useful tool to rule out false-positive diagnoses. The RLS-DI is a 10-item questionnaire designed to improve diagnostic decision making in suspected cases of RLS.[79]

As in non-MS patients, conservative and pharmacologic treatments are available for the treatment of RLS and PLMD, and they can be considered alone or in combination. Conservative approaches are a reasonable first step, and they include the removal of various agents known to exacerbate these conditions. An evaluation of the patient's medication list is recommended, with reduction or discontinuation of medications that can cause or worsen RLS or PLMD and may be used often in MS (dopamine antagonists, lithium, selective serotonin reuptake inhibitors, tricyclic antidepressants). Iron supplementation with vitamin C should be implemented for ferritin level less than 50 ng/mL. Dopaminergic agents such as pramipexole, ropinirole, and rotigotine, along with the alpha-2-delta ligand gabapentin enacarbil, have been approved by the FDA for moderate to severe RLS, and all may be considered in patients with MS. If side effects of dopaminergic agents are poorly tolerated by the patient (nausea, hypotension, hallucinations, dyskinesia, or risk-taking behavior) or the treatment is ineffective, agents such as the alpha-2 delta ligands (gabapentin, pregabalin) or carbamazepine can be useful alternatives that may serve a dual role in treating concomitant neuropathic pain apart from RLS. Benzodiazepines are a third alternative, although less attractive due to potential complications of daytime sleepiness and exacerbation of sleep-related breathing disorders. Lastly, opiates are considered for select cases, but due to tolerance, potential dependence, and problematic side effects, their use in patients with MS is usually discouraged.

Narcolepsy

Narcolepsy is characterized by abnormal manifestations of REM sleep intrusion into wakefulness. An irrepressible need to sleep is an essential symptom, often accompanied by cataplexy, sleep paralysis, and hypnagogic hallucinations. PSG with multiple sleep latency testing (MSLT) is required to establish a diagnosis. While narcolepsy is estimated to affect 0.02–0.05% of the general population, the overall prevalence of narcolepsy among persons with MS is unknown.[80]

Two subtypes of primary narcolepsy exist. Narcolepsy type 1, distinguished by the presence of hypocretin deficiency, can be established clinically by the presence of cataplexy (a reliable clinical marker for hypocretin deficiency), or with CSF studies (hypocretin ≤ 110 pg/dL or one third of normal mean values); and narcolepsy type 2, which is associated with normal hypocretin levels.

Despite the low prevalence of narcolepsy, an intriguing potential nexus of genetic risk has been identified with respect to MS. Type 1 narcolepsy has been associated with immune-mediated degradation of hypocretin-secreting cell bodies within the hypothalamus and is associated with the human leukocyte antigen DQB1*0602 allele.[81] Nearly all patients with narcolepsy and hypocretin deficiency have this allele. Moreover, the preponderance of genetic risk for MS is associated with the HLA-DRB1*1501 allele, found in high linkage disequilibrium with DQB1*0602.[82] Conditions that may affect the hypothalamus, including MS, may also cause secondary forms of narcolepsy, or narcolepsy related to a medical condition.[83,84] A previous report suggests that MS may be one of the most common causes of secondary narcolepsy.[85]

Although specific guidelines for management of secondary narcolepsy due to MS have yet to be formulated, management strategies used in the general population that may also be considered for MS patients include wake-promoting agents (modafinil and armodafinil), stimulants, or sodium oxybate for hypersomnia, and sodium oxybate or REM-suppressing antidepressants for cataplexy.[86–89] New hypothalamic lesions leading to secondary narcolepsy may be treated with high-dose steroids.

Rapid Eye Movement Sleep Behavior Disorder

RBD is parasomnia characterized by a loss of motor inhibition during REM sleep, resulting in excessive and sometimes violent nocturnal vocal or motor activity, and dream enactment. Both idiopathic (primary) and secondary forms exist.

Idiopathic RBD is most common in male patients age 50 years or older, and is thought to be a *form fruste* of α-synucleinopathies due to the fact that the majority of patients who present with idiopathic RBD will eventually develop signs of a synuclein-mediated neurodegenerative disease.[90,91] Occurring in less than 1% of the population, typical RBD case descriptions involve individuals without any significant past neurological or psychiatric history who are found, by a spouse or partner, to physically perform dreamed activities.[92–94] Secondary forms of RDB are most commonly associated with conditions that affect pontine REM generators, including MS (Fig. 9.2A–B).[95] One study reported PSG-confirmed RBD in 3 of 135 patients with MS (2 of whom were on antidepressants), but population-based data are lacking and the prevalence of secondary RBD in MS is still unknown.

Interestingly, case reports have shown that RBD may be the first clinical manifestation of MS. RBD was noted as the presenting symptom of a 25-year-old woman who sought medical attention after a 6 month history of awakening in terror from violent dreams, sometimes having injured her limbs or head by thrashing about during sleep. She was found to have an inflammatory demyelinating lesion in the pons, and further evaluation ultimately established a diagnosis of MS. RBD ceased with administration of corticosteroids.[96] Another case report described a 51-year-old woman with a 24-year history of MS whose husband worried that she began to have nightly episodes of screaming, flailing, and punching; she was diagnosed with

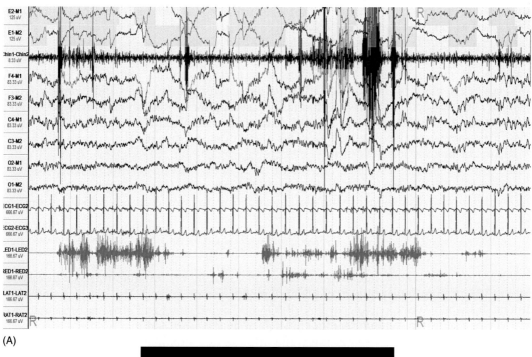

(A)

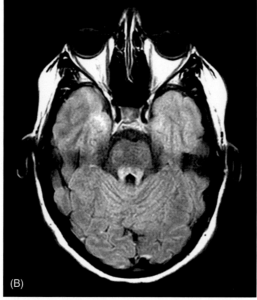

(B)

FIGURE 9.2 (A) Polysomnography showing loss of REM atonia. (B) MRI of the brain (axial FLAIR image) showing T2 hyperintense lesion in the posterior pons.

RBD, verified by PSG, proximate to a clinical relapse. A correlated demyelinating lesion was discovered in the dorsal pons. She was treated with corticosteroids acutely then prescribed clonazepam for prevention, resulting in clinical improvement.[95]

Overnight PSG is required to confirm loss of REM atonia, and rule out other conditions that may exacerbate or mimic RBD, such as sleep apnea, nocturnal seizures, or other parasomnias. Given that RBD is exceptionally rare in otherwise young adults and young women in particular, such individuals who present with symptoms of RBD should undergo a complete neurological workup with brain MRI. Patients with known MS who present with new-onset RBD should also be evaluated for signs of radiographic progression.

Treatment options for RBD include clonazepam, melatonin, pramipexole (though with mixed results in pertinent studies), and modification of the sleeping environment for safety.[97] For secondary RBD as seen in some patients with MS, concurrent management of the underlying condition with disease-modifying therapy and in acute cases, corticosteroids, is essential.

CONCLUSIONS

MS is a common chronic neurological disease of young adults. Fatigue and sleep disorders among MS patients impair quality of life and are often undiagnosed. The less common conditions described—CSA, narcolepsy, and RBD—all have distinct and striking presentations with fewer diagnostic confounders for clinicians than the more commonly seen sleep disorders, presenting an opportunity to intervene in these rare situations. Commonly occurring sleep disorders such as insomnia, OSA, and RLS, however, are more frequently encountered in patients with MS, and their prompt treatment enhances health and quality of life for these patients. When initial efforts to manage fatigue and daytime sleepiness fail to alleviate symptoms, it is incumbent upon the clinician to consider the spectrum of sleep disorders and seek appropriate referral to a sleep specialist if necessary.

References

1. Hirtz D, Thurman DJ, Gwinn-Hardy K, Mohamed M, Chaudhuri AR, Zalutsky R. How common are the "common" neurologic disorders?. *Neurology*. 2007;68:326–337.
2. Polman CH, Reingold SC, Edan G, et al. Diagnostic criteria for multiple sclerosis: 2005 revisions to the "McDonald Criteria". *Ann Neurol*. 2005;58:840–846.
3. Polman CH, Reingold SC, Banwell B, et al. Diagnostic criteria for multiple sclerosis: 2010 revisions to the McDonald criteria. *Ann Neurol*. 2011;69:292–302.
4. Lublin FD, Reingold SC, Cohen JA, et al. Defining the clinical course of multiple sclerosis: the 2013 revisions. *Neurology*. 2014;83:278–286.
5. Crespy L, Zaaraoui W, Lemaire M, et al. Prevalence of grey matter pathology in early multiple sclerosis assessed by magnetization transfer ratio imaging. *PloS One*. 2011;6:e24969.
6. Schlaeger R, Papinutto N, Panara V, et al. Spinal cord gray matter atrophy correlates with multiple sclerosis disability. *Ann Neurol*. 2014;76:568–580.
7. Fjaer S, Bo L, Lundervold A, et al. Deep gray matter demyelination detected by magnetization transfer ratio in the cuprizone model. *PloS One*. 2013;8:e84162.
8. Filippi M, Preziosa P, Copetti M, et al. Gray matter damage predicts the accumulation of disability 13 years later in MS. *Neurology*. 2013;81:1759–1767.
9. Johnson KP, Nelson BJ. Multiple sclerosis: diagnostic usefulness of cerebrospinal fluid. *Ann Neurol*. 1977;2:425–431.

10. Housley WJ, Pitt D, Hafler DA. Biomarkers in multiple sclerosis. *Clin Immunol.* 2015;161:51–58.

11. Irani DN. *Cerebrospinal fluid in clinical practice.* Philiedelphia, PA: Saunders Elsevier; 2009.

12. Sorensen PS, Blinkenberg M. The potential role for ocrelizumab in the treatment of multiple sclerosis: current evidence and future prospects. *Ther Adv Neurol Disord.* 2016;9:44–52.

13. Shirani A, Okuda DT, Stuve O. Therapeutic advances and future prospects in progressive forms of multiple sclerosis. *Neurotherapeutics.* 2016;13:58–69.

14. Ontaneda D, Fox RJ. Progressive multiple sclerosis. *Curr Opin Neurol.* 2015;28:237–243.

15. Sedel F, Papeix C, Bellanger A, et al. High doses of biotin in chronic progressive multiple sclerosis: a pilot study. *Mult Scler Relat Disord.* 2015;4:159–169.

16. Krupp LB, Alvarez LA, LaRocca NG, Scheinberg LC. Fatigue in multiple sclerosis. *Arch Neurol.* 1988;45:435–437.

17. Krupp L. Fatigue is intrinsic to multiple sclerosis (MS) and is the most commonly reported symptom of the disease. *Mult Scler.* 2006;12:367–368.

18. Lerdal A, Celius EG, Krupp L, Dahl AA. A prospective study of patterns of fatigue in multiple sclerosis. *Eur J Neurol.* 2007;14:1338–1343.

19. Janardhan V, Bakshi R. Quality of life in patients with multiple sclerosis: the impact of fatigue and depression. *J Neurol Sci.* 2002;205:51–58.

20. Flachenecker P, Bihler I, Weber F, Gottschalk M, Toyka KV, Rieckmann P. Cytokine mRNA expression in patients with multiple sclerosis and fatigue. *Mult Scler.* 2004;10:165–169.

21. Tartaglia MC, Narayanan S, Francis SJ, et al. The relationship between diffuse axonal damage and fatigue in multiple sclerosis. *Arch Neurol.* 2004;61:201–207.

22. Tellez N, Alonso J, Rio J, et al. The basal ganglia: a substrate for fatigue in multiple sclerosis. *Neuroradiology.* 2008;50:17–23.

23. Veauthier C, Paul F. Fatigue in multiple sclerosis: which patient should be referred to a sleep specialist?. *Mult Scler.* 2012;18:248–249.

24. Braley TJ, Segal BM, Chervin RD. Sleep-disordered breathing in multiple sclerosis. *Neurology.* 2012;79:929–936.

25. Brass SD, Li CS, Auerbach S. The underdiagnosis of sleep disorders in patients with multiple sclerosis. *J Clin Sleep Med.* 2014;10:1025–1031.

26. Braley TJ, Segal BM, Chervin RD. Obstructive sleep apnea is an under-recognized and consequential morbidity in multiple sclerosis. *J Clin Sleep Med.* 2014;10:709–710.

27. Cote I, Trojan DA, Kaminska M, et al. Impact of sleep disorder treatment on fatigue in multiple sclerosis. *Mult Scler.* 2013;19:480–489.

28. Veauthier C, Gaede G, Radbruch H, Gottschalk S, Wernecke KD, Paul F. Treatment of sleep disorders may improve fatigue in multiple sclerosis. *Clin Neurol Neurosurg.* 2013;115:1826–1830.

29. Chervin RD. Sleepiness, fatigue, tiredness, and lack of energy in obstructive sleep apnea. *Chest.* 2000;118:372–379.

30. Krupp LB, LaRocca NG, Muir-Nash J, Steinberg AD. The fatigue severity scale. Application to patients with multiple sclerosis and systemic lupus erythematosus. *Arch Neurol.* 1989;46:1121–1123.

31. Iriarte J, Katsamakis G, de Castro P. The Fatigue Descriptive Scale (FDS): a useful tool to evaluate fatigue in multiple sclerosis. *Mult Scler.* 1999;5:10–16.

32. Flachenecker P, Kümpfel T, Kallmann B, et al. Fatigue in multiple sclerosis: a comparison of different rating scales and correlation to clinical parameters. *Mult Scler.* 2002;8:523–526.

33. Induruwa I, Constantinescu CS, Gran B. Fatigue in multiple sclerosis—a brief review. *J Neurol Sci.* 2012;323: 9–15.

34. Mills RJ, Young CA, Pallant JF, Tennant A. Development of a patient reported outcome scale for fatigue in multiple sclerosis: The Neurological Fatigue Index (NFI-MS). *Health Qual Life Outcomes.* 2010;8:22.

35. Veauthier C, Radbruch H, Gaede G, et al. Fatigue in multiple sclerosis is closely related to sleep disorders: a polysomnographic cross-sectional study. *Mult Scler.* 2011;17:613–622.

36. Kaminska M, Kimoff RJ, Benedetti A, et al. Obstructive sleep apnea is associated with fatigue in multiple sclerosis. *Mult Scler.* 2012;18:1159–1169.

37. Boe Lunde HM, Aae TF, Indrevag W, et al. Poor sleep in patients with multiple sclerosis. *PloS One.* 2012;7:e49996.

38. Neau JP, Paquereau J, Auche V, et al. Sleep disorders and multiple sclerosis: a clinical and polysomnography study. *Eur J Neurol.* 2012;68:8–15.

39. Merlino G, Fratticci L, Lenchig C, et al. Prevalence of 'poor sleep' among patients with multiple sclerosis: an independent predictor of mental and physical status. *Sleep Med.* 2009;10:26–34.

40. Tachibana N, Howard RS, Hirsch NP, Miller DH, Moseley IF, Fish D. Sleep problems in multiple sclerosis. *Eur J Neurol*. 1994;34:320–323.

41. American Academy of Sleep Medicine. *International Classification of Sleep Disorders (ICSD-3): Diagnostic and Coding Manual*. 3rd ed. Darien, IL: American Academy of Sleep Medicine; 2014.

42. Stanton BR, Barnes F, Silber E. Sleep and fatigue in multiple sclerosis. *Mult Scler*. 2006;12:481–486.

43. Morin CM, Belleville G, Belanger L, Ivers H. The Insomnia Severity Index: psychometric indicators to detect insomnia cases and evaluate treatment response. *Sleep*. 2011;34:601–608.

44. Figved N, Klevan G, Myhr KM, et al. Neuropsychiatric symptoms in patients with multiple sclerosis. *Acta Psychiat Scand*. 2005;112:463–468.

45. Baron KG, Corden M, Jin L, Mohr DC. Impact of psychotherapy on insomnia symptoms in patients with depression and multiple sclerosis. *J Behav Med*. 2011;34:92–101.

46. Veauthier C. Hypnotic use and multiple sclerosis related fatigue: a forgotten confounder. *Sleep Med*. 2015;16(3):319.

47. Braley TJ, Segal BM, Chervin RD. Hypnotic use and fatigue in multiple sclerosis. *Sleep Med*. 2015;16:131–137.

48. Peppard PE, Young T, Palta M, Dempsey J, Skatrud J. Longitudinal study of moderate weight change and sleep-disordered breathing. *JAMA*. 2000;284:3015–3021.

49. Braley TJ, Segal BM, Chervin RD. Obstructive sleep apnea and fatigue in patients with multiple sclerosis. *J Clin Sleep Med*. 2014;10:155–162.

50. Chung F, Subramanyam R, Liao P, Sasaki E, Shapiro C, Sun Y. High STOP-Bang score indicates a high probability of obstructive sleep apnoea. *Brit J Anaesth*. 2012;108:768–775.

51. Fogel RB, Trinder J, White DP, et al. The effect of sleep onset on upper airway muscle activity in patients with sleep apnoea versus controls. *J Physiol*. 2005;564:549–562.

52. Jordan AS, White DP. Pharyngeal motor control and the pathogenesis of obstructive sleep apnea. *Respir Physiol Neurobiol*. 2008;160:1–7.

53. Losurdo A, Testani E, Scarano E, Massimi L, Della Marca G. What causes sleep-disordered breathing in Chiari I malformation? Comment on: "MRI findings and sleep apnea in children with Chiari I malformation". *Pediatr Neurol*. 2013;49:e11–e13.

54. Shochat T, Pillar G. Sleep apnoea in the older adult : pathophysiology, epidemiology, consequences and management. *Drugs Aging*. 2003;20:551–560.

55. Auer RN, Rowlands CG, Perry SF, Remmers JE. Multiple sclerosis with medullary plaques and fatal sleep apnea (Ondine's curse). *Clin Neuropathol*. 1996;15:101–105.

56. Severinghaus JW, Mitchell RA. Ondine's curse—failure of respiratory center automaticity while awake. *Clin Res*. 1962;10:122.

57. Chung F, Subramanyam R, Liao P, Sasaki E, Shapiro C, Sun Y. High STOP-Bang score indicates a high probability of obstructive sleep apnoea. *Brit J Anaesth*. 2012;108:768–775.

58. Dias RA, Hardin KA, Rose H, Agius MA, Apperson ML, Brass SD. Sleepiness, fatigue, and risk of obstructive sleep apnea using the STOP-BANG questionnaire in multiple sclerosis: a pilot study. *Sleep & breathing = Schlaf & Atmung*. 2012;16:1255–1265.

59. Ramar K, Dort LC, Katz SG, et al. Clinical practice guideline for the treatment of obstructive sleep apnea and snoring with oral appliance therapy: an update for 2015. *J Clin Sleep Med*. 2015;11:773–827.

60. Ferini-Strambi L. Sleep disorders in multiple sclerosis. *Handb Clin Neurol*. 2011;99:1139–1146.

61. Ferini-Strambi L, Filippi M, Martinelli V, et al. Nocturnal sleep study in multiple sclerosis: correlations with clinical and brain magnetic resonance imaging findings. *J Neurol Sci*. 1994;125:194–197.

62. Manconi M, Ferini-Strambi L, Filippi M, et al. Multicenter case-control study on restless legs syndrome in multiple sclerosis: the REMS study. *Sleep*. 2008;31:944–952.

63. Ekbom KA. Restless legs syndrome. *Neurology*. 1960;10:868–873.

64. Manconi M, Fabbrini M, Bonanni E, et al. High prevalence of restless legs syndrome in multiple sclerosis. *Eur J Neurol*. 2007;14:534–539.

65. Miri S, Rohani M, Sahraian MA, et al. Restless legs syndrome in Iranian patients with multiple sclerosis. *Neurol Sci*. 2013;34:1105–1108.

66. Li Y, Munger KL, Batool-Anwar S, De Vito K, Ascherio A, Gao X. Association of multiple sclerosis with restless legs syndrome and other sleep disorders in women. *Neurology*. 2012;78:1500–1506.

67. Manconi M, Rocca MA, Ferini-Strambi L, et al. Restless legs syndrome is a common finding in multiple sclerosis and correlates with cervical cord damage. *Mult Scler*. 2008;14:86–93.

68. Montplaisir J, Boucher S, Poirier G, Lavigne G, Lapierre O, Lesperance P. Clinical, polysomnographic, and genetic characteristics of restless legs syndrome: a study of 133 patients diagnosed with new standard criteria. *Mov Disord*. 1997;12:61–65.

69. Veauthier C, Gaede G, Radbruch H, Sieb JP, Wernecke KD, Paul F. Periodic limb movements during REM sleep in multiple sclerosis: a previously undescribed entity. *Neuropsychiatr Dis Treat*. 2015;11:2323–2329.

70. Kurtzke JF. Rating neurologic impairment in multiple sclerosis: An expanded disability status scale (EDSS). *Neurology*. 1983;33:1444.

71. Winkelmann J, Schormair B, Lichtner P, et al. Genome-wide association study of restless legs syndrome identifies common variants in three genomic regions. *Nat Genet*. 2007;39:1000–1006.

72. Pichler I, Hicks AA, Pramstaller PP. Restless legs syndrome: an update on genetics and future perspectives. *Clin Genet*. 2008;73:297–305.

73. Cervenka S, Palhagen SE, Comley RA, et al. Support for dopaminergic hypoactivity in restless legs syndrome: a PET study on D2-receptor binding. *Brain*. 2006;129:2017–2028.

74. Earley CJ, Connor JR, Beard JL, Clardy SL, Allen RP. Ferritin levels in the cerebrospinal fluid and restless legs syndrome: effects of different clinical phenotypes. *Sleep*. 2005;28:1069–1075.

75. Frauscher B, Loscher W, Hogl B, Poewe W, Kofler M. Auditory startle reaction is disinhibited in idiopathic restless legs syndrome. *Sleep*. 2007;30:489–493.

76. Allen RP, Picchietti D, Hening WA, et al. Restless legs syndrome: diagnostic criteria, special considerations, and epidemiology. A report from the restless legs syndrome diagnosis and epidemiology workshop at the National Institutes of Health. *Sleep Med*. 2003;4:101–119.

77. Bassetti CL, Mauerhofer D, Gugger M, Mathis J, Hess CW. Restless legs syndrome: a clinical study of 55 patients. *Eur J Neurol*. 2001;45:67–74.

78. Hening WA, Allen RP, Washburn M, Lesage SR, Earley CJ. The four diagnostic criteria for Restless Legs Syndrome are unable to exclude confounding conditions ("mimics"). *Sleep Med*. 2009;10:976–981.

79. Benes H, von Eye A, Kohnen R. Empirical evaluation of the accuracy of diagnostic criteria for restless legs syndrome. *Sleep Med*. 2009;10:524–530.

80. Ohayon MM, Ferini-Strambi L, Plazzi G, Smirne S, Castronovo V. Frequency of narcolepsy symptoms and other sleep disorders in narcoleptic patients and their first-degree relatives. *J Sleep Res*. 2005;14:437–445.

81. De la Herran-Arita AK, Garcia-Garcia F. Narcolepsy as an immune-mediated disease. *Sleep Disord*. 2014;2014:792687.

82. Patsopoulos NA, Barcellos LF, Hintzen RQ, et al. Fine-mapping the genetic association of the major histocompatibility complex in multiple sclerosis: HLA and non-HLA effects. *PLoS Genet*. 2013;9:e1003926.

83. Aldrich MS, Naylor MW. Narcolepsy associated with lesions of the diencephalon. *Neurology*. 1989;39:1505–1508.

84. Vetrugno R, Stecchi S, Plazzi G, et al. Narcolepsy-like syndrome in multiple sclerosis. *Sleep Med*. 2009;10:389–391.

85. Nishino S, Kanbayashi T. Symptomatic narcolepsy, cataplexy and hypersomnia, and their implications in the hypothalamic hypocretin/orexin system. *Sleep Med Rev*. 2005;9:269–310.

86. Wise MS, Arand DL, Auger RR, Brooks SN, Watson NF. Treatment of narcolepsy and other hypersomnias of central origin. *Sleep*. 2007;30:1712–1727.

87. Xyrem International Study GroupA double-blind, placebo-controlled study demonstrates sodium oxybate is effective for the treatment of excessive daytime sleepiness in narcolepsy. *J Clin Sleep Med*. 2005;1:391–397.

88. Xyrem International Study GroupFurther evidence supporting the use of sodium oxybate for the treatment of cataplexy: a double-blind, placebo-controlled study in 228 patients. *Sleep Med*. 2005;6:415–421.

89. Morgenthaler TI, Kapur VK, Brown T, et al. Practice parameters for the treatment of narcolepsy and other hypersomnias of central origin: an American Academy of Sleep Medicine Report. *Sleep*. 2007;30:1705–1711.

90. Iranzo A, Molinuevo JL, Santamaria J, et al. Rapid-eye-movement sleep behaviour disorder as an early marker for a neurodegenerative disorder: a descriptive study. *Lancet Neurol*. 2006;5:572–577.

91. Postuma RB, Gagnon JF, Vendette M, Montplaisir JY. Markers of neurodegeneration in idiopathic rapid eye movement sleep behaviour disorder and Parkinson's disease. *Brain*. 2009;132:3298–3307.

92. Schenck CH, Bundlie SR, Ettinger MG, Mahowald MW. Chronic behavioral disorders of human REM sleep: a new category of parasomnia. *Sleep*. 1986;9:293–308.

93. Schenck CH, Bundlie SR, Patterson AL, Mahowald MW. Rapid eye movement sleep behavior disorder. A treatable parasomnia affecting older adults. *JAMA*. 1987;257:1786–1789.

94. Mahowald MW, Schenck CH. REM sleep behaviour disorder: a window on the sleeping brain. *Brain*. 2015;138:1131–1133.

95. Tippmann-Peikert M, Boeve BF, Keegan BM. REM sleep behavior disorder initiated by acute brainstem multiple sclerosis. *Neurology*. 2006;66:1277–1279.

96. Plazzi G, Montagna P, Remitting REM. Sleep behavior disorder as the initial sign of multiple sclerosis. *Sleep Med*. 2002;3:437–439.

97. Aurora RN, Zak RS, Standards of Practice Committeeet al. Best practice guide for the treatment of REM sleep behavior disorder (RBD). *J Clin Sleep Med*. 2010;6:85–95.

Sleep and Neuromuscular Disease

S. Sakamuri, J.W. Day

Department of Neurology and Neurological Sciences, Palo Alto, CA, USA

O U T L I N E

Sleep and Neurologic Disease. http://dx.doi.org/10.1016/B978-0-12-804074-4.00010-8

INTRODUCTION

Neuromuscular diseases (NMD) by definition involve the muscle, neuromuscular junction, nerve, plexus, nerve root, or anterior horn cells. Depending on the pathophysiologic mechanism, neuromuscular diseases may involve the corticospinal tracts, upper motor neurons, and other aspects of the central nervous system (CNS).

An incomplete summary of NMD is included in Table 10.1. These are a heterogenous group, ranging from acute-onset conditions to chronic, slowly progressive disorders. They may appear at birth, in childhood, or later in life. Their etiologies vary, and include developmental, degenerative, metabolic, immune-mediated, infectious, traumatic, and toxic causes. Some are treatable, while others are irreversible. Weakness may be intermittent, fluctuating, static, or progressive.

TABLE 10.1   Differential of Neuromuscular Diseases

Muscle
Polymyositis
Dermatomyositis
Inclusion body myositis
Duchenne muscular dystrophy
Becker muscular dystrophy
Nemaline myopathy
Critical illness myopathy
Statin myopathy
McArdle disease
Pompe disease
Neuromuscular junction
Myasthenia gravis
Lambert–Eaton myasthenic syndrome
Congenital myasthenic syndrome
Botulinum toxicity
Nerve
Guillain–Barre syndrome
Chronic inflammatory demyelinating polyneuropathy
Inherited neuropathy (Charcot–Marie–Tooth)
Phrenic neuropathy (traumatic, diabetic, etc.)
Vasculitic neuropathy
Critical illness polyneuropathy
Plexus
Parsonage–Turner syndrome
Diabetic amyotrophy
Tumor infiltration
Trauma
Root, anterior horn cell, cord
Amyotrophic lateral sclerosis
Spinobulbar muscular atrophy
Spinal muscular atrophy
Poliomyelitis and enterovirus-related syndromes
Spinal cord injury

It is important to note that essentially any process causing respiratory muscle weakness can cause sleep-disordered breathing.

Evaluation for these conditions begins with a thorough general, neurological, and neuromuscular examination. Based on this evaluation, further testing may be indicated, such as serum laboratory evaluations (e.g., creatine kinase, acetylcholine receptor antibodies), electromyography and nerve conduction studies, muscle and nerve biopsy, brain and lung imaging, and genetic testing.

THE PHYSIOLOGY OF NORMAL BREATHING

The major muscle of inspiration is the diaphragm, a large band of muscle with a central fibrous tendon positioned between the xiphoid, costal margins, rib cage, and upper lumbar vertebrae. It is innervated by the phrenic nerve, which arises from the C3–C5 nerve roots. Inspiration also involves external intercostal muscles (supplied by the intercostal nerves that arise from corresponding thoracic roots), and the scalene and sternocleidomastoid "accessory" muscles, which are innervated by the C3–C6 nerve roots and spinal accessory nerve (cranial nerve XI), respectively.

Exhalation is typically mediated by the passive elastic recoil of the thoracic wall and rib cage.[1] The internal intercostal muscles (innervated by the intercostal nerves), rectus abdominus (innervated by the thoracoabdominal nerves), and other abdominal musculature also play a role, particularly when forceful exhalation is required for a cough.

Patency of the upper airway is necessary to allow for adequate intake of oxygenated air.[2] The upper airway structures include the lips, tongue, soft palate, epiglottis and larynx, supplied by cranial nerves VII, IX, X, and XII. During inspiration, the diaphragm contracts and moves downward, pushing the abdominal contents downward and outward and creating negative intrathoracic pressure. Air travels through the upper airway to fill this vacuum. During exhalation, the diaphragm relaxes, and elastic recoil of the lungs and chest wall forces air outward. Control of respiratory function is largely automatic and driven by brainstem centers. Impulses are relayed from medullary respiratory neurons to lower motor neurons of the phrenic and intercostal nerves.

Adjustments to the respiratory drive are mediated, in part, by the central and peripheral chemoreceptors in the brainstem and carotid sinus, respectively.[3] The central chemoreceptors in the ventral medulla stimulate ventilation in response to reductions in cerebrospinal fluid (CSF) pH caused by hypercapnia. The peripheral chemoreceptors in the carotid and aortic bodies similarly respond to reductions in pH and elevations in CO_2, as well as changes in the partial pressure of O_2 in arterial blood. The CNS plays an important role in regulating ventilation, particularly in patients with chronic NMD.

MECHANISMS OF SLEEP DISTURBANCES IN NEUROMUSCULAR DISEASE

NMDs affect sleep in a number of ways:

- restrictive pattern of ventilation due to inspiratory muscle weakness
- airway obstruction due to pharyngeal weakness

- secondary central hypoventilation syndrome due to chronic hypercapnia
- primary central disturbances of sleep regulation
- restless legs syndrome (RLS)
- neuropathic pain

Hypoventilation due to Compromise of the Respiratory Pump

When neuromuscular weakness involves the respiratory musculature, reduced ventilation results in inadequate delivery of oxygen to lung capillaries (hypoxemia) and inadequate clearance of carbon dioxide (hypercapnia). Hypoventilation occurs in two ways:

- "Obstruction" of air flow is caused by weakness in the upper airway.
- "Restriction" of air flow is caused by reduced diaphragm and chest wall excursion due to muscle weakness or fibrosis.

Patients are more vulnerable to the effects of neuromuscular weakness during sleep for several reasons:

- In the recumbent position, diaphragmatic excursion is not aided by gravity, resulting in decreased inspiration.
- In rapid eye movement (REM) sleep, atonia of the respiratory accessory muscles further reduces inspiratory effort.
- Weakness and hypotonia of the upper airway increase the risk of airway obstruction.
- In sleep, there is less central chemoreceptor sensitivity and less response to hypoxemia and hypercapnia than there is during wakefulness.

In patients with neuromuscular ventilatory insufficiency and disrupted sleep, hypoxemia and hypercapnia intermittently trigger increases in respiratory effort. This results in arousals from sleep and a majority of the sleep period spent in stage 1 and 2 non-REM (NREM) sleep. Patients or their sleep partners may not be aware of these involuntary arousals; instead, patients may endorse symptoms of daytime fatigue and sleepiness, morning headaches, and cognitive difficulties.

As noted previously, several factors result in more shallow breathing during sleep, typically before wakeful breathing is affected. As such, nocturnal hypoventilation often precedes daytime hypoventilation. In disorders that primarily affect the diaphragm, sleep-related hypoventilation may be the only sign of neuromuscular impairment. The sleep clinician thus may be the first to identify respiratory muscle involvement by a NMD.

Other Structural Changes that Affect the Respiratory Pump

Congenital or early-onset NMDs are often associated with deformities of craniofacial structures due to weakness during development. Examination reveals a high, narrow palate, narrow face, and micrognathia (Fig. 10.1). Conditions such as Pompe disease and Duchenne muscular dystrophy (DMD) are associated with macroglossia. These changes may in turn contribute to upper airway obstruction.

These patients are also prone to paraspinal muscle dysfunction, resulting in thoracic cage deformities (pectus excavatum or pectus carinatum) and scoliosis (Fig. 10.2).[4] This chronic remodeling limits chest wall mobility and may contribute to ventilatory restriction. Reduced

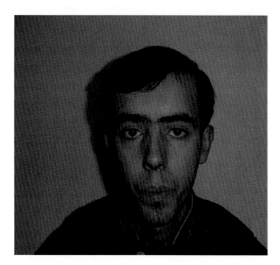

FIGURE 10.1   **Elongated face, ptosis, temporal wasting, frontal balding, and micrognathia in a patient with myotonic dystrophy type 1.** *Source: Figure 1, Image B only, from Kurt S et al. Combination of myotonic dystrophy and hereditary motor and sensory neuropathy. J Neurol Sci. 2010;288(1–2):197–199.*

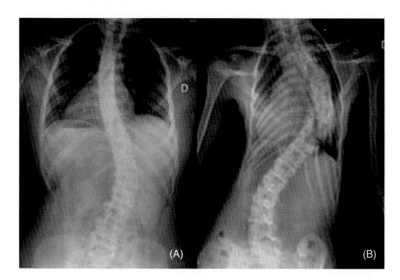

FIGURE 10.2   **Vertebral column X-rays indicating severe dorso-lumbar scoliosis in two patients with myopathies due to ryanodine receptor (*RYR1*) gene mutations.** *Source: Figure 1, Images A and B only, Rocha J et al. Ryanodine myopathies without central cores–clinical, histopathologic, and genetic description of three cases. Pediatr Neurol. 2014;51(2):275–278.*

chest wall mobility can also occur in late-onset diseases such as amyotrophic lateral sclerosis (ALS), where chronic respiratory weakness can lead to reduced chest wall compliance.

Truncal obesity and increased abdominal content (e.g., due to constipation, aerophagia, or organomegaly) occur in conditions such as Duchenne muscular dystrophy. This further predisposes patients to reduced vital capacity via restrictive mechanisms.[5]

ASSESSMENTS OF HYPOVENTILATION AND SLEEP DISTURBANCE IN NEUROMUSCULAR DISEASE

The astute clinician can identify this complication during the routine neurological history and examination, even before sleep studies are performed.

Symptoms

The symptoms of nocturnal hypoventilation in sleep are listed in Table 10.2. Patients will often dismiss these symptoms as normal. More severe neuromuscular respiratory weakness may result in overt dyspnea or orthopnea during activities of daily living.

A number of clinical scales can be used to screen for poor sleep quality and ventilatory impairment. These include the Epworth sleepiness scale (ESS), the Borg dyspnea scale, and the Fatigue and Daytime Sleepiness Scale (FDSS).

Clinical Signs

Proximal weakness often correlates with respiratory muscle weakness. Neck flexion strength is most accurately tested with the patient in the supine position. One can also observe for orthopnea, accessory muscle use, and paradoxical breathing. Paradoxical breathing (or thoracoabdominal asynchrony) takes two forms:

- Inward movement of the abdominal wall on inspiration due to diaphragm weakness, which is common in neuromuscular disorders.
- Inward movement of the chest wall on inspiration due to poor thoracic cage elasticity. This is seen in children with spinal muscular atrophy (SMA), where intercostal weakness exceeds diaphragm weakness.[6]

A thorough neuromuscular exam should include evaluation of ocular and bulbar function, proximal and distal strength, reflexes, sensation, and gait, screening for muscle fatigability, myotonia, atrophy, fasciculations, craniofacial deformities including temporal wasting, and

TABLE 10.2   Symptoms of Nocturnal Hypoventilation

Daytime symptoms
Fatigue
Excessive daytime sleepiness
Morning headaches
Fatigue upon awakening
Reliance on an alarm
Sleeping in on weekends

Nighttime symptoms
Snoring
Apneas
Restlessness
Frequent awakenings
Nocturia

other musculoskeletal abnormalities. Disease-specific nonmusculoskeletal features may include frontal balding, neurofibromas, cataracts, hypogonadism, and gynecomastia.

As a proxy measure of vital capacity, patients may be asked to count out loud to the highest number possible in one exhalation, enunciating at a standardized rate of two numbers each second. The forced vital capacity (FVC) is approximately 100 mL for each number counted, and a proposed normal cutoff is 25 numbers.[7] In individuals with nasopharyngeal weakness, the nostrils should be closed with a clip during this assessment to improve its accuracy.

Tongue and pharyngeal weakness can contribute to upper airway obstruction and collapse.[8] Symptoms of bulbar dysfunction may include difficulty chewing and swallowing, ineffective cough due to poor glottal closure, trouble managing secretions, and reduced speech volume and rapidity. The examination should include an assessment of speech quality (nasal or spastic speech), tongue strength and speed, palatal elevation, gag reflex, and ability to swallow and cough.

Laboratory Testing

In addition to the exam, patients should undergo basic vital signs including respiratory rate, pulse oximetry, and, when possible, end-tidal CO_2 (though this may underestimate arterial pCO_2).[9] These measures serve as a gross screen for significant respiratory dysfunction. It should be emphasized again that signs of nocturnal hypercapnia will often precede such markers of daytime hypercapnia.[10]

Patients with neuromuscular disorders should also undergo pulmonary function tests (PFTs), even in the absence of sleep symptoms or daytime hypoventilation. Abnormalities on these measures may predict nocturnal desaturations and hypercapnia,[11,12] and provide a useful measure of disease progression or response to treatment. Basic PFTs include forced vital capacity (FVC), slow vital capacity (SVC), and forced expiratory volume in the first minute (FEV_1). Maximum inspiratory and expiratory pressures (MIP and MEP) and sniff nasal inspiratory pressure (SNIP) also measure inspiratory and expiratory muscle strength. Note that all PFTs can be confounded by the presence of bulbar or pseudobulbar weakness or limited patient effort.[13]

In many patients with neuromuscular disorders, a polysomnogram (PSG) is not required for initiation of noninvasive ventilation (NIV). Any patient with a neuromuscular diagnosis and FVC <50% of predicted, MIP <60 cm H_2O, or $PaCO_2$ >45 mmHg will meet current Medicare criteria for use of nocturnal NIV without a diagnostic PSG (Table 10.3). On the other hand, PSG remains an essential diagnostic tool in patients without daytime hypoventilation, as criteria for NIV also include O_2 saturation <88% for greater than five consecutive minutes while asleep. PSG is also valuable to optimize fittings and settings for all neuromuscular patients who require nocturnal NIV. Lastly, despite the Medicare criteria, PSG is often required to support the clinician's decision to implement NIV.

TABLE 10.3   Medicare Criteria for Coverage of Noninvasive Ventilation in Patients with Neuromuscular Disease

One of the following:
- FVC <50% predicted FVC
- MIP <60 cm H_2O
- $PaCO_2$ >45 mmHg
- SaO_2 <88% for >5 consecutive minutes on PSG

There are several findings on PSG that may be identified in patients with NMDs. Obstructive apneas appear when upper airway collapse results in reduced airflow despite normal respiratory effort. Nonobstructive apneas occur when limited chest wall movement results in poor airflow and hypoxia, without observable obstruction. This limited movement is due to muscle weakness, contracture of the thoracic cage, or severe upper airway collapse. Note that this finding may be termed a "central apnea." A diagnostic pitfall occurs when nonobstructive apneas are incorrectly attributed to a primary deficit in the central respiratory drive, leading to unnecessary investigations for CNS diseases, such as brainstem stroke or *PHOX2B* mutations. Thus, more accurate terms are "nonobstructive" or "pseudocentral" apneas. Lastly, paradoxical chest and abdominal wall movements may also confound identification of obstructive or central apneas.[6] To avoid misdiagnosis, NMD should be considered in the differential for all patients with a significant number of nonobstructive apneas (more than 5 per hour) during sleep.

Imaging such as chest X-ray and sniff fluoroscopy (SNIF) can evaluate for diaphragmatic weakness. On chest X-ray, the paralyzed diaphragm may be abnormally elevated. During the inspiratory phase of SNIF testing, the muscle does not demonstrate normal downward excursion, and can even have paradoxical elevation. These imaging studies are well suited for detection of hemidiaphragmatic weakness (as seen in cases of phrenic nerve paralyses due to trauma, tumor invasion, diabetic nerve palsies, and brachial plexitis). However, in most NMD the diaphragm weakness is bilateral, causing the chest X-ray and SNIF to appear "pseudo-normal" due to the symmetric diaphragm appearance.

The phrenic nerve conduction study is being explored in the research setting as a direct marker of cervical motor neuron, phrenic nerve, and diaphragm function in patients with ALS and Charcot–Marie–Tooth disease (CMT). This noninvasive test is not confounded by the low patient effort and bulbar weakness that can affect traditional PFTs.[13,14]

Ultrasound of the phrenic nerve and diaphragm are also being explored to measure neuromuscular structure and function. Phrenic nerve enlargement has been demonstrated in some forms of CMT.[14]

TREATMENT OF NEUROMUSCULAR HYPOVENTILATION

The mainstay of treatment is the use of NIV to maintain adequate air exchange while keeping the upper airway patent. This intervention compensates for multiple mechanisms of neuromuscular weakness. NIV options include pressure-based devices, such as bilevel positive airway pressure, and advanced devices that offer volume-targeted modes to maintain adequate minute ventilation, such as average volume assured pressure support (AVAPS). A backup rate may be utilized if there is evidence of reduced ventilatory drive or if the patient is unable to maintain adequate respiratory effort.

The choice of device is often made in conjunction with a neuromuscular specialist and pulmonologist. Pressure-based devices allow for easier detection of air leaks, but volume-based devices may be more effective due to guaranteed volume delivery. This has been demonstrated in the ALS population, though aerophagia contributes to lower tolerance.[15] For insurance purposes, NIV is divided into two categories: respiratory assist devices and portable

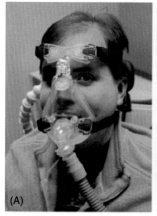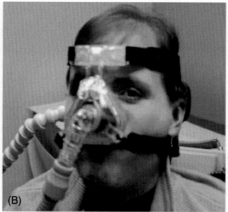

FIGURE 10.3 Different means of delivering noninvasive positive-pressure ventilation: (A) full mask, (B) nasal mask, (C) mouthpiece. *Source: Figure 977*, Murray and Nadel's Textbook of Respiratory Medicine. *6th ed. Vol. 2. 2016;1691–1706.e4.*

ventilators, based on the sophistication of the device. Disease severity often affects what type of equipment is authorized by the insurance provider.

Continuous positive airway pressure (CPAP) is commonly used to treat obstructive sleep apnea in the general population. In neuromuscular patients with impaired expiratory strength, CPAP can impede exhalation and reduce tidal volume, thereby exaggerating hypoventilation and hypoxemia. As a result, bilevel and AVAPS are preferable[16,17] and likely safer. CPAP is occasionally used in neuromuscular patients who have obstructive apneas due to isolated pharyngeal weakness, but have preserved inspiratory and expiratory muscle function. Outside of this scenario, CPAP is not considered a respiratory assist device in this population.

The interface for NIV is chosen based on the patient's comfort and degree of handicap. Options include the nasal mask, nasal pillows, and full-face mask interfaces (Fig. 10.3). The input of respiratory and occupational therapists can be invaluable, especially in patients with coexistent hand weakness who have trouble applying or removing their NIV device.

NIV settings are typically based on calculation of tidal volume (using ideal body weight) and patient tolerance, and later adjusted based on the degree of ventilatory impairment or gas exchange abnormalities. Device titration can be optimized with in-laboratory titration PSG. However, patients with neuromuscular weakness may not be able to undergo reliable overnight PSG testing due to logistical challenges. In those cases, home titration of portable ventilators by respiratory therapists may be an acceptable alternative.

NIV can be utilized in any situation where nocturnal hypoventilation is impacting sleep, regardless of disease duration. This includes reversible or episodic conditions [i.e., Guillain–Barre syndrome (GBS), myasthenia gravis] and progressive diseases (i.e., ALS, Duchenne muscular dystrophy). This is especially true in myasthenia gravis, where acute to subacute onset of nocturnal hypoventilation can result in sleep deprivation and exaggerated daytime fatigue,[18] both of which can be remedied quickly with NIV while awaiting the drug response of immunomodulatory treatment.[19]

Phrenic Nerve Stimulation and Diaphragmatic Conditioning

Patients with high spinal cord injury or a congenital central hypoventilation syndrome (*PHOX2B* mutation) have intact neuromuscular structures, including phrenic nerves, diaphragm, and intercostal muscles. Phrenic nerve and diaphragm stimulators utilize electrical stimulation to directly generate a normal respiratory rate and volume, allowing some patients to wean off of invasive mechanical ventilation, despite the loss of CNS input.[20]

In patients with ALS, a device to condition the denervated diaphragm and potentially improve muscle resilience has been explored. This diaphragm stimulator provides low intensity current to the muscle, without the effect of simulating typical respiration rate and volume. Although there has been suggestion of improvement in some measures of sleep, reports of improved survival have not yet been validated.[21] Investigators in one randomized controlled trial found increased mortality in the implanted group, and recommended against its use; additional trials are ongoing.[22]

Other Measures to Minimize Hypoventilation

Daytime measures that improve vital capacity, improve cough, and clear secretions may also have positive impacts on nocturnal ventilation. These include:

- daytime use of NIV
- mechanical insufflator–exsufflator (CoughAssist), a device to aid in airway clearance
- chest percussion devices to loosen secretions
- nebulizer treatments and other forms of pulmonary toilette.[23]

CENTRAL HYPOVENTILATION SYNDROME DUE TO CHRONIC HYPERCAPNIA

Patients with longstanding neuromuscular weakness and chronic hypoventilation often develop chronic hypercapnia. The central chemoreceptors become desensitized to elevated CO_2 and the ventilatory drive becomes dependent on hypoxemia alone. This is termed "central hypoventilation," as the central respiratory effort does not increase to manage the rising CO_2. It bears reminding that this "central" process develops due to peripheral neuromuscular weakness; as such, it is treated by using NIV to correct hypoventilation.

In the awake patient, the presence of hypercapnia on end-tidal CO_2, pulse CO_2, or blood gas indicates chronic hypercapnia. Other symptoms are described previously (Table 10.2). More severe hypercapnia may result in cognitive slowing or obtundation. In sleep, the ventilatory drive is naturally depressed, further increasing hypercapnia. This becomes even more apparent during REM sleep, when skeletal muscle tone is suppressed and the accessory muscles of respiration are not activated to assist the diaphragm.

The PSG may be difficult to interpret when there is primary weakness and secondary central hypoventilation. The underlying obstructive and/or restrictive neuromuscular process leads to apneic events. These may be detected as obstructive or nonobstructive apneas ("pseudocentral," in that reduced ventilatory effort due to muscle weakness is misinterpreted as poor central respiratory drive). True central apneas due to CNS desensitization to hypercapnia may also be present, but are difficult to distinguish on routine PSG.[24] If the chronic hypercapnia is not taken into account, then a congenital or acquired central hypoventilation syndrome (i.e., brainstem stroke, *PHOX2B* mutation) may be incorrectly diagnosed while the underlying NMD is overlooked. Thus, PSG findings should always be interpreted in the context of clinical findings and tests of respiratory function.[25]

Oxygen supplementation is occasionally initiated during the PSG to correct persistent hypoxemia, often when the clinician or technician is unaware of the underlying diagnosis. In the setting of chronic hypoventilation, correction of hypoxemia results in further depression of respiratory function. In this instance oxygen supplementation can reduce respiratory drive, heighten hypercapnia and paradoxically result in greater hypoxemia. For these reasons, oxygen supplementation is not recommended during the routine assessment of central hypoventilation syndrome due to the risk of producing an exaggerated respiratory depression.

Treatment of Central Nocturnal Hypoventilation

The sleep clinician may be the first to recognize the effects of chronic hypercapnia and initiate investigations for underlying neuromuscular respiratory weakness. NIV is used to treat the primary obstructive and/or restrictive airway disease causing the cascade of hypoventilation, chronic hypercapnia, and central respiratory depression.

While oxygen supplementation in the absence of ventilatory support can be harmful, oxygen is occasionally used in conjunction with NIV when hypoxemia is severe. It is essential that the underlying disease physiology is recognized and managed, and hypoxemia is not treated in isolation.

CENTRAL SLEEP DYSFUNCTION IN NEUROMUSCULAR DISEASE

Some NMD are associated with central sleep disturbances that appear independent of respiratory weakness. In addition, it is postulated that sleep architecture in neuromuscular patients naturally evolves to reduce REM-associated hypoventilation, resulting in reduced time spent in REM, and loss of REM atonia (also termed EMG augmentation).[24]

Myotonic Dystrophy

Myotonic dystrophy types 1 (DM1) and 2 (DM2) are autosomal dominant disorders associated with myotonia (delayed muscle relaxation), bulbar and limb weakness.[26] In DM1 in particular, widespread effects on splicing regulation result in CNS, endocrine, and other systemic manifestations.

Multiple peripheral and central mechanisms converge to cause sleep disturbances in DM1. These manifest as excessive daytime sleepiness, nocturnal hypoventilation, sleep-onset REM periods, sleep fragmentation, and periodic limb movements of sleep (PLMS).[27] Excessive daytime sleepiness appears early and is out of proportion to the degree of respiratory muscle and craniofacial weakness.[28,29] This implies other regulatory processes in the sleep-wake cycle, including the loss of serotonergic neurons in the brainstem raphe nuclei,[30] disturbances in growth hormone secretion,[31] and persistence of neonatal patterns of sleep due to alternative splicing in CNS circuits.[32] In DM2, alternative splicing in dopaminergic and other pathways has been proposed as a mechanism for the increased incidence of RLS in these patients.[33]

The hypothalamic neurotransmitter hypocretin (also known as orexin) may be involved in the pathophysiology of DM1. Central regulation of sleep is mediated in part by hypocretin signaling, and reduced levels have been associated with narcolepsy[34]. A small group of patients with DM1 were found to have a significant decrease in CSF hypocretin-1, whereas a larger group did not display significant abnormalities.[32,35] Thus the contribution of hypocretin deficiency to hypersomnia in DM1 is not yet clear.

In addition to PSG, the Rasch-built FDSS is being explored as a tool for the evaluation of sleep in DM1.[36]

Guillain–Barre Syndrome

This acute-onset, immune-mediated polyradiculoneuropathy often occurs in the context of *Campylobacter jejuni* infection. Symptoms may include neuropathic weakness, sensory changes, and autonomic instability. In addition, many patients report fatigue,[37] and disordered sleep has been described with sleep-onset REM and loss of REM atonia.[38]

In some patients with GBS, CSF studies have revealed absence of hypocretin, similar to what is seen in narcolepsy;[39] patients may experience persistent sleep disturbances months after the onset of GBS.[40] This raises the possibility of immune-mediated hypothalamic dysfunction and subsequent hypocretin deficiency leading to chronic sleep-related disturbances and fatigue. Interestingly, a small group of patients with a related condition, chronic inflammatory demyelinating polyneuropathy (CIDP), were not found to have reduced CSF hypocretin levels.[40]

UPPER MOTOR NEURON DYSFUNCTION

Involvement of the upper motor neurons (also known as corticospinal or pyramidal dysfunction) is predominant in a number of neurological diseases, including primary lateral sclerosis and hereditary spastic paraparesis. Primary CNS impairment of the respiratory drive in sleep has been postulated in such patients where muscle weakness is minimal, but sleep-disordered breathing is significant.[41] The precise mechanism is unknown.

RESTLESS LEGS SYNDROME IN NEUROMUSCULAR DISEASE

RLS, or Willis–Ekbom disease, is defined by an unpleasant urge to move the legs and/or arms, relief with movement, and a predominance of symptoms during evening hours and during periods of inactivity. The etiology is complex and likely multifactorial. Postulated mechanisms include genetic predisposition, decreased iron stores, thalamic dysfunction, and dysregulation of dopamine and other neurotransmitters.[42] PLMS may also be present. Peripheral neuropathies are associated with an increased incidence of RLS and PLMS. These include diabetic polyneuropathy, inherited polyneuropathies (CMT), and CIDP.[43–46] The pathophysiology may be related to axonal dysfunction or lack of spinal inhibitory input.[47,48]

Frequent symptoms may require daily use of dopamine agonists or calcium channel alpha-2-delta ligands, such as gabapentin. Serum ferritin should always be checked and levels

<50 should be treated with iron and vitamin C supplementation. It should be noted that all of the medications used in the treatment of RLS may exacerbate excessive daytime sleepiness in patients with neuromuscular hypoventilation. Opiates, including methadone, should be avoided in these patients as they can exacerbate sleep disordered breathing and nocturnal hypoventilation. For a complete review of RLS and its treatment, the reader is directed to the movement disorders chapter in this text.

PAIN SECONDARY TO NEUROMUSCULAR DISEASE

Pain is common in NMD; the causes may include neuropathic pain from sensory neuropathy, muscle pain from myositis, and other musculoskeletal pain due to weakness, inability to move, and orthopedic problems. These symptoms should be discussed at each visit. The mainstay of pharmacologic treatment is neuropathic pain medications (i.e., gabapentin, pregabalin, nortriptyline) in low doses. While not as sedating as opiates, the patient should still be queried on follow up visits for any symptoms of respiratory depression. Difficulties in positioning may be addressed with a hospital bed or alternating pressure mattress cover. An occupational therapist can provide valuable input on equipment.

EVIDENCE FOR TREATMENT OF SLEEP DISTURBANCES IN NEUROMUSCULAR DISEASE

As a general rule, treatment with NIV is recommended in any patient with NMD and nocturnal hypoventilation. In many conditions, NIV is associated with improved quality of life and even extended duration of life.

Amyotrophic Lateral Sclerosis

ALS is characterized by progressive degeneration of the upper and lower motor neurons. Respiratory failure is common, and typically results in death within 3–5 years of disease onset. Sleep-disordered breathing is a common feature due to weakness of the respiratory musculature. A number of studies have demonstrated that NIV improves sleep architecture, nocturnal hypoxia, and hypercapnia, even in patients with bulbar weakness (in whom mask fit and titration are assumed to be more challenging).[49,50,51]

The benefit of NIV in ALS goes beyond sleep because it unequivocally prolongs survival.[52] In a randomized controlled trial of 41 ALS patients with either orthopnea and MIP <60 cm H_2O, or with symptomatic daytime hypercapnia, patients with moderate or no bulbar dysfunction who used NIV had prolonged survival by 205 days, and also reported large benefits in quality of life on several indices.[53] The sleep clinician may be an important advocate in helping ALS patients avail themselves of this life-altering treatment.

Myotonic Dystrophy

As in other NMDs, the use of NIV in DM improves nocturnal hypoventilation on PSG and symptoms of daytime sleepiness.[54] In addition, there is limited data to support the use of stimulant medications for symptomatic treatment of excessive daytime sleepiness in patients with DM1.[55] In a small crossover trial, methylphenidate 20 mg daily significantly reduced excessive daytime sleepiness as measured by ESS.[56] However, a small randomized controlled trial of modafinil 300 mg daily was ineffective in improving results on maintenance of wakefulness test scores or self-reports of sleepiness as measured by a global self-assessment scale.[57] Of note, in a nonrandomized observational study of 145 patients, many patients and caregivers reported meaningful benefit from modafinil on open-ended surveys (the ESS and other measures were not used).[58] Further stimulant trials are ongoing.

Duchenne Muscular Dystrophy

Duchenne Muscular Dystrophy (DMD) is an X-linked disorder that results in mutations to the dystrophin gene, which is essential to muscle membrane and cytoskeleton stability. In early childhood, boys manifest proximal greater than distal weakness and Gower's sign, a maneuver in which the arms are used to push up to a standing position. Over time, restrictive ventilatory disease develops due to thoracic wall weakness and scoliosis. Standard of care includes corticosteroids to slow disease progression; however, these medications can result in obesity and further impact respiratory function. Disease-associated cardiomyopathy also compounds efforts to improve sleep-related breathing, and it is important to recognize that untreated ventilatory insufficiency can contribute to cardiac disease.

PSG in DMD reveals obstructive and restrictive components.[59] As in other NMDs, morning headaches and daytime fatigue may precede frank respiratory symptoms.[60] Predictors of sleep-disordered breathing include a higher body mass index (BMI) and decreased abdominal wall strength.[61,62] Since the oral and pharyngeal musculature is less affected than truncal, ventilatory, and limb musculature in Duchenne patients, NIV can typically be used well into the course of the disease without resorting to tracheostomy to maintain adequate daytime and nighttime ventilation.

In DMD, the use of NIV improves quality of sleep, quality of life, and survival.[63] DMD-specific guidelines support nocturnal NIV if the following abnormalities are noted:[64]

- signs or symptoms of hypoventilation,
- awake baseline SpO_2 <95% and/or blood or end-tidal CO_2 >45 mmHg,
- apnea–hypopnea index (AHI) >10 per hour; or four or more episodes of SpO_2 <92%; or desaturations of at least 4% per hour of sleep during PSG.

Spinal Muscular Atrophy

Mutations in the survival motor neuron 1 (*SMN1*) gene result in this autosomal recessive anterior horn cell disease. Residual function of the *SMN2* gene allows for some modulation of disease severity. Patients develop proximal greater than distal weakness with lower motor neuron features. Onset ranges from childhood to adulthood, and earlier symptoms are associated with a more severe and potentially fatal course.

Restrictive lung disease is common, in large part due to intercostal muscle weakness, although the diaphragm is also involved to a lesser degree. As a result, contrary to what is seen in other NMDs, patients with SMA often breathe better when recumbent than when upright. Paradoxical breathing is common, with inward movement of the chest wall during inspiration.

Nocturnal hypoventilation has been well documented in the infant SMA population, and certainly occurs in adult patients with SMA.[65] Daytime fatigue due to increasing nocturnal hypoventilation might be erroneously attributed to disease progression in the skeletal muscles, when in fact the worsening hypoventilation is quite treatable.[66] Treatment with NIV significantly improves the subjective and objective quality of sleep in infants and children, and incontrovertibly improves survival.[65,67] There is limited evidence to suggest that a CNS process in infants with SMA may reduce arousal thresholds during sleep.[68] If so, this may further impact the quality of sleep, independent of nocturnal hypoventilation.

Pompe Disease

This autosomal recessive metabolic disorder is caused by mutations in lysosomal acid alpha-glucosidase (acid maltase), leading to pathologic intracellular accumulations of glycogen and exaggerated lysosomal function. Ultimately, the disease manifests in infants as severe hypotonia and cardiomyopathy, or presents later in life with progressive skeletal myopathy and hypoventilation. The diaphragm is often involved.

Respiratory muscle weakness and nocturnal hypoventilation are significant in both infantile- and adult-onset forms of the disease.[69,70] Treatment with NIV improves gas exchange during sleep and reduces symptoms of daytime sleepiness and fatigue.[71] There is evidence that respiratory muscle training improves strength during inspiration and expiration, and also improves performance on PFT parameters; this could theoretically improve nocturnal hypoventilation as well.[72] Enzyme replacement therapy may also stabilize the progression of sleep disturbances.[73]

CONCLUSIONS

NMDs affect sleep via several mechanisms. Respiratory and bulbar weakness and abnormal skeletal anatomy result in hypoventilation. Over time, secondary hypercapnia results in desensitization of central chemoreceptors and diminished respiratory drive. In certain conditions, additional central changes in sleep patterns can occur independent of respiratory weakness. PSG may reveal obstructive and nonobstructive (pseudocentral) apneas, and hypoventilation can become severe during REM sleep. NIV is the mainstay of treatment of hypoventilation. Supplemental oxygen should be used with extreme caution, if at all. Stimulants and other treatments may be used in certain conditions.

The astute sleep clinician may be the first to recognize hypoventilation and respiratory muscle weakness due to NMD. Early and ongoing treatment of these patients can significantly improve sleep, quality of life, and survival.

Acknowledgment

The authors thank Michelle Cao, DO, for her contributions to this chapter.

References

1. Molkov YI, Shevtsova NA, Park C, Ben-Tal A, Smith JC, Rubin JE, Rybak IA. A closed-loop model of the respiratory system: focus on hypercapnia and active expiration. *PLoS One*. 2014;9(10):e109894.
2. van Lunteren E, Strohl KP. The muscles of the upper airways. *Clin Chest Med*. 1986;7(2):171–188.
3. Guyenet PG, Bayliss DA. Neural control of breathing and CO_2 homeostasis. *Neuron*. 2015;87(5):946–961.
4. Halawi MJ, Lark RK, Fitch RD. Neuromuscular scoliosis: current concepts. *Orthopedics*. 2015;38(6):e452–e456.
5. Canapari CA, Barrowman N, Hoey L, Walker SW, Townsend E, Tseng BS, Katz SL. Truncal fat distribution correlates with decreased vital capacity in Duchenne muscular dystrophy. *Pediatr Pulmonol*. 2015;50(1):63–70.
6. Testa MB, Pavone M, Bertini E, Petrone A, Pagani M, Cutrera R. Sleep-disordered breathing in spinal muscular atrophy types 1 and 2. *Am J Phys Med Rehabil*. 2005;84(9):666–670.
7. Elsheikh B, Arnold WD, Gharibshahi S, Reynolds J, Freimer M, Kissel JT. Correlation of single-breath count test and neck flexor muscle strength with spirometry in myasthenia gravis. *Muscle Nerve*. 2016;53(1):134–136.
8. Aboussouan LS, Lewis RA, Shy ME. Disorders of pulmonary function, sleep, and the upper airway in Charcot–Marie–Tooth disease. *Lung*. 2007;185(1):1–7.
9. Won YH, Choi WA, Lee JW, Bach JR, Park J, Kang SW. Sleep transcutaneous vs. end-tidal CO_2 monitoring for patients with neuromuscular disease. *Am J Phys Med Rehabil*. 2016;95(2):91–95.
10. Bersanini C, Khirani S, Ramirez A, Lofaso F, Aubertin G, Beydon N, Mayer M, Maincent K, Boule M, Fauroux B. Nocturnal hypoxaemia and hypercapnia in children with neuromuscular disorders. *Eur Respir J*. 2012;39(5):1206–1212.
11. Ragette R, Mellies U, Schwake C, Voit T, Teschler H. Patterns and predictors of sleep disordered breathing in primary myopathies. *Thorax*. 2002;57(8):724–728.
12. Mellies U, Ragette R, Schwake C, Boehm H, Voit T, Teschler H. Daytime predictors of sleep disordered breathing in children and adolescents with neuromuscular disorders. *Neuromuscul Disord*. 2003;13(2):123–128.
13. Jenkins JA, Sakamuri S, Katz JS, Forshew DA, Guion L, Moore D, Miller RG. Phrenic nerve conduction studies as a biomarker of respiratory insufficiency in amyotrophic lateral sclerosis. *Amyotroph Lateral Scler Frontotemporal Degener*. 2016;17(3–4):213–220.
14. de Carvalho Alcantara M, Nogueira-Barbosa MH, Fernandes RM, da Silva GA, Sander HH, Lourenco CM. Respiratory dysfunction in Charcot-Marie-Tooth disease type 1A. *J Neurol*. 2015;262(5):1164–1171.
15. Sancho J, Servera E, Morelot-Panzini C, Salachas F, Similowski T, Gonzalez-Bermejo J. Non-invasive ventilation effectiveness and the effect of ventilatory mode on survival in ALS patients. *Amyotroph Lateral Scler Frontotemporal Degener*. 2014;15(1–2):55–61.
16. Kushida CA, Littner MR, Hirshkowitz M, et al. Practice parameters for the use of continuous and bilevel positive airway pressure devices to treat adult patients with sleep-related breathing disorders. *Sleep*. 2006;29(3):375–380.
17. Perrin C, D'Ambrosio C, White A, Hill NS. Sleep in restrictive and neuromuscular respiratory disorders. *Semin Respir Crit Care Med*. 2005;26(1):117–130.
18. Quera-Salva MA, Guilleminault C, Chevret S, et al. Breathing disorders during sleep in myasthenia gravis. *Ann Neurol*. 1992;31(1):86–92.
19. Rabinstein AA. Acute neuromuscular respiratory failure. *Continuum (Minneap Minn)*. 2015;21:1324–1345:5 Neurocritical Care.
20. Gonzalez-Bermejo J, LLontop C, Similowski T, Morelot-Panzini C. Respiratory neuromodulation in patients with neurological pathologies: for whom and how?. *Ann Phys Rehabil Med*.. 2015;58(4):238–244.
21. Scherer K, Bedlack RS. Diaphragm pacing in amyotrophic lateral sclerosis: a literature review. *Muscle Nerve*. 2012;46(1):1–8.
22. Di PWC, Di PSGC, McDermott CJ, et al. Safety and efficacy of diaphragm pacing in patients with respiratory insufficiency due to amyotrophic lateral sclerosis (DiPALS): a multicentre, open-label, randomised controlled trial. *Lancet Neurol*. 2015;14(9):883–892.
23. Ambrosino N, Carpene N, Gherardi M. Chronic respiratory care for neuromuscular diseases in adults. *Eur Respir J*. 2009;34(2):444–451.
24. Aboussouan LS. Sleep-disordered breathing in neuromuscular disease. *Am J Respir Crit Care Med*. 2015;191(9):979–989.
25. Caruana-Montaldo B, Gleeson K, Zwillich CW. The control of breathing in clinical practice. *Chest*. 2000;117(1):205–225.
26. Harley HG, Brook JD, Rundle SA, et al. Expansion of an unstable DNA region and phenotypic variation in myotonic dystrophy. *Nature*. 1992;355(6360):545–546.
27. Laberge L, Gagnon C, Dauvilliers Y. Daytime sleepiness and myotonic dystrophy. *Curr Neurol Neurosci Rep*. 2013;13(4):340.

28. Meola G, Sansone V. Cerebral involvement in myotonic dystrophies. *Muscle Nerve*. 2007;36(3):294–306.
29. Kalkman JS, Schillings ML, van der Werf SP, et al. Experienced fatigue in facioscapulohumeral dystrophy, myotonic dystrophy, and HMSN-I. *J Neurol Neurosurg Psychiatry*. 2005;76(10):1406–1409.
30. Ono S, Takahashi K, Jinnai K, et al. Loss of serotonin-containing neurons in the raphe of patients with myotonic dystrophy: a quantitative immunohistochemical study and relation to hypersomnia. *Neurology*. 1998;50(2):535–538.
31. Culebras A, Podolsky S, Leopold NA. Absence of sleep-related growth hormone elevations in myotonic dystrophy. *Neurology*. 1977;27(2):165–167.
32. Ciafaloni E, Mignot E, Sansone V, et al. The hypocretin neurotransmission system in myotonic dystrophy type 1. *Neurology*. 2008;70(3):226–230.
33. Lam EM, Shepard PW, St Louis EK, et al. Restless legs syndrome and daytime sleepiness are prominent in myotonic dystrophy type 2. *Neurology*. 2013;81(2):157–164.
34. Nishino S, Ripley B, Overeem S, Lammers GJ, Mignot E. Hypocretin (orexin) deficiency in human narcolepsy. *Lancet*. 2000;355(9197):39–40.
35. Martinez-Rodriguez JE, Lin L, Iranzo A, et al. Decreased hypocretin-1 (Orexin-A) levels in the cerebrospinal fluid of patients with myotonic dystrophy and excessive daytime sleepiness. *Sleep*. 2003;26(3):287–290.
36. Hermans MC, Merkies IS, Laberge L, Blom EW, Tennant A, Faber CG. Fatigue and daytime sleepiness scale in myotonic dystrophy type 1. *Muscle Nerve*. 2013;47(1):89–95.
37. Merkies IS, Faber CG. Fatigue in immune-mediated neuropathies. *Neuromuscul Disord*. 2012;22(suppl 3):S203–S207.
38. Cochen V, Arnulf I, Demeret S, et al. Vivid dreams, hallucinations, psychosis and REM sleep in Guillain–Barre syndrome. *Brain*. 2005;128(Pt 11):2535–2545.
39. Fronczek R, Baumann CR, Lammers GJ, Bassetti CL, Overeem S. Hypocretin/orexin disturbances in neurological disorders. *Sleep Med Rev*. 2009;13(1):9–22.
40. Nishino S, Kanbayashi T, Fujiki N, et al. CSF hypocretin levels in Guillain-Barre syndrome and other inflammatory neuropathies. *Neurology*. 2003;61(6):823–825.
41. Gouveia RG, Pinto A, Evangelista T, Atalaia A, Conceicao I, de Carvalho M. Evidence for central abnormality in respiratory control in primary lateral sclerosis. *Amyotroph Lateral Scler*. 2006;7(1):57–60.
42. Wijemanne S, Jankovic J. Restless legs syndrome: clinical presentation diagnosis and treatment. *Sleep Med*. 2015;16(6):678–690.
43. Rajabally YA, Shah RS. Restless legs syndrome in chronic inflammatory demyelinating polyneuropathy. *Muscle Nerve*. 2010;42(2):252–256.
44. Shah RS, Rajabally YA. Restless legs syndrome in sensory axonal neuropathy: a case-control study. *Rev Neurol (Paris)*. 2013;169(3):228–233.
45. Nineb A, Rosso C, Dumurgier J, Nordine T, Lefaucheur JP, Creange A. Restless legs syndrome is frequently overlooked in patients being evaluated for polyneuropathies. *Eur J Neurol*. 2007;14(7):788–792.
46. Boentert M, Knop K, Schuhmacher C, Gess B, Okegwo A, Young P. Sleep disorders in Charcot-Marie-Tooth disease type 1. *J Neurol Neurosurg Psychiatry*. 2014;85(3):319–325.
47. Gemignani F, Brindani F, Negrotti A, Vitetta F, Alfieri S, Marbini A. Restless legs syndrome and polyneuropathy. *Mov Disord*. 2006;21(8):1254–1257.
48. Stiasny-Kolster K, Pfau DB, Oertel WH, Treede RD, Magerl W. Hyperalgesia and functional sensory loss in restless legs syndrome. *Pain*. 2013;154(8):1457–1463.
49. Boentert M, Brenscheidt I, Glatz C, Young P. Effects of non-invasive ventilation on objective sleep and nocturnal respiration in patients with amyotrophic lateral sclerosis. *J Neurol*. 2015;262(9):2073–2082.
50. Katzberg HD, Selegiman A, Guion L, et al. Effects of noninvasive ventilation on sleep outcomes in amyotrophic lateral sclerosis. *J Clin Sleep Med*. 2013;9(4):345–351.
51. Vrijsen B, Buyse B, Belge C, et al. Noninvasive ventilation improves sleep in amyotrophic lateral sclerosis: a prospective polysomnographic study. *J Clin Sleep Med*. 2015;11(5):559–566.
52. Miller RG, Jackson CE, Kasarskis EJ, et al. Practice parameter update: the care of the patient with amyotrophic lateral sclerosis: drug, nutritional, and respiratory therapies (an evidence-based review): report of the Quality Standards Subcommittee of the American Academy of Neurology. *Neurology*. 2009;73(15):1218–1226.
53. Bourke SC, Tomlinson M, Williams TL, Bullock RE, Shaw PJ, Gibson GJ. Effects of non-invasive ventilation on survival and quality of life in patients with amyotrophic lateral sclerosis: a randomised controlled trial. *Lancet Neurol*. 2006;5(2):140–147.
54. Monteiro R, Bento J, Goncalves MR, Pinto T, Winck JC. Genetics correlates with lung function and nocturnal ventilation in myotonic dystrophy. *Sleep Breath*. 2013;17(3):1087–1092.

55. Annane D, Moore DH, Barnes PR, Miller RG. Psychostimulants for hypersomnia (excessive daytime sleepiness) in myotonic dystrophy. *Cochrane Database Syst Rev*. 2006;19(3):CD003218.

56. Puymirat J, Bouchard JP, Mathieu J. Efficacy and tolerability of a 20-mg dose of methylphenidate for the treatment of daytime sleepiness in adult patients with myotonic dystrophy type 1: a 2-center, randomized, double-blind, placebo-controlled, 3-week crossover trial. *Clin Ther*. 2012;34(5):1103–1111.

57. Orlikowski D, Chevret S, Quera-Salva MA, et al. Modafinil for the treatment of hypersomnia associated with myotonic muscular dystrophy in adults: a multicenter, prospective, randomized, double-blind, placebo-controlled, 4-week trial. *Clin Ther*. 2009;31(8):1765–1773.

58. Hilton-Jones D, Bowler M, Lochmueller H, et al. Modafinil for excessive daytime sleepiness in myotonic dystrophy type 1—the patients' perspective. *Neuromuscul Disord*. 2012;22(7):597–603.

59. Suresh S, Wales P, Dakin C, Harris MA, Cooper DG. Sleep-related breathing disorder in Duchenne muscular dystrophy: disease spectrum in the paediatric population. *J Paediatr Child Health*. 2005;41(9–10):500–503.

60. Nozoe KT, Moreira GA, Tolino JR, Pradella-Hallinan M, Tufik S, Andersen ML. The sleep characteristics in symptomatic patients with Duchenne muscular dystrophy. *Sleep Breath*. 2015;19(3):1051–1056.

61. Sawnani H, Thampratankul L, Szczesniak RD, Fenchel MC, Simakajornboon N. Sleep disordered breathing in young boys with Duchenne muscular dystrophy. *J Pediatr*. 2015;166(3):640–645:e641.

62. Romei M, D'Angelo MG, LoMauro A, et al. Low abdominal contribution to breathing as daytime predictor of nocturnal desaturation in adolescents and young adults with Duchenne muscular dystrophy. *Respir Med*. 2012;106(2):276–283.

63. Finder JD, Birnkrant D, Carl J, et al. Respiratory care of the patient with Duchenne muscular dystrophy: ATS consensus statement. *Am J Respir Crit Care Med*. 2004;170(4):456–465.

64. Bushby K, Finkel R, Birnkrant DJ, et al. Diagnosis and management of Duchenne muscular dystrophy, part 2: implementation of multidisciplinary care. *Lancet Neurol*. 2010;9(2):177–189.

65. Mellies U, Dohna-Schwake C, Stehling F, Voit T. Sleep disordered breathing in spinal muscular atrophy. *Neuromuscul Disord*. 2004;14(12):797–803.

66. Puruckherr M, Mehta JB, Girish MR, Byrd Jr RP, Roy TM. Severe obstructive sleep apnea in a patient with spinal muscle atrophy. *Chest*. 2004;126(5):1705–1707.

67. Lemoine TJ, Swoboda KJ, Bratton SL, Holubkov R, Mundorff M, Srivastava R. Spinal muscular atrophy type 1: are proactive respiratory interventions associated with longer survival?. *Pediatr Crit Care Med*. 2012;13(3):e161–e165.

68. Verrillo E, Bruni O, Pavone M, et al. Sleep architecture in infants with spinal muscular atrophy type 1. *Sleep Med*. 2014;15(10):1246–1250.

69. Kansagra S, Austin S, DeArmey S, Kishnani PS, Kravitz RM. Polysomnographic findings in infantile Pompe disease. *Am J Med Genet A*. 2013;161A(12):3196–3200.

70. Boentert M, Karabul N, Wenninger S, et al. Sleep-related symptoms and sleep-disordered breathing in adult Pompe disease. *Eur J Neurol*. 2015;22(2):369–376:e327.

71. Mellies U, Stehling F, Dohna-Schwake C, Ragette R, Teschler H, Voit T. Respiratory failure in Pompe disease: treatment with noninvasive ventilation. *Neurology*. 2005;64(8):1465–1467.

72. Jones HN, Crisp KD, Robey RR, Case LE, Kravitz RM, Kishnani PS. Respiratory muscle training (RMT) in late-onset Pompe disease (LOPD): effects of training and detraining. *Mol Genet Metab*. 2015;117(2):120–128.

73. Kansagra S, Austin S, DeArmey S, Kazi Z, Kravitz RM, Kishnani PS. Longitudinal polysomnographic findings in infantile Pompe disease. *Am J Med Genet A*. 2015;167A(4):858–861.

Sleep and Headache

M. O'Hare, R.P. Cowan

Department of Neurology and Neurological Sciences, Palo Alto,
CA, USA

O U T L I N E

Sleep and Neurologic Disease. http://dx.doi.org/10.1016/B978-0-12-804074-4.00011-X

INTRODUCTION

Sleep and headache share a well-recognized, bidirectional relationship, with complex and incompletely understood interactions. The physiology of sleep shares many features with the pathophysiology of headache disorders, both in terms of the neuroanatomical pathways and the neurotransmitters that are involved. This may explain features of primary headache disorders like migraine, cluster headache and hypnic headache; all conditions whose clinical phenotype is intrinsically related to sleep.

Moreover, the painful experience of headache itself disrupts sleep, potentially creating a vicious circle of reinforcement. Both sleep disturbance and chronic pain also greatly increase the risk of depression and anxiety, further affecting the complex relationship between sleep and headache. Similarly, sleep disorders such as sleep apnea can lead to secondary headaches, and may in turn affect the expression of various primary headache disorders.

THE CLINICAL RELATIONSHIP BETWEEN SLEEP AND HEADACHE DISORDERS

Certain primary headache disorders are closely related to sleep by their phenotype, including migraine, cluster headache, and hypnic headache. The clinical features of these particular disorders suggest a relationship to sleep pathology. Shared neuroanatomical and neurotransmitter pathways may explain the mechanisms underlying this relationship.

Migraine is Clinically Related to Sleep

Migraine is a recurring headache disorder characterized by attacks of unilateral, pulsating head pain, that is, moderate to severe in intensity, with associated photophobia and phonophobia, and/or nausea and vomiting.[1] The complex relationship between migraine and sleep was noted as early as 1873, when Liveing described the therapeutic effects of sleep on a migraine attack—"the pain is [...] relieved [...] by sleep when the sufferer is fortunate enough to procure it." Liveing also described the propensity for migraine attacks to begin on waking and to be precipitated by disturbances of sleep patterns.[2]

This phenomenon is still observed, with sleep reported by many migraineurs as a reliable abortive therapy for acute migraine. There is also clear evidence that changes in sleep patterns (e.g., decreased sleep, increased sleep, or changing time zones) can precipitate migraine attacks.[3-7] In addition, sleep disorders contribute to the evolution of episodic migraine into its chronic form.[8] The use of behavioral sleep modification techniques may result in an improvement in migraine frequency, and may be an effective adjunctive therapy in reverting chronic migraine back to its episodic form.[3,4]

Sleep complaints are prevalent among migraineurs, with difficulty initiating or maintaining sleep reported by over half of patients, and frequent difficulty reported by a third of patients.[9] This is particularly striking in chronic (transformed) migraine, where patients almost invariably report modifiable poor sleep habits, nonrestorative sleep, and shorter sleep durations.[9,10] Evidence from a case-control study suggests that excessive daytime sleepiness is more common in migraineurs than controls, with an odds ratio of 3.1.[11] However, a more recent study did not support this, suggesting that although migraineurs are more likely to report sleepiness, they do not score higher on the Epworth sleepiness scale (ESS), a common objective measurement of recent sleepiness.[12]

Migraine attacks exhibit clear circadian timing, with attacks more likely to occur between the hours of 4 a.m. and 9 a.m.[13] A disturbance of circadian rhythms in migraine is also supported by "circa-septan" and circannual patterns that vary according to localization in the northern or southern hemisphere.[14] Parasomnias such as sleep terrors, somnambulism, sleep bruxism, restless legs syndrome (RLS), and nocturnal enuresis are also seen more commonly in migraineurs.[15] The mechanisms underlying the association of these disorders with migraine are poorly understood. European case-control observational studies have estimated the prevalence of RLS in migraineurs at 22–26%, compared with a prevalence of 5–8% in the general population.[16,17] The comorbidity of migraine and RLS appears to be associated with higher migraine attack frequency, greater migraine-related disability, and increasing impairment in subjective sleep quality.[18] The mechanism underlying this association remains obscure, and pathology relating to the hypothalamus or to dopaminergic signaling has been proposed.[18,19] Migraine-induced sleep disruption may also play a role in RLS severity. For a complete review of the subject see the review by Mitsikostas et al.[20]

Migraine attacks may be related to specific sleep stages, although the exact mechanisms remain unclear. Nocturnal arousal from sleep with migraine may be more likely during REM sleep.[21,22] In addition, migraine appears more likely to occur following a night's sleep with excessive durations of slow wave and REM sleep.[23] Some patients report that daytime napping can trigger migraine attacks, and this has been reported to occur following naps in which slow wave or REM sleep is obtained.[23] However, these findings were not supported by more recent polysomnographic (PSG) studies of migraineurs by Goder et al.[22] Findings consistent with reduced cortical activation were observed the night preceding a migraine (reduced number of arousals, reduced beta power in slow wave sleep and reduced REM density).[22] Similarly, studies of migraineurs during pain-free periods have demonstrated reduced cortical arousal.[24,25]

Engstrom et al. compared sleep patterns in migraineurs during both interictal and peri-ictal phases to healthy controls.[12] Migraineurs reported greater subjective sleep complaints such as insomnia, tiredness or pain-related sleep difficulties. However, the difference between the migraine population and healthy controls was much less significant when looking at PSG data, which demonstrated increased amounts of slow wave sleep and reduced fast arousals in the interictal migraine group.

Cluster Headache is Clinically Related to Sleep

Cluster headache (CH) is a distinct primary headache disorder characterized by attacks of excruciating pain, typically in a unilateral retro-orbital distribution, with associated ipsilateral autonomic features.[1] It occurs more commonly in males.[26] In its typical episodic form, patients experience "clusters" of headaches occurring up to 8 times per day, with periods of remission in between, each lasting between 20 and 180 minutes.[27] CH is a disorder intricately associated with sleep, often with a predictable chronobiological pattern.[28] CH often awakens the patient at the same time each night (circadian rhythmicity), and tends to occur in clusters in the same season each year (circannual periodicity).[29]

A causal relationship between REM sleep and cluster attacks has been postulated. Headache waking CH patients from sleep occurred more frequently during REM sleep in a number of case series.[30–32] However, this data has since been challenged by more recent studies that have reported no such relationship.[33,34] In fact, these more recent case series demonstrate headache onset occurring more predominantly during stage 2 non-REM (NREM) sleep in both episodic and chronic CH patients. There also appears to be a reduction in the duration of slow wave sleep and REM sleep duration in these cases.[34] Overall, it appears that a specific sleep stage abnormality is insufficient in explaining CH attack onset,[33] and the relationship is more complicated than previously hypothesized.

CH is also strongly associated with obstructive sleep apnea (OSA). OSA prevalence of up to 80% has been reported in a CH population.[35,36] This association will be further discussed later in this chapter, along with neuroimaging and neuroendocrine evidence supporting a pathophysiological link between sleep and CH.

Hypnic Headache is Clinically Related to Sleep

Hypnic headache is the only primary headache disorder, that is, defined specifically in relation to sleep.[37] The most recent International Classification of Headache Disorders (ICHD 3-beta) diagnostic criteria require that hypnic headaches occur only during sleep, and result in awakening from sleep.[1] The presence of cranial autonomic features or restlessness exclude the diagnosis according to ICHD 3-beta criteria, and are more suggestive of a trigeminal autonomic cephalgia.[1] However, some authors dispute this, reporting "mild" autonomic features in several cases.[38–41]

Hypnic headache is a rare condition, accounting for only 0.07–0.35% of headaches seen in specialist clinics.[42,43] The underlying pathophysiology remains poorly understood.[44] It has been proposed that hypnic headache may in fact represent an age-related phenotypic change of another sleep-related primary headache disorder such as migraine.[14,45] Hypnic headache may also coexist with other primary headache conditions such as migraine or tension-type headache.[38,39,44] In a recent case series of 23 patients, 16 had a preceding diagnosis of migraine.[38]

Hypnic headaches typically begin in the 5th decade of life, more often in females, and are generally moderate to severe in intensity and bilateral in distribution.[38,39,44] Migrainous features have been reported.[38] Given the advanced age at headache onset, the diagnosis of hypnic headache should only be made after a careful evaluation for other causes of nocturnal headache such as elevated intracranial pressure, space-occupying lesion, arterial hypertension, and OSA.[39,44] However, hypnic headache may also coexist with OSA, thus the presence of a sleep disorder does not exclude the primary headache diagnosis.[1,44] The prevalence of OSA is high in hypnic headache patients; however, this is reflective of the age distribution, and OSA prevalence is normal when compared to age-matched controls.[44]

Hypnic headaches tend to occur at a specific time each night, earning the nickname "alarm clock headache."[46] This points strongly toward an underlying disorder of chronobiology.[39] Hypnic headache may be associated with age-related alterations in sleep physiology.[14,37] There is a reduction in slow wave sleep seen in the elderly,[37] as well as increasingly fragmented sleep, with increased daytime napping and increased nocturnal waking,[47] and this may predispose patients to this disorder. Hypnic headache does not appear to be strictly associated with a specific sleep phase, and attacks can occur in both REM and non-REM sleep, even within the same individual.[48,49]

Structural neuroanatomical changes reported in hypnic headache patients[40] further support a circadian disorder, with decreased gray matter volume in the posterior hypothalamus,[40] a key structure in sleep physiology. Similar structural changes have been reported in narcolepsy patients.[50] Age-related melatonin deficiency has also been proposed as a pathophysiological mechanism in this condition;[37] the possible role of melatonin will be discussed later in this chapter

Summary of Headache Disorders and Their Associated Sleep Disorders

Headache disorder	Associated sleep disorders
Migraine	• Insomnia[9] • Parasomnias (sleep terrors, somnambulism, sleep bruxism, restless leg syndrome)[15]
Cluster headache	• Obstructive and central sleep apnea[35,36]
Hypnic headache	• Associated with disturbed sleep by definition[1,51]
Tension-type headache	• Insomnia[52-55]
Chronic daily headache	• Insomnia[9,10,56,57] • Obstructive sleep apnea;[58] snoring[59]

THE SHARED NEUROBIOLOGICAL SUBSTRATE OF SLEEP AND HEADACHE

The transition between wake and sleep is controlled by the interaction between arousal and sleep-promoting pathways in the brain. Many of the structures, pathways, and neurotransmitters involved in this process are also implicated in headache pathophysiology. The convergence between sleep and headache pathways primarily localizes to diencephalic and brainstem structures (Fig. 11.1).[60]

Sensory afferents mediating headache pain converge on the trigeminal nucleus caudalis (TNC) of the brainstem, which then project to the ventral posteromedial nucleus of the thalamus. These TNC projections interact with numerous brainstem and diencephalic nuclei as they ascend through the midbrain,[60] including the periacqueductal grey matter (PAG), locus coeruleus, hypothalamus, limbic system, and autonomic brainstem centers such as the solitary nucleus and the parabrachial nucleus.[14,60,61]

Of equal importance in headache pathophysiology are the descending pathways that modulate nociceptive pathways in the TNC. Brainstem structures including the rostroventromedial medulla, nucleus raphe magnus, locus coeruleus, and PAG modulate the trigeminovascular pathway and are influenced by descending inputs from the thalamus and hypothalamus.[14,61]

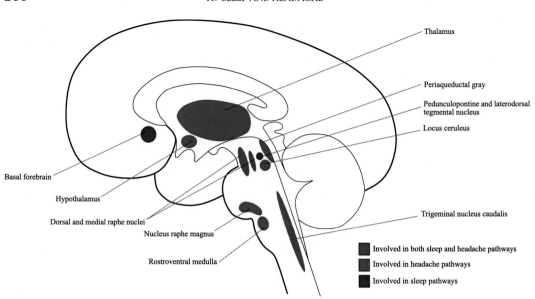

FIGURE 11.1 **Key structures involved in sleep and headache pathways.** Key structures involved in both sleep and headache localize to the midbrain and diencephalon.[60] Wakefulness is promoted by ascending activating pathways projecting from the brainstem to the thalamus and basal forebrain, from structures that include the locus ceruleus (LC), periacqueductal gray matter (PAG), dorsal and medial raphe nuclei, and the tuberomammillary nuclei of the hypothalamus.[66] Orexin-containing neurons in the lateral hypothalamus maintain stability of the waking state.[66] Inhibition of these nuclei by the VLPO promotes sleep.[66] Pain afferents ascend from the trigeminal nucleus caudalis to the ventral posteromedial nucleus of the thalamus, interacting with multiple nuclei as they ascend (including the LC, PAG and hypothalamus).[60] Descending pathways that modulate nociception involve the thalamus, hypothalamus, LC, PAG, rostroventral medulla and nucleus raphe magnus.[14,61]

The Hypothalamus is a Key Structure Linking Headache and Sleep

Hypothalamic structures are vital for maintaining wakefulness, and significant lesions of the posterolateral hypothalamus result in an unremitting sleep-like state.[62] The importance of the hypothalamus in this regard was first recognized by von Economo, who described how inflammation in this region, "encephalitis lethargica," produced profound disturbances of sleep/wake control.[63,64] In particular, the posterior hypothalamus is a critical wake-promoting area.[65] Histaminergic neurons of the tuberomammilary nuclei project widely to cortical and diencephalic structures to promote wakefulness.

In the lateral hypothalamus, a specific population of orexin-containing neurons reciprocally excite many activating monoaminergic systems, and also project to the basal forebrain and cortex.[62,66] These orexinergic projections are thought to have a specific role in coordinating the activity of other activating systems, maintaining the stability of the waking state, and preventing inappropriate sudden onset of sleep. Orexin-containing neurons are deficient in patients with narcolepsy, a condition characterized by inappropriate sleep onset as well as intrusion of a REM-like-state into waking hours.[62]

In migraine, hypothalamic involvement has long been hypothesized, as premonitory symptoms (e.g., changes in appetite or mood, yawning, and autonomic features) suggest localization to this region.[37] Functional imaging studies support this hypothesis, demonstrating activation of the hypothalamus, midbrain PAG, and dorsal pons during the early premonitory phase of migraine.[67] In addition, the comorbidity of migraine and nocturnal enuresis suggests a shared hypothalamic pathological substrate[68] because the hypothalamus plays an important role in micturition. Adolescents with migraine are more likely to report a history of nocturnal enuresis (41% in episodic and 49% in chronic migraine compared to 12% of controls), and also achieve bladder control at a later age than controls.[68]

The distinctive rhythmicity of CH and hypnic headache is also highly suggestive of hypothalamic involvement. In addition, secondary CH has been shown to arise from suprasellar invasion of a pituitary adenoma,[69] further supporting a role for the hypothalamus in the development of CH. Neuroendocrine abnormalities provided the first objective evidence supporting hypothalamic dysfunction in CH. The pronounced gender discrepancy in CH patients, a disorder with a male to female ratio of 9:1, prompted sex hormone studies. These revealed lower testosterone levels in CH patients, with altered rhythmicity of secretion.[70,71] In addition, abnormalities in the secretion of prolactin, gonadotrophins, growth hormone, thyroid-stimulating hormone, and cortisol have also been reported.[72] Challenge testing of the hypothalamic-pituitary axis (with the insulin tolerance test) results in blunted responses, consistent with hypothalamic dysfunction—even in comparison to a population of chronic back pain patients.[73,74] Similarly, studies of hypothalamic function in chronic migraine patients have revealed blunted profiles of prolactin secretion, and increased cortisol levels. Measurements of growth hormone secretion were normal in this population.[75]

Radiological evidence also supports hypothalamic dysfunction in CH, with positron emission tomography (PET) scans demonstrating increased regional cerebral blood flow in the ipsilateral posterior hypothalamus during cluster attacks,[27,76,77] as well as other less specific pain-processing areas such as the anterior cingulate cortex. This suggests a role for the hypothalamus as a generator of CH. Hypothalamic activation does not occur after injection of capsaicin into the forehead, suggesting a CH-specific association.[27] MRI scans of episodic and chronic CH patients show increased volume of the inferior posterior hypothalamus bilaterally, in the region correlating to the area of specific activation seen on PET imaging.[78] In addition, proton MR spectroscopy studies in CH patients demonstrate abnormal metabolite ratios consistent with loss or dysfunction of neurons and myelin in the hypothalamus.[79]

Functional MRI (fMRI) studies confirm ipsilateral activation of the posterior hypothalamus in the cluster headache ictal phase.[80] Resting state fMRI studies also demonstrate abnormal connectivity during cluster attacks between the hypothalamus and other brain regions typically involved in the "pain matrix" (such as the anterior and posterior cingulate cortex).[26] Abnormal connectivity between the hypothalamus and regions of the frontal, parietal, and temporal cortex persists interictally.[26] Hypothalamic activation was initially thought to be specific to CH genesis, but has also been demonstrated on PET scanning of spontaneous migraine attacks.[81] Similarly, abnormal functional connectivity of the hypothalamus in migraineurs has been reported, particularly to regions such as the locus coeruleus and other areas controlling autonomic tone.[82]

In addition, hypothalamic activation has been demonstrated in various other trigeminal autonomic cephalgias (TACs), including hemicranias continua, short-lasting unilateral neuralgiform headache with conjunctival injection and tearing (SUNCT), and paroxysmal hemicranias.[83–85] The pronounced circadian rhythmicity of CH is absent in these other TACs, and the persistence of posterior hypothalamic activation suggests that hypothalamic activation may in fact be more indicative of pain than specific circadian circuitry.[14]

Deep brain stimulation (DBS) of the posterior inferior hypothalamus has been effective in intractable cases of chronic CH,[86] with improvement in sleep architecture and quality after treatment, as well as pain relief.[87] However, there have been two case reports associating posterior hypothalamus DBS with disruption of sleep patterns.[88] In both cases, patients with intractable CH and SUNCT as well as comorbid sleep disorders (REM behavioral disorder in one; non-REM parasomnia in the other) underwent DBS implantation of the posterior hypothalamus. In both cases, headache control improved, but this correlated temporally with both objective and subjective evidence of worsening sleep disturbance.

In summary, there is an abundance of structural and functional neuroimaging data supporting a role for the hypothalamus in headache pathogenesis. Dysfunction of the hypothalamus may result in disinhibition of the PAG and TNC, resulting in uncontrolled trigeminovascular activation.[79]

The Brainstem Contains Key Structures Linking Headache and Sleep

Key structures in the overlap between headache and sleep localize to the brainstem, including the locus coeruleus, ventral PAG, and dorsal raphe nucleus. These cell bodies send monoaminergic projections to thalamic nuclei, the lateral hypothalamus, basal forebrain, and cortex. They are important in maintaining arousal, and are highly active during wakeful hours.[14,66,89] In addition, the ventrolateral PAG (vlPAG) plays a specific role in switching off REM sleep (i.e., it is a REM-off zone), and is innervated by orexinergic neurons of the lateral hypothalamus.[61] As well as their role in arousal, these brainstem structures also play an important role in the modulation of headache perception, acting as an endogenous "antinociceptive system."[37]

Abnormal Serotonin Signaling in the Brainstem may Link Headache and Sleep

The main CNS serotoninergic nucleus is the dorsal raphe nucleus.[90] This is a key antinociceptive structure[91] that is also involved in promoting wakefulness. A significant role for serotonin in migraine pathophysiology has long been suspected.[92,93] This theory is supported by a wealth of indirect evidence. Serotonin metabolism in migraineurs is abnormal, both during and between attacks. Plasma serotonin is abnormally low in the interictal phase in migraineurs, and increases during attacks, while the inverse is observed in plasma measures of plasma 5-HIAA, the principle metabolite of serotonin.[92] CSF 5-HIAA is increased (which may reflect increased central serotonin turnover),[94] and the capacity for serotonin synthesis is increased in migraineurs.[95]

In addition, the mechanism of action of many pharmacologic agents involved in provoking, aborting, and preventing migraine attacks relates to serotonin (e.g., reserpine, triptans, selective serotonin reuptake inhibitors).[92,93] One unifying theory explaining the altered serotonin metabolism observed in migraineurs proposes that migraine is a condition of central serotonin dysregulation, leading to a cycle of recurrent headaches and reduced pain threshold.[95]

Serotoninergic activity of the dorsal raphe nucleus is characterized by circadian rhythmicity. Waking hours are associated with tonic activity of the serotoninergic neurons, which decreases during slow wave sleep, becoming almost totally quiescent during REM sleep.[96] These fluctuations in serotoninergic activity may also contribute to the relationship between migraine and sleep.

The Homeostatic and Circadian Drives to Sleep

Many ascending activating pathways projecting from the brainstem to the thalamus and basal forebrain promote a state of wakefulness.[66] These arousal systems are counterbalanced in a controlled manner by an endogenous drive to sleep, which leads to arousal pathways being "switched off" by inhibition arising from the ventrolateral preoptic nucleus (VLPO).[62,66] GABAergic inhibitory neurons of the VLPO project to the tuberomammillary nucleus, locus coeruleus, and the dorsal and median raphe nuclei.[66] Bilateral lesions of the VLPO in animal models result in pronounced insomnia, with a dramatic reduction particularly of non-REM sleep.[97]

These opposing systems are controlled by two separate processes, resulting in a consistent sleep–wake cycle.[98] The first process, the homeostatic sleep drive, increases the longer while one is awake and decreases during sleep. The molecular basis of this "sleep homeostat" model is the accumulation of adenosine and other "somnogens," metabolic by-products over the course of the waking hours.[62] The existence of such a molecule was hypothesized as early as 1913, when Pieron reported inducing sleep in dogs by the intraventricular injection of CSF from sleep-deprived dogs.[99] He explained this phenomenon by hypothesizing a "fatigue substance" or "hypnotoxin" produced by physiological exertion. Adenosine is the molecule now understood to fill that role, and is considered an endogenous somnogen,[100] that is, thought to activate sleep-promoting pathways in the VLPO and inhibit wakefulness-promoting pathways in the basal forebrain.[101] The soporific effects of adenosine are counteracted by caffeine, an antagonist of adenosine receptors.[100]

The second process is the circadian rhythm. This daily rhythm is established by a "circadian pacemaker" in the suprachiasmatic nucleus (SCN) of the anterior hypothalamus, which is entrained to light-dark periodicity by inputs from photosensitive melanopsin-containing retinal ganglion cells.[66,89,102] The SCN predominantly projects to the subparaventricular zone, which in turn projects to the dorsal medial hypothalamus (DMH). GABAergic neurons in the DMH project to the lateral hypothalamus while glutamatergic neurons project to the VLPO, thus promoting the sleep state. The multiple relays within the hypothalamus may explain how factors other than simple light–dark entrainment of the SCN can influence circadian sleep regulation, including other environmental stimuli such as food availability.[66,89]

The SCN also projects to the paraventricular hypothalamic nucleus, controlling melatonin release from the pineal gland. Melatonin may then feedback and modulate SCN activity, as evidenced by the ability of melatonin to entrain circadian rhythms in blind patients, with animal models demonstrating this effect only in the presence of an intact SCN. Other SCN inputs that can modulate the circadian rhythm include those related to locomotor activity and sleep states.[66]

The Role of Adenosine in Primary Headache Pathophysiology

Adenosine plays a role in both pain and sleep, acting primarily on A1 and A2A receptors in the CNS. The pain-related effects of adenosine vary in a receptor-dependent manner, with A1 stimulation having antinociceptive effects, and A2A stimulation resulting in pain.[103] The sleep-promoting effects of adenosine are primarily mediated via A1 receptor activation, with some contribution from A2A receptor activation.[104] Administration of adenosine can cause migraine in susceptible patients,[105] and elevated circulating levels of adenosine have been recorded during migraine attacks.[106] Adenosine also facilitates the effects of vasoactive intestinal peptide (VIP) and calcitonin-related gene peptide (CGRP), molecules known to be important in migraine pathophysiology.[107]

Adenosine exerts its somnogenic effect via inactivation of the $Ca_V2.1$ channel, which is mutated in a form of familial hemiplegic migraine, the CACNA1A sub-type.[108] In addition, a specific haplotype of the A2A receptor has been identified, that is, associated with the development of migraine with aura.[107] In addition to a genetic predisposition to migraine, CACNA1A mutant mice have altered sleep phenotypes, with more rapid adjustment of circadian rhythm to phase shifts of the light–dark cycle.[109] They are also less susceptible to the somnogenic effects of adenosine.[108]

Given the association between adenosine, sleep, and nociception, therapeutic applications of both adenosine agonists and antagonists have been investigated in primary headache disorders. Selective A1 receptor agonists reduce CGRP release in response to dural stimulation and reduce trigeminovascular activation, and are of interest in the development of novel treatments for migraine and CH.[103,110] Caffeine, widely used as a neurostimulant drug, acts as an antagonist at both A1 and A2A receptor types, and is also useful as a treatment adjunct in headaches.[111] Interestingly, caffeine is often therapeutically effective in the treatment of hypnic headache,[44] and caffeine withdrawal itself may also cause headache.[111]

The Role of Melatonin in Headache Pathophysiology

Melatonin secretion from the pineal gland is regulated by the SCN and begins at dim-light onset, with hypnotic effects.[112] Melatonin may also have analgesic effects, and disordered melatonin secretion may result in headache. Mechanisms underlying the interaction between melatonin and headache may include potentiation of GABAergic inhibition of pain pathways, modulation of 5-HT signaling, reduced production of proinflammatory cytokines, inhibition of nitric oxide synthase, antioxidant effects, and the induction of cytokines acting at opioid receptors (melatonin-induced-opioids).[37,113]

Pineal cysts are common incidental findings on imaging,[114] and are associated with an increased incidence of headache, with a migrainous phenotype predominating.[115] Disordered melatonin secretion, rather than mass effect, is thought to underlie this phenomenon because the association is unrelated to cyst size.[115] Abnormal melatonin secretion profiles have been described as a result of structural pineal disease associated with headache, with resolution of headache and normalization of melatonin secretion after surgical resection.[116]

Melatonin secretion profiles are also abnormal in primary headache disorders. However, it should be noted that this finding lacks specificity, and has also been demonstrated in conditions such as alcohol ingestion, fibromyalgia, and depression.[117–119] Initial studies by

Claustrat et al. indicated that plasma melatonin levels were lower in migraine patients when compared to controls, a difference that was more pronounced in those with comorbid depression.[120] Subsequent studies of urinary melatonin concentrations demonstrated lower levels in migraineurs during the interictal period, with more pronounced decreases during migraine attacks.[121,122] In patients with chronic migraine, nocturnal melatonin secretion was not found to be significantly different than healthy controls, however there was a significant phase delay in the peak melatonin concentration, similar to that seen in patients with delayed sleep phase syndrome. Those with chronic migraine and comorbid insomnia exhibited significantly lower melatonin levels in addition to the phase delay.[75]

Nocturnal plasma melatonin levels in CH are abnormally low in comparison to controls,[123] particularly in smokers,[124] and are lower during cluster attacks than in the interictal phase.[125] Melatonin levels in these patients are persistently low throughout the calendar year, with no clear circannual or seasonal rhythm identified.[124] Melatonin secretion declines with age, and because hypnic headache is a disease of the middle-aged to elderly population, melatonin deficiency has been proposed as a pathophysiological explanation.[37] The administration of exogenous melatonin is therefore of interest as a therapeutic option in a range of primary headache disorders.

Evidence for the use of melatonin in the treatment of migraine is limited, with conflicting reports. It is regarded as "Level U" quality evidence.[126] An open label study of melatonin (3mg) in 34 episodic migraine patients reported an improvement in headache frequency, severity, and duration with treatment.[127] In a subsequent randomized controlled trial of 179 episodic migraine patients, the response to melatonin (3mg) exceeded the response to placebo and was equivalent to amitriptyline therapy, although with better tolerability.[128] However, another randomized controlled trial of melatonin therapy (2 mg) in episodic migraine ($n = 48$) showed no improvement in attack frequency or sleep quality compared to treatment with placebo.[129] The migraine population included in this study had measures of sleep quality that were almost normal at baseline, and the possibility remains that melatonin may be helpful in selected migraineurs with comorbid sleep disorders and abnormal melatonin production.[130] It should be noted that melatonin therapy is typically extremely well tolerated[131] and that higher doses of up to 10 mg have been evaluated in nonheadache disorders.[132]

A small randomized controlled trial investigating the efficacy of melatonin in CH prophylaxis also demonstrated a benefit, with a significant reduction in the number of daily attacks.[133] Of the 10 treated patients, 5 patients (all episodic) had a clear response. The two patients in the group with chronic CH had no response. Subsequent anecdotal evidence has suggested a therapeutic effect of melatonin in chronic CH.[134] Another trial investigating the use of melatonin as an adjunctive therapy in nine patients with episodic and chronic CH showed no improvement in attack frequency or analgesic use;[135] however, the dose used (2 mg) was much lower than that used in the earlier positive trial (10 mg).

Agomelatine is an antidepressant drug acting as an agonist at the MT1 and MT2 melatonin receptor and as a selective 5-HT2C receptor antagonist.[136] Its use in the treatment of patients with migraine and comorbid depression has been reported in a case series with encouraging results.[136] Lithium, which is used in the effective treatment of CH and hypnic headaches, may exert its effect in part by increasing melatonin concentrations.[37] Melatonin has been anecdotally effective in treatment of hypnic headache.[137] Ramelteon, a selective melatonin MT1/MT2 receptor agonist with a longer half-life than melatonin, is also anecdotally effective.[138]

Hemicrania continua (HC) is defined in part by its responsiveness to indomethacin,[1] however the tolerability of long-term indomethacin therapy can be problematic. The benign side-effect profile of melatonin is very appealing in these instances. The efficacy of melatonin in HC treatment has been suggested by a number of case reports.[139–141] Tapering indomethacin therapy following the introduction of melatonin was attempted in a recent study of 11 HC patients, with two patients achieving total freedom from pain on melatonin alone, three partial responders who were able to reduce their indomethacin dose, and no response to melatonin in six cases.[142]

The Role of Orexin in Headache and Sleep

The orexin system of the hypothalamus is activated in response to stimuli such as emotion and stress via inputs from the insular cortex and amygdala, the biological clock via inputs from the dorsomedial nucleus of the hypothalamus, and sleep–wake transitions via inputs from the VLPO and dorsal raphe nucleus. Activation of the orexinergic system results in behavioral changes including increased wakefulness, feeding, and sympathetic tone.[143]

Orexinergic neurons from the posterolateral hypothalamus interact in a complex manner with the TNC, and may facilitate or inhibit TNC nociception by receptor-specific pathways.[144,145] Manipulation of orexinergic signaling has been evaluated for its therapeutic implications in migraine as well as in sleep disorders.[144,146] Orexin signaling is also involved in appetite regulation, reward pathways, and even the pathology of anxiety and depression.[146] Orexinergic pathways are known to promote feeding behaviors, and dysregulation of this system may contribute to the increased prevalence of obesity in migraineurs.[147,148]

Evidence from animal migraine models suggests that systemic treatment with a dual antagonist of orexin receptors 1 and 2 acts both peripherally and centrally to reduce neurogenic dural vasodilation, cortical spreading depression, and TNC activation.[144] This suggests a primarily facilitatory role of orexin in TNC nociception. Orexin levels are increased in the initial stages of a migraine attack, and CSF orexin concentrations are higher in patients with chronic migraine or medication overuse headache compared to controls.[149] Although a recent randomized controlled trial of a dual orexin antagonist as a migraine prophylactic agent was negative,[150] future avenues for clinical trials may include the study of these agents in CH.[151]

There is an increased incidence of migraine in narcolepsy patients, a population known to be deficient in orexinergic neurons of the hypothalamus.[152,153] However, this relationship remains controversial. A case-control study failed to demonstrate any relationship between migraine and narcolepsy, instead demonstrating a higher prevalence of "unspecific" headaches in narcolepsy patients, predominantly tension-type.[154]

Cortical Spreading Depression may Link Headache and Sleep

Cortical spreading depression (CSD) is a spreading wave of neuronal and glial depolarization that is thought to correlate with the aura preceding migraine headache in a subset of patients.[155] It results in neurogenic inflammation,[156] as well as activation of meningeal nociceptors[157] and the TNC.[156] CSD also affects the physiology of subsequent sleep, with pronounced changes persisting for hours. As the percentage of non-REM sleep increases, REM sleep decreases, and slow-wave activity is increased ipsilateral to the side of CSD stimulus.[158]

These changes may be related to CSD-induced alteration of ionic conductance, or to CSD-induced disruption of cortical cholinergic transmission.[158] CSD completely silences wakefulness-promoting basal forebrain cholinergic projections[159] and may also promote increased non-REM sleep via induction of COX-2 activity and increased prostaglandin production.[160]

A genetic model of the relationship between sleep disorders and migraine has been identified in one family with familial hemiplegic migraine (FHM) and familial advanced sleep phase syndrome (FASPS). A missense mutation was identified in the gene encoding CK1 delta, a kinase known to be involved in circadian rhythm control. Knock-out mice for this mutation exhibited a markedly lower threshold for CSD initiation, and greater dilation of pial arterioles during CSD,[161] thus acting similarly to transgenic models of familial hemiplegic migraine.[162] Nitroglycerin-induced migraine models in these mice also demonstrate an increased propensity to nociception and TNC activation.[161]

Given this information, screening for circadian rhythm sleep disorders (CRSD) has been proposed in migraine patients.[163] CRSD are conditions in which there is a failure of synchronicity between the external day/night and the internal biological clock, and include delayed sleep phase syndrome, "jet lag," and shift worker syndrome.[51] Timing of the circadian system can be evaluated with salivary measurements of the onset of melatonin secretion in the evening, which may guide the timing of exogenous melatonin therapy in treating comorbid headache and CRSD.[163] Stronger evidence in the form of a randomized-controlled trial is needed to adequately assess the clinical utility of routine screening and treatment of CRSD in the migraine population.

HEADACHE ITSELF CAN RESULT IN SLEEP DISTURBANCES

Sleep disturbance is increasingly prevalent with increasing headache frequency,[53] and can be predicted by other headache-related factors including headache severity and headache-related disability.[164] Pain makes restful sleep very difficult because it is associated with a state of stress and heightened alertness.[165] However, painful stimuli alone do not explain the sleep disturbance reported by patients with pain disorders, as painful stimuli during sleep in healthy subjects result in only brief cortical arousals and postural adjustments without significant sleep disruption or recollection of disrupted sleep the following day.[166] Chronic pain, however, has been associated with increased activity of systems modulating ascending activation as well as nociception, such as the raphe magnus.[165,166] Pain has been shown experimentally to alter sleep microarchitecture, with an overall arousal effect of decreased delta activity and increased high frequency activity.[167]

Behavioral factors may also play a role and chronic pain may be considered to condition a patient toward insomnia by maladaptive strategies such as increased daytime napping and spending increased amounts of time in the bedroom while alert and in pain.[165,168] A bidirectional relationship between sleep and pain has been illustrated in female fibromyalgia patients, with a poor night's sleep predictive of a bad pain day, and vice versa.[169] It is difficult to establish direction or causality in the relationship between sleep and headache, and both processes are very likely to feed into one another.[170] The relationship between pain and sleep is complex, and while increasing severity of pain correlates with worsening sleep, the association is even stronger for cognitive factors such as low mood and rumination.[165]

Psychiatric Comorbidities are Common in Headache Patients, and may Further Disturb Sleep

Depression and anxiety disorders are highly prevalent in headache patient populations, and can further disrupt sleep. However, the association of sleep disturbance in migraineurs persists despite correction for psychiatric comorbidities, and cannot be entirely explained by psychological factors.[171]

A recent cross-sectional population-based Danish study[172] investigated the epidemiology of headache and sleep disorders. The percentage of individuals reporting both complaints was 18.1%. The presence of severe comorbid headache and sleep disorders was strongly associated with stress, depression, and anxiety, as well as significantly impaired quality of life. Engstrom et al. demonstrated that increasing levels of anxiety were associated with more superficial sleep in both migraineurs and healthy controls. However, migraineurs appeared to be more vulnerable to this effect, with reduced slow wave sleep and higher reported insomnia associated with anxiety.[12]

SLEEP DISORDERS CAN CAUSE OR EXACERBATE HEADACHE

Typically healthy people who do not experience regular headaches will develop sleep deprivation headaches when provoked; these tend to be dull, frontal, and aching in nature.[173] Paiva et al. found that patients presenting with early morning headache were more likely to have an underlying sleep disorder, and that treatment of the latter resulted in improvement in headache.[174] Patients with insomnia have a 3.2-fold increased risk of reporting migraine, and a 2.3-fold increased risk of reporting tension type headache (TTH), after adjusting for anxiety and depression.[52] The presence of sleep disorders other than sleep disordered breathing, such as insomnia, was associated with an increased incidence of new-onset migraine in a large longitudinal population-based study.[175]

Sleep disturbances are also strongly suspected to be a risk factor for the development of chronic daily headache (CDH), although a causal relationship has not been demonstrated.[56] Patients with CDH are more likely to report sleep complaints than those with episodic headache or the general population.[57]

A correlation between sleep disturbance and pain is not only seen in headache, but has also been reported in many diverse chronic pain conditions.[165] Evidence suggesting that sleep disturbance contributes to pain initially came from animal studies demonstrating that sleep deprivation resulted in a reduced pain threshold.[176] In humans, experimentally interfering with slow wave sleep results in increased musculoskeletal tenderness[177] reminiscent of fibromyalgia, as well as a reduced pain threshold.[178]

Impaired sleep may result in disordered pain processing.[170] Specifically, disrupted REM sleep has been shown to impair the antinociceptive effect of the endogenous opioid system in animal models.[179] Reduced serotoninergic signaling after sleep deprivation may also contribute to a state of hyperalgesia.[170] Lack of sleep has been reported as a precipitating factor for both migraine and TTH, and has been reported to worsen existing TTH.[54,55] Patients with migraine and to a lesser extent TTH are significantly more likely to complain of severe sleep disturbance than headache free patients, with just a marginal reduction in the strength of association after controlling for depression and anxiety.[53] Two consecutive nights of self-reported poor sleep in chronic migraine patients is highly predictive of increased headache activity,[180] and reduced sleep duration is associated with increased headache severity. This effect appears to be synergistic with that of high stress levels,[180] and may suggest a protective effect of a regular sleep schedule during periods of high stress.[181]

The Relationship Between Sleep Apnea and Headache

The prevalence of OSA has dramatically increased over the last two to three decades, in tandem with the rising prevalence of obesity.[182] Recent estimates of prevalence suggest that polysomnographic evidence of a significant number of apnea/hypnoea episodes is found in 33.9% of males and 17.4% of females in the 30–70 age group; with evidence of moderate to severe disease in 13 and 5.6%, respectively.[182]

Sleep apnea and sleep-disordered breathing may interact with headache disorders in a number of ways.

According to ICHD-3 beta criteria, the diagnosis of sleep apnea-related headache requires the presence of headache on awakening, in association with a diagnosis of sleep apnea. Evidence of causality is required, either by the headache developing soon after the onset of OSA, or by the headaches improving or worsening in parallel with OSA.[1] The nature of the headache is described as typically bilateral, lasting less than 4 h after waking, and lacking in migrainous features.[1] However, this clearly defined syndrome does not describe the various other ways in which sleep apnea may modulate other primary headache disorders.

Could Sleep Apnea Cause or Exacerbate Primary Headache Disorders?

The relationship between OSA and primary headache disorders was examined in a large cross-sectional population-based Norwegian study. This data did not support a relationship between migraine and OSA in the general population,[183] and furthermore did not demonstrate any dose-response relationship between severe OSA and migraine severity. Similarly, no relationship between TTH and OSA was identified.[184] This is consistent with previous studies reporting the prevalence of OSA in a headache population to be similar to that of the general population.[185] Therefore, while the evidence may not support an increased prevalence of primary headache disorders in OSA patients, the question of whether the frequency and severity of headaches may be influenced by the presence of OSA remains to be answered.

Refractory CDH may be associated with OSA. In one study, patients attending a headache clinic were referred for polysomnography if they were experiencing CDH that was refractory to standard treatment, and in addition reported one of the following: headaches predominantly occurring during sleep or in the morning, snoring, or daytime somnolence.[58] Of the 72 patients studied, 29.2% received a diagnosis of OSA (13.7% of women and 66.6% of men).[58] Patients without OSA who simply snore habitually also appear to be at increased risk of CDH.[59]

Milder Forms of Sleep-disordered Breathing may Also be Associated with Headache

Morning headache may be a relatively nonspecific finding for sleep apnea.[186,187] In one study of patients undergoing PSG for snoring, the prevalence and characteristics of headache (including morning headache and an ICHD diagnosis of migraine) did not differ significantly between the group ultimately diagnosed as having sleep apnea and those who were just "snorers". Of note, headache was strongly correlated with measures of depression in both groups. Despite this lack of specificity, headaches in the group with sleep apnea did respond to treatment with continuous positive airways pressure (CPAP).[186] This is at odds with the recent Norwegian epidemiological data, which demonstrates that the presence of morning headache is predictive of OSA in the general population, and was present in 11.8% of PSG-confirmed OSA cases.[188] The relationship persisted after correction for potential confounders, including depression and BMI.

The Pathophysiology of Headaches Related to Sleep-disordered Breathing is Unclear

Episodic hypoxia is one hypothesis that has been proposed to explain OSA-related headache.[14] Experimental induction of arterial oxygen desaturations to 75–80% have been demonstrated to trigger migraine in migraineurs,[189] and headache is highly prevalent in those ascending to high altitudes.[190] However, the duration or severity of oxygen desaturation in OSA patients studied by Kristiansen et al. was not predictive of the presence of OSA-related headache.[188] Similarly, a study by Sand et al. did not identify a dose-response relationship between the severity of apnea, oxygen desaturation, and headache.[187]

Alternatively, hypercapnia secondary to respiratory insufficiency may result in altered pH, vasodilation and trigeminal activation.[14] "Turtle headaches" are described as bilateral, generalized headaches occurring in a patient who wakes up, pulls the bedclothes over their head, and goes back to sleep again.[191] The proposed mechanism in this instance is the induction of mild hypoxia and hypercapnia.[192,193]

CPAP Therapy may Improve Headache Associated with Sleep-disordered Breathing

In one study of 11 patients with comorbid sleep apnea and migraine, treatment with CPAP resulted in a marked improvement in migraine frequency and a significant reduction in medication use.[194] Similarly, nonspecific headache in a sleep apnea population is more likely to improve in those who are compliant with CPAP therapy.[195] However, Mitsikostas et al. reported an improvement in headache frequency with CPAP therapy in just 23.8% of patients with comorbid OSA and CDH, with a worsening of headache frequency in 66.6%.[58] Of note, this particular population was one of refractory CDH, with patients having already failed to respond to at least two different prophylactic pharmacotherapies.

Cluster Headache is Strongly Associated with OSA

OSA has been reported in up to 80% of the CH population.[35,36] The association is strengthened by evidence demonstrating that in the cluster-free period CH patients have significantly fewer apneic and desaturation episodes, and the diagnosis of OSA is less likely to be made.[196] Interestingly, both central and obstructive apneas have been observed during cluster attacks, even though central apneas were not present outside the cluster period.[196]

The mechanism underlying this relationship is poorly understood. The two processes may be associated with a shared neurobiological substrate in the hypothalamus. Alternatively, a causal relationship may exist, possibly secondary to hypoxia.[35] This latter theory is suggested by the therapeutic effect of oxygen administration in the abortion of a cluster headache attack. However, central hypothalamic dysregulation of both processes appears more likely in light of the recently reported central apnea occurring in association with cluster attacks.[196] Evidence that CPAP therapy improves CH is limited and conflicting, with a lack of prospective studies.[36,58,196,197]

CONCLUSIONS

Sleep and headache disorders are intricately related, both in clinical presentation and physiology. Shared anatomical pathways may help to explain this relationship, and recent functional imaging studies suggest that the hypothalamus, long suspected on clinical grounds, may be involved in both processes. The roles of shared neurotransmitters such as serotonin, adenosine, melatonin, and orexin lend further support to this relationship.

These physiological intersections not only provide insights into the underlying headache pathology but also suggest novel therapeutic approaches. Of particular interest for future research is the potential role of melatonin in the treatment of selected headache patients. Given its benign side-effect profile, even modestly positive effects in this medication-sensitive population would be welcomed. Future research may focus on the therapeutic effect of melatonin in migraine patients stratified by sleep quality indices. Screening selected patients for circadian rhythm sleep disorders (as proposed by Rovers et al.),[163] to further personalize melatonin therapy is also appealing; however, further evidence is needed before widespread adoption of this approach.

Sleep and headache disorders cause or exacerbate each other in a complex, bidirectional manner, perhaps as a reflection of their shared neurobiological substrate. The common comorbidities of depression and anxiety further modify and complicate this relationship, and are associated with lower quality of life indices.[172] The importance of these relationships may be overlooked or underestimated in clinical practice, resulting in potentially avoidable distress.

Rains et al. propose general guidelines for approaching comorbid sleep and headache disorders.[198] These include taking a careful history to clarify the way in which headache and sleep interact in an individual. Standardized sleep questionnaires have also been suggested as a screening tool in all headache patients.[14] Sleep apnea should be suspected and screened for, not only in those with morning headaches or with signs and symptoms of OSA but also in those with specific headache diagnoses that may be associated with higher risk of comorbid OSA (i.e., CH, hypnic headache, chronic migraine, and chronic TTH). Screening for depression and anxiety is also recommended.[198]

The presence of comorbid sleep disorders in a headache patient, with or without coexisting OSA, depression, or anxiety, will influence the choice of pharmacotherapy in that individual. Bidirectional treatment with a single agent targeting both depression and headache may be helpful in some patients.[198]

Impressive results in reducing headache frequency have been reported with behavioral sleep modifications in a migraine population.[4] The behavioral intervention used in this study consisted of five components: scheduling a consistent bedtime allowing for 8 h in bed; elimination of TV, reading, or music in bed; the use of a visualization technique to shorten time to sleep onset; moving the evening meal ≥4 h before bedtime and limiting fluid intake within 2 h of bedtime; and the discontinuation of naps.[4] At a minimum, patients should be counselled on pragmatic lifestyle modifications including adherence to a consistent sleep/wake schedule, avoiding "screen time" before going to sleep, and reducing caffeine and alcohol intake.[199] There is no evidence at present to support a reduction in headache frequency in insomniac migraineurs treated with hypnotic agents.[200]

The intersection between sleep and headache is clinically complex, remains incompletely understood, and provides many pathophysiological insights and novel therapeutic targets. In addition, it affords the conscientious physician an opportunity to consider his or her patient in a holistic light, bearing in mind the high prevalence of comorbid headache, sleep and psychiatric disorders, and how these factors may influence one another.

References

1. Headache Classification Committee of the International Headache Society. The International Classification of Headache Disorders, 3rd edition (beta version). *Cephalalgia*. 2013;33(9):629–808.
2. Liveing E. *On Megrim, Sick-Headache and Some Allied Disorders: A Contribution to the Pathology of Nerve-storms*. London: J. and A. Churchill. 1873.
3. Bruni O, Galli F, Guidetti V. Sleep hygiene and migraine in children and adolescents. *Cephalalgia*. 1999;19(suppl 25):57–59.
4. Calhoun AH, Ford S. Behavioral sleep modification may revert transformed migraine to episodic migraine. *Headache V* 47. 2007;(8):1178–1183.
5. Blau JN. Resolution of migraine attacks: sleep and the recovery phase. *J Neurol Neurosurg Psychiatry*. 1982;45(3):223–226.
6. Robbins L. Precipitating factors in migraine: a retrospective review of 494 patients. *Headache*. 1994;34(4):214–216.
7. Peres M, Stiles M, Siow H, Dogramji K, Silberstein S, Cipolla-Neto J. Chronobiological features in episodic and chronic migraine. *Cephalalgia*. 2003;23(7):590–591.
8. Bigal ME, Lipton RB. Modifiable risk factors for migraine progression. *Headache*. 2006;46(9):1334–1343.
9. Kelman L, Rains JC. Headache and sleep: examination of sleep patterns and complaints in a large clinical sample of migraineurs. *Headache*. 2005;45(7):904–910.
10. Calhoun AH, Ford S, Finkel AG, Kahn KA, Mann JD. The prevalence and spectrum of sleep problems in women with transformed migraine. *Headache*. 2006;46(4):604–610.
11. Barbanti P, Fabbrini G, Aurilia C, Vanacore N, Cruccu G. A case-control study on excessive daytime sleepiness in episodic migraine. *Cephalalgia*. 2007;27(10):1115–1119.
12. Engstrom M, Hagen K, Bjork MH, et al. Sleep quality, arousal and pain thresholds in migraineurs: a blinded controlled polysomnographic study. *J Headache Pain*. 2013;14:12.
13. Fox AW, Davis RL. Migraine chronobiology. *Headache*. 1998;38(6):436–441.
14. Brennan KC, Charles A. Sleep and headache. *Semin Neurol*. 2009;29(4):406–418.
15. Guidetti V, Dosi C, Bruni O. The relationship between sleep and headache in children: implications for treatment. *Cephalalgia*. 2014;34(10):767–776.
16. Seidel S, Bock A, Schlegel W, et al. Increased RLS prevalence in children and adolescents with migraine: a case-control study. *Cephalalgia*. 2012;32(9):693–699.
17. d'Onofrio F, Bussone G, Cologno D, et al. Restless legs syndrome and primary headaches: a clinical study. *Neurol Sci*. 2008;29(suppl 1):S169–S172.
18. Lucchesi C, Bonanni E, Maestri M, Siciliano G, Murri L, Gori S. Evidence of increased restless legs syndrome occurrence in chronic and highly disabling migraine. *Funct Neurol*. 2012;27(2):91–94.
19. Chen PK, Fuh JL, Wang SJ. Bidirectional triggering association between migraine and restless legs syndrome: a diary study. *Cephalalgia*. 2016;36(5):431–436.

20. Mitsikostas DD, Viskos A, Papadopoulos D. Sleep and headache: the clinical relationship. *Headache.* 2010;50(7):1233–1245.

21. Dexter JD, Weitzman ED. The relationship of nocturnal headaches to sleep stage patterns. *Neurology.* 1970;20(5):513–518.

22. Goder R, Fritzer G, Kapsokalyvas A, et al. Polysomnographic findings in nights preceding a migraine attack. *Cephalalgia.* 2001;21(1):31–37.

23. Dexter JD. The relationship between Stage III + IV + REM sleep and arousals with migraine. *Headache.* 1979;19(7):364–369.

24. Della Marca G, Vollono C, Rubino M, Di Trapani G, Mariotti P, Tonali PA. Dysfunction of arousal systems in sleep-related migraine without aura. *Cephalalgia.* 2006;26(7):857–864.

25. Nayak C, Sinha S, Nagappa M, et al. Study of sleep microstructure in patients of migraine without aura. *Sleep Breath.* 2015;20(1):263–269.

26. Qiu E, Wang Y, Ma L, et al. Abnormal brain functional connectivity of the hypothalamus in cluster headaches. *PloS One.* 2013;8(2):e57896.

27. May A, Bahra A, Buchel C, Frackowiak RS, Goadsby PJ. PET and MRA findings in cluster headache and MRA in experimental pain. *Neurology.* 2000;55(9):1328–1335.

28. Barloese MCJ. Neurobiology and sleep disorders in cluster headache. *J Headache Pain.* 2015;16:78.

29. Pringsheim T. Cluster headache: evidence for a disorder of circadian rhythm and hypothalamic function. *Canadian J Neurol Sci.* 2002;29(1):33–40.

30. Nobre ME, Leal AJ, Filho PMF. Investigation into sleep disturbance of patients suffering from cluster headache. *Cephalalgia.* 2005;25(7):488–492.

31. Kudrow L, McGinty DJ, Phillips ER, Stevenson M. Sleep apnea in cluster headache. *Cephalalgia.* 1984;4(1):33–38.

32. Della Marca G, Vollono C, Rubino M, Capuano A, Di Trapani G, Mariotti P. A sleep study in cluster headache. *Cephalalgia.* 2006;26(3):290–294.

33. Terzaghi M, Ghiotto N, Sances G, Rustioni V, Nappi G, Manni R. Episodic cluster headache: NREM prevalence of nocturnal attacks. Time to look beyond macrostructural analysis? *Headache.* 2010;50(6):1050–1054.

34. Zaremba S, Holle D, Wessendorf TE, Diener HC, Katsarava Z, Obermann M. Cluster headache shows no association with rapid eye movement sleep. *Cephalalgia.* 2012;32(4):289–296.

35. Graff-Radford SB, Newman A. Obstructive sleep apnea and cluster headache. *Headache.* 2004;44(6):607–610.

36. Chervin RD, Zallek SN, Lin X, Hall JM, Sharma N, Hedger KM. Sleep disordered breathing in patients with cluster headache. *Neurology.* 2000;54(12):2302–2306.

37. Dodick DW, Eross EJ, Parish JM, Silber M. Clinical, anatomical, and physiologic relationship between sleep and headache. *Headache.* 2003;43(3):282–292.

38. Ruiz M, Mulero P, Pedraza MI, et al. From wakefulness to sleep: migraine and hypnic headache association in a series of 23 patients. *Headache.* 2015;55(1):167–173.

39. Manni R, Ghiotto N, Hypnic headache. Michael J, Aminoff FB, Dick FS, eds. *Handbook of Clinical Neurology,* vol. 97. The Netherlands: Elsevier; 2010:469–472:Chapter 41.

40. Holle D, Naegel S, Krebs S, et al. Hypothalamic gray matter volume loss in hypnic headache. *Ann Neurol.* 2011;69(3):533–539.

41. Liang J-F, Wang S-J. Hypnic headache: a review of clinical features, therapeutic options and outcomes. *Cephalalgia.* 2014;34(10):795–805.

42. Lisotto C, Mainardi F, Maggioni F, Zanchin G. Episodic hypnic headache? *Cephalalgia.* 2004;24(8):681–685.

43. Silva-Néto RP, Almeida KJ. Hypnic headache: a descriptive study of 25 new cases in Brazil. *J Neurol Sci.* 2014;338(1–2):166–168.

44. Holle D, Naegel S, Obermann M. Pathophysiology of hypnic headache. *Cephalalgia.* 2014;34(10):806–812.

45. Ghiotto N, Sances G, Di Lorenzo G, et al. Report of eight new cases of hypnic headache and mini-review of the literature. *Funct Neurol.* 2002;17(4):211–219.

46. Dodick DW, Mosek AC, Campbell JK. The hypnic ("alarm clock") headache syndrome. *Cephalalgia.* 1998;18(3):152–156.

47. Edwards BA, O'Driscoll DM, Ali A, Jordan AS, Trinder J, Malhotra A. Aging and sleep: physiology and pathophysiology. *Sem Respi Crit Care.* 2010;31(5):618–633.

48. Holle D, Naegel S, Obermann M. Hypnic headache. *Cephalalgia.* 2013;33(16):1349–1357.

49. Holle D, Wessendorf TE, Zaremba S, et al. Serial polysomnography in hypnic headache. *Cephalalgia.* 2011;31(3):286–290.

50. Draganski B, Geisler P, Hajak G, et al. Hypothalamic gray matter changes in narcoleptic patients. *Nat Med.* 2002;8(11):1186–1188.

51. American Academy of Sleep Medicine. *International Classification of Sleep Disorders.* 3rd ed. Darien, IL, Chicago: American Academy of Sleep Medicine; 2014.

52. Yeung W-F, Chung K-F, Wong C-Y. Relationship between insomnia and headache in community-based middle-aged Hong Kong Chinese women. *J Headache Pain.* 2010;11(3):187–195.

53. Odegard SS, Engstrom M, Sand T, Stovner LJ, Zwart JA, Hagen K. Associations between sleep disturbance and primary headaches: the third Nord-Trondelag Health Study. *J Headache Pain.* 2010;11(3):197–206.

54. Spierings ELH, Ranke AH, Honkoop PC. Precipitating and aggravating factors of migraine versus tension-type headache. *Headache.* 2001;41(6):554–558.

55. Langemark M, Olesen J, Poulsen DL, Bech P. Clinical characterization of patients with chronic tension headache. *Headache.* 1988;28(9):590–596.

56. Cho SJ, Chu MK. Risk factors of chronic daily headache or chronic migraine. *Curr Pain Headache Rep.* 2015;19(1):465.

57. Rueda-Sanchez M, Diaz-Martinez LA. Prevalence and associated factors for episodic and chronic daily headache in the Colombian population. *Cephalalgia.* 2008;28(3):216–225.

58. Mitsikostas DD, Vikelis M, Viskos A. Refractory chronic headache associated with obstructive sleep apnoea syndrome. *Cephalalgia.* 2008;28(2):139–143.

59. Scher AI, Lipton RB, Stewart WF. Habitual snoring as a risk factor for chronic daily headache. *Neurology.* 2003;60(8):1366–1368.

60. Evers S. Sleep and headache: the biological basis. *Headache.* 2010;50(7):1246–1251.

61. Holland PR. Headache and sleep: shared pathophysiological mechanisms. *Cephalalgia.* 2014;34(10):725–744.

62. Rosenwasser AM. Functional neuroanatomy of sleep and circadian rhythms. *Brain Res Rev.* 2009;61(2):281–306.

63. Economo C. Sleep as a problem of localization. *J Nerv Ment Dis.* 1930;71(3):249–259.

64. Saper CB, Scammell TE, Lu J. Hypothalamic regulation of sleep and circadian rhythms. *Nature.* 2005;437(7063):1257–1263.

65. Brown RE, Basheer R, McKenna JT, Strecker RE, McCarley RW. Control of sleep and wakefulness. *Physiol Rev.* 2012;92(3):1087–1187.

66. Fuller PM, Gooley JJ, Saper CB. Neurobiology of the sleep-wake cycle: sleep architecture, circadian regulation, and regulatory feedback. *J Biol Rhythms.* 2006;21(6):482–493.

67. Maniyar FH, Sprenger T, Monteith T, Schankin C, Goadsby PJ. Brain activations in the premonitory phase of nitroglycerin-triggered migraine attacks. *Brain.* 2014;137(Pt 1):232–241.

68. Lin J, Rodrigues Masruha M, Prieto Peres MF, et al. Nocturnal enuresis antecedent is common in adolescents with migraine. *Eur Neurol.* 2012;67(6):354–359.

69. Tfelt-Hansen P, Paulson OB, Krabbe AA. Invasive adenoma of the pituitary gland and chronic migrainous neuralgia. A rare coincidence or a causal relationship? *Cephalalgia.* 1982;2(1):25–28.

70. Facchinetti F, Nappi G, Cicoli C, et al. Reduced testosterone levels in cluster headache: a stress-related phenomenon? *Cephalalgia.* 1986;6(1):29–34.

71. Kudrow L. Plasma testosterone levels in cluster headache preliminary results. *Headache.* 1976;16(1):28–31.

72. Leone M, Bussone G. A review of hormonal findings in cluster headache. Evidence for hypothalamic involvement. *Cephalalgia.* 1993;13(5):309–317.

73. Leone M, Zappacosta BM, Valentini S, Colangelo AM, Bussone G. The insulin tolerance test and the ovine corticotrophin-releasing hormone test in episodic cluster headache. *Cephalalgia.* 1991;11(6):269–274.

74. Leone M, Maltempo C, Gritti A, Bussone G. The insulin tolerance test and ovine corticotrophin-releasing-hormone test in episodic cluster headache. II: Comparison with low back pain patients. *Cephalalgia.* 1994;14(5):357–364.

75. Peres MF, Sanchez del Rio M, Seabra ML, et al. Hypothalamic involvement in chronic migraine. *J Neurol Neurosurg Psychiatry.* 2001;71(6):747–751.

76. May A, Bahra A, Büchel C, Frackowiak RSJ, Goadsby PJ. Hypothalamic activation in cluster headache attacks. *Lancet.* 1998;352(9124):275–278.

77. Sprenger T, Boecker H, Tolle TR, Bussone G, May A, Leone M. Specific hypothalamic activation during a spontaneous cluster headache attack. *Neurology.* 2004;62(3):516–517.

78. May A, Ashburner J, Buchel C, et al. Correlation between structural and functional changes in brain in an idiopathic headache syndrome. *Nat Med.* 1999;5(7):836–838.

79. Wang SJ, Lirng JF, Fuh JL, Chen JJ. Reduction in hypothalamic (1)H-MRS metabolite ratios in patients with cluster headache. *J Neurol Neurosurg Psychiatry.* 2006;77(5):622–625.

80. Morelli N, Pesaresi I, Cafforio G, et al. Functional magnetic resonance imaging in episodic cluster headache. *J Headache Pain.* 2009;10(1):11–14.

81. Denuelle M, Fabre N, Payoux P, Chollet F, Geraud G. Hypothalamic activation in spontaneous migraine attacks. *Headache.* 2007;47(10):1418–1426.

82. Moulton EA, Becerra L, Johnson A, Burstein R, Borsook D. Altered hypothalamic functional connectivity with autonomic circuits and the locus coeruleus in migraine. *PloS One.* 2014;9(4):e95508.

83. Matharu MS, Cohen AS, McGonigle DJ, Ward N, Frackowiak RS, Goadsby PJ. Posterior hypothalamic and brainstem activation in hemicrania continua. *Headache.* 2004;44(8):747–761.

84. May A, Bahra A, Büchel C, Turner R, Goadsby PJ. Functional magnetic resonance imaging in spontaneous attacks of SUNCT: short-lasting neuralgiform headache with conjunctival injection and tearing. *Ann Neurol.* 1999;46(5):791–794.

85. Matharu MS, Cohen AS, Frackowiak RSJ, Goadsby PJ. Posterior hypothalamic activation in paroxysmal hemicrania. *Ann Neurol.* 2006;59(3):535–545.

86. Leone M, Franzini A, Broggi G, May A, Bussone G. Long-term follow-up of bilateral hypothalamic stimulation for intractable cluster headache. *Brain.* 2004;127(10):2259–2264.

87. Vetrugno R, Pierangeli G, Leone M, et al. Effect on sleep of posterior hypothalamus stimulation in cluster headache. *Headache.* 2007;47(7):1085–1090.

88. Kovac S, Wright M-A, Eriksson SH, Zrinzo L, Matharu M, Walker MC. The effect of posterior hypothalamus region deep brain stimulation on sleep. *Cephalalgia.* 2014;34(3):219–223.

89. Saper CB, Cano G, Scammell TE. Homeostatic, circadian, and emotional regulation of sleep. *J Comp Neurol.* 2005;493(1):92–98.

90. Frazer AHJ. Serotonin. In: Siegel GJAB, Albers RW, eds. *Basic Neurochemistry: Molecular, Cellular and Medical Aspects.* Philadelphia: Lippincott-Raven; 1999.

91. Wang QP, Nakai Y. The dorsal raphe: an important nucleus in pain modulation. *Brain Res Bull.* 1994;34(6):575–585.

92. Ferrari MD, Saxena PR. On serotonin and migraine: a clinical and pharmacological review. *Cephalalgia.* 1993;13(3):151–165.

93. Silberstein SD. Serotonin (5-HT) and migraine. *Headache.* 1994;34(7):408–417.

94. Kovács K, Bors L, Tóthfalusi L, et al. Cerebrospinal fluid (CSF) investigations in migraine. *Cephalalgia.* 1989;9(1):53–57.

95. Chugani DC, Niimura K, Chaturvedi S, et al. Increased brain serotonin synthesis in migraine. *Neurology.* 1999;53(7):1473–1479.

96. Gervasoni D, Peyron C, Rampon C, et al. Role and origin of the GABAergic innervation of dorsal raphe serotonergic neurons. *J Neurosci.* 2000;20(11):4217–4225.

97. Lu J, Greco MA, Shiromani P, Saper CB. Effect of lesions of the ventrolateral preoptic nucleus on NREM and REM sleep. *J Neurosci.* 2000;20(10):3830–3842.

98. Borbely AA, Achermann P. Sleep homeostasis and models of sleep regulation. *J Biol Rhythms.* 1999;14(6):557–568.

99. Pappenheimer J, Miller T, Goodrich C. Sleep-promoting effects of cerebrospinal fluid from sleep-deprived goats. *Proc Natl Acad Sci USA.* 1967;58(2):513.

100. Huang Z-L, Zhang Z, Qu W-M, Chapter fourteen: roles of adenosine and its receptors in sleep–wake regulation. Akihisa M, ed. *International Review of Neurobiology,* 119. Academic Press; 2014:349–371.

101. Pierre-Herve Luppi PF. Neuroanatomy and physiology of sleep and wakefulness. In: Eric Nofzinger PM, Michael J, Thorpy, eds. *Neuroimaging of Sleep and Sleep Disorders.* Cambridge: Cambridge University Press; 2013.

102. Gooley JJ, Lu J, Fischer D, Saper CB. A broad role for melanopsin in nonvisual photoreception. *J Neurosci.* 2003;23(18):7093–7106.

103. Giffin NJ, Kowacs F, Libri V, Williams P, Goadsby PJ, Kaube H. Effect of the adenosine A1 receptor agonist GR79236 on trigeminal nociception with blink reflex recordings in healthy human subjects. *Cephalalgia.* 2003;23(4):287–292.

104. Landolt H-P. Sleep homeostasis: a role for adenosine in humans? *Biochemi Pharmacol.* 2008;75(11):2070–2079.

105. Brown S, Waterer GW. Migraine precipitated by adenosine. *Med J Aust.* 1995;162(7):389, 391.

106. Guieu R, Devaux C, Henry H, et al. Adenosine and migraine. *Can J Neurol Sci.* 1998;25(1):55–58.

107. Hohoff C, Marziniak M, Lesch K-P, Deckert J, Sommer C, Mössner R. An adenosine A2A receptor gene haplotype is associated with migraine with aura. *Cephalalgia*. 2007;27(2):177–181.

108. Deboer T, van Diepen HC, Ferrari MD, Van den Maagdenberg AM, Meijer JH. Reduced sleep and low adenosinergic sensitivity in cacna1a R192Q mutant mice. *Sleep*. 2013;36(1):127–136.

109. van Oosterhout F, Michel S, Deboer T, et al. Enhanced circadian phase resetting in R192Q Cav2.1 calcium channel migraine mice. *Ann Neurol*. 2008;64(3):315–324.

110. Goadsby PJ, Hoskin KL, Storer RJ, Edvinsson L, Connor HE. Adenosine A1 receptor agonists inhibit trigeminovascular nociceptive transmission. *Brain*. 2002;125(6):1392–1401.

111. Shapiro R. Caffeine and headaches. *Curr Pain Headache Rep*. 2008;12(4):311–315.

112. Brzezinski A. Melatonin in humans. *N Engl J Med*. 1997;336(3):186–195.

113. Peres M. Melatonin, the pineal gland and their implications for headache disorders. *Cephalalgia*. 2005;25(6):403–411.

114. Sawamura Y, Ikeda J, Ozawa M, Minoshima Y, Saito H, Abe H. Magnetic resonance images reveal a high incidence of asymptomatic pineal cysts in young women. *Neurosurgery*. 1995;37(1):11–15.

115. Seifert CL, Woeller A, Valet M, et al. Headaches and pineal cyst: a case-control study. *Headache*. 2008;48(3):448–452.

116. Meyer S, Oberkircher N, Boing A, Larsen A, Eymann R, Kutschke G. Disturbance in melatonin metabolism as a causative factor for recurrent headaches in a girl with a pineal cyst? *Acta Paediatr*. 2013;102(2):e51–e52.

117. Germain A, Kupfer DJ. Circadian rhythm disturbances in depression. *Hum Psychopharmacol*. 2008;23(7):571–585.

118. Ekman AC, Leppaluoto J, Huttunen P, Aranko K, Vakkuri O. Ethanol inhibits melatonin secretion in healthy volunteers in a dose-dependent randomized double blind cross-over study. *J Clin Endocrinol Metab*. 1993;77(3):780–783.

119. Wikner J, Hirsch U, Wetterberg L, Rojdmark S. Fibromyalgia–a syndrome associated with decreased nocturnal melatonin secretion. *Clin Endocrinol*. 1998;49(2):179–183.

120. Claustrat B, Loisy C, Brun J, Beorchia S, Arnaud JL, Chazot G. Nocturnal plasma melatonin levels in migraine: a preliminary report. *Headache*. 1989;29(4):242–245.

121. Murialdo G, Fonzi S, Costelli P, et al. Urinary melatonin excretion throughout the ovarian cycle in menstrually related migraine. *Cephalalgia*. 1994;14(3):205–209.

122. Brun J, Claustrat B, Saddier P, Chazot G. Nocturnal melatonin excretion is decreased in patients with migraine without aura attacks associated with menses. *Cephalalgia*. 1995;15(2):136–139.

123. Chazot G, Claustrat B, Brun J, Jordan D, Sassolas G, Schott B. A chronobiological study of melatonin, cortisol growth hormone and prolactin secretion in cluster headache. *Cephalalgia*. 1984;4(4):213–220.

124. Waldenlind E, Ekbom K, Wetterberg L, et al. Lowered circannual urinary melatonin concentrations in episodic cluster headache. *Cephalalgia*. 1994;14(3):199–204.

125. Waldenlind E, Gustafsson SA, Ekbom K, Wetterberg L. Circadian secretion of cortisol and melatonin in cluster headache during active cluster periods and remission. *J Neurol Neurosurg Psychiatry*. 1987;50(2):207–213.

126. Tepper SJ. Nutraceutical and other modalities for the treatment of headache. *Continuum (Minneapolis Minn)*. 2015;21(4 Headache):1018–1031.

127. Peres MF, Zukerman E, da Cunha Tanuri F, Moreira FR, Cipolla-Neto J. Melatonin, 3 mg, is effective for migraine prevention. *Neurology*. 2004;63(4):757.

128. Peres MGA. Double-blind, placebo controlled, randomized clinical trial comparing melatonin 3 mg, amitriptyline 25 mg and placebo for migraine prevention. *Neurology*. 2013;80(suppl):S40.005.

129. Alstadhaug KB, Odeh F, Salvesen R, Bekkelund SI. Prophylaxis of migraine with melatonin: a randomized controlled trial. *Neurology*. 2010;75(17):1527–1532.

130. Nagtegaal JE, Smits MG, Swart AC, Kerkhof GA, van der Meer YG. Melatonin-responsive headache in delayed sleep phase syndrome: preliminary observations. *Headache*. 1998;38(4):303–307.

131. Buscemi N, Vandermeer B, Hooton N, et al. The efficacy and safety of exogenous melatonin for primary sleep disorders. *J Gen Intern Med*. 2005;20(12):1151–1158.

132. Liira J, Verbeek JH, Costa G, et al. Pharmacological interventions for sleepiness and sleep disturbances caused by shift work. *Cochrane Database Syst Rev*. 2014;8:Cd009776.

133. Leone M, D'Amico D, Moschiano F, Fraschini F, Bussone G. Melatonin versus placebo in the prophylaxis of cluster headache: a double-blind pilot study with parallel groups. *Cephalalgia*. 1996;16(7):494–496.

134. Peres MF, Rozen TD. Melatonin in the preventive treatment of chronic cluster headache. *Cephalalgia*. 2001;21(10):993–995.

135. Pringsheim T, Magnoux E, Dobson CF, Hamel E, Aube M. Melatonin as adjunctive therapy in the prophylaxis of cluster headache: a pilot study. *Headache*. 2002;42(8):787–792.

136. Plasencia-García BO, Romero-Guillena SL, Quirós-López A, Ruiz-Doblado S. Agomelatine and migraine management: a successfully treated case series. *Ther Adv Psychopharmacol*. 2015;5(4):243–245.

137. Dodick DW. Polysomnography in hypnic headache syndrome. *Headache*. 2000;40(9):748–752.

138. Arai M. A case of unilateral hypnic headache: rapid response to ramelteon, a selective melatonin MT1/MT2 receptor agonist. *Headache*. 2015;55(7):1010–1011.

139. Rozen TD. Melatonin responsive hemicrania continua. *Headache*. 2006;46(7):1203–1204.

140. Hollingworth M, Young TM. Melatonin responsive hemicrania continua in which indomethacin was associated with contralateral headache. *Headache*. 2014;54(5):916–919.

141. Spears RC. Hemicrania continua: a case in which a patient experienced complete relief on melatonin. *Headache*. 2006;46(3):524–527.

142. Rozen TD. How effective is melatonin as a preventive treatment for hemicrania continua? A clinic-based study. *Headache*. 2015;55(3):430–436.

143. Scammell TE, Winrow CJ. Orexin receptors: pharmacology and therapeutic opportunities. *Annu Rev Pharmacol Toxicol*. 2011;51:243–266.

144. Hoffmann J, Supronsinchai W, Akerman S, et al. Evidence for orexinergic mechanisms in migraine. *Neurobiol Dis*. 2015;74:137–143.

145. Bartsch T, Levy MJ, Knight YE, Goadsby PJ. Differential modulation of nociceptive dural input to [hypocretin] orexin A and B receptor activation in the posterior hypothalamic area. *Pain*. 2004;109(3):367–378.

146. Gotter AL, Roecker AJ, Hargreaves R, Coleman PJ, Winrow CJ, Renger JJ. Orexin receptors as therapeutic drug targets. *Prog Brain Res*. 2012;198:163–188.

147. Peterlin BL, Rapoport AM, Kurth T. Migraine and obesity: epidemiology, mechanisms, and implications. *Headache*. 2010;50(4):631–648.

148. Holland P, Goadsby PJ. The hypothalamic orexinergic system: pain and primary headaches. *Headache*. 2007;47(6):951–962.

149. Sarchielli P, Rainero I, Coppola F, et al. Involvement of corticotrophin-releasing factor and orexin-a in chronic migraine and medication-overuse headache: findings from cerebrospinal fluid. *Cephalalgia*. 2008;28(7):714–722.

150. Chabi A, Zhang Y, Jackson S, et al. Randomized controlled trial of the orexin receptor antagonist filorexant for migraine prophylaxis. *Cephalalgia*. 2015;35(5):379–388.

151. Goadsby PJ. Putting migraine to sleep: rexants as a preventive strategy. *Cephalalgia*. 2015;35(5):377–378.

152. Dahmen N, Querings K, Grun B, Bierbrauer J. Increased frequency of migraine in narcoleptic patients. *Neurology*. 1999;52(6):1291–1293.

153. Dahmen N, Kasten M, Wieczorek S, Gencik M, Epplen J, Ullrich B. Increased frequency of migraine in narcoleptic patients: a confirmatory study. *Cephalalgia*. 2003;23(1):14–19.

154. Evers S, Group TDS. Migraine and idiopathic narcolepsy: a case-control study. *Cephalalgia*. 2003;23(8):786–789.

155. Ayata C, Lauritzen M. Spreading depression, spreading depolarizations, and the cerebral vasculature. *Physiol Rev*. 2015;95(3):953–993.

156. Bolay H, Reuter U, Dunn AK, Huang Z, Boas DA, Moskowitz MA. Intrinsic brain activity triggers trigeminal meningeal afferents in a migraine model. *Nat Med*. 2002;8(2):136–142.

157. Zhang X, Levy D, Noseda R, Kainz V, Jakubowski M, Burstein R. Activation of meningeal nociceptors by cortical spreading depression: implications for migraine with aura. *J Neurosci*. 2010;30(26):8807–8814.

158. Faraguna U, Nelson A, Vyazovskiy VV, Cirelli C, Tononi G. Unilateral cortical spreading depression affects sleep need and induces molecular and electrophysiological signs of synaptic potentiation in vivo. *Cereb Cortex*. 2010;20(12):2939–2947.

159. Szentgyorgyi V, Balatoni B, Toth A, Detari L. Effect of cortical spreading depression on basal forebrain neurons. *Exp Brain Res*. 2006;169(2):261–265.

160. Cui Y, Kataoka Y, Inui T, et al. Up-regulated neuronal COX-2 expression after cortical spreading depression is involved in non-REM sleep induction in rats. *J Neurosci Res*. 2008;86(4):929–936.

161. Brennan KC, Bates EA, Shapiro RE, et al. Casein kinase Iδ mutations in familial migraine and advanced sleep phase. *Sci Transl Med*. 2013;5(183):1–11.

162. van den Maagdenberg AMJM, Pietrobon D, Pizzorusso T, et al. A Cacna1a knockin migraine mouse model with increased susceptibility to cortical spreading depression. *Neuron*. 2004;41(5):701–710.

163. Rovers J, Smits M, Duffy JF. Headache and sleep: also assess circadian rhythm sleep disorders. *Headache*. 2014;54(1):175–177.

164. Heyer GL, Rose SC, Merison K, Perkins SQ, Lee JE. Specific headache factors predict sleep disturbances among youth with migraine. *Pediatr Neurol*. 2014;51(4):489–493.

165. Smith MT, Haythornthwaite JA. How do sleep disturbance and chronic pain inter-relate? Insights from the longitudinal and cognitive-behavioral clinical trials literature. *Sleep Med Rev*. 2004;8(2):119–132.

166. Foo H, Mason P. Brainstem modulation of pain during sleep and waking. *Sleep Med Rev*. 2003;7(2):145–154.

167. Drewes AM, Nielsen KD, Arendt-Nielsen L, Birket-Smith L, Hansen LM. The effect of cutaneous and deep pain on the electroencephalogram during sleep—an experimental study. *Sleep*. 1997;20(8):632–640.

168. Spielman AJ, Caruso LS, Glovinsky PB. A behavioral perspective on insomnia treatment. *Psychiatr Clin North Am*. 1987;10(4):541–553.

169. Affleck G, Urrows S, Tennen H, Higgins P, Abeles M. Sequential daily relations of sleep, pain intensity, and attention to pain among women with fibromyalgia. *Pain*. 1996;68(2–3):363–368.

170. Lautenbacher S, Kundermann B, Krieg JC. Sleep deprivation and pain perception. *Sleep Med Rev*. 2006;10(5):357–369.

171. Seidel S, Hartl T, Weber M, et al. Quality of sleep, fatigue and daytime sleepiness in migraine—a controlled study. *Cephalalgia*. 2009;29(6):662–669.

172. Lund N, Westergaard ML, Barloese M, Glumer C, Jensen RH. Epidemiology of concurrent headache and sleep problems in Denmark. *Cephalalgia*. 2014;34(10):833–845.

173. Blau JN. Sleep deprivation headache. *Cephalalgia*. 1990;10(4):157–160.

174. Paiva T, Batista A, Martins P, Martins A. The relationship between headaches and sleep disturbances. *Headache*. 1995;35(10):590–596.

175. Harnod T, Wang YC, Kao CH. Association of migraine and sleep-related breathing disorder: a population-based cohort study. *Medicine*. 2015;94(36):e1506.

176. Cooperman NR, Mullin FJ, Kleitman N. Studies on the physiology of sleep. XI. Further observations on the effects of prolonged sleeplessness. *Am J Physiol*. 1934;107:589–593.

177. Moldofsky H, Scarisbrick P. Induction of neurasthenic musculoskeletal pain syndrome by selective sleep stage deprivation. *Psychosom Med*. 1976;38(1):35–44.

178. Lentz MJ, Landis CA, Rothermel J, Shaver JL. Effects of selective slow wave sleep disruption on musculoskeletal pain and fatigue in middle aged women. *J Rheumatol*. 1999;26(7):1586–1592.

179. Ukponmwan OE, Rupreht J, Dzoljic MR. REM sleep deprivation decreases the antinociceptive property of enkephalinase-inhibition, morphine and cold-water-swim. *Gen Pharmacol*. 1984;15(3):255–258.

180. Houle TT, Butschek RA, Turner DP, Smitherman TA, Rains JC, Penzien DB. Stress and sleep duration predict headache severity in chronic headache sufferers. *Pain*. 2012;153(12):2432–2440.

181. Rains JC, Davis RE, Smitherman TA. Tension-type headache and sleep. *Curr Neurol Neurosci Rep*. 2015;15(2):520.

182. Peppard PE, Young T, Barnet JH, Palta M, Hagen EW, Hla KM. Increased prevalence of sleep-disordered breathing in adults. *Am J Epidemiol*. 2013;177(9):1006–1014.

183. Kristiansen HA, Kværner KJ, Akre H, Øverland B, Russell MB. Migraine and sleep apnea in the general population. *J Headache Pain*. 2011;12(1):55–61.

184. Kristiansen HA, Kvaerner KJ, Akre H, Overland B, Russell MB. Tension-type headache and sleep apnea in the general population. *J Headache Pain*. 2011;12(1):63–69.

185. Jensen R, Olsborg C, Salvesen R, Torbergsen T, Bekkelund SI. Is obstructive sleep apnea syndrome associated with headache? *Acta Neurol Scand*. 2004;109(3):180–184.

186. Neau J-P, Paquereau J, Bailbe M, Meurice J-C, Ingrand P, Gil R. Relationship between sleep apnoea syndrome, snoring and headaches. *Cephalalgia*. 2002;22(5):333–339.

187. Sand T, Hagen K, Schrader H. Sleep apnoea and chronic headache. *Cephalalgia*. 2003;23(2):90–95.

188. Kristiansen HA, Kvaerner KJ, Akre H, Overland B, Sandvik L, Russell MB. Sleep apnoea headache in the general population. *Cephalalgia*. 2012;32(6):451–458.

189. Schoonman G, Sándor P, Agosti R, et al. Normobaric hypoxia and nitroglycerin as trigger factors for migraine. *Cephalalgia*. 2006;26(7):816–819.

190. Marmura M, Hernandez P. High-altitude headache. *Curr Pain Headache Rep*. 2015;19(5):1–7.

191. Gilbert GJ. Turtle headaches. *JAMA*. 1982;248(8):921.

192. King AB. Hypoxia and hypercapnia from bedcovers. *JAMA*. 1972;220(13):1745.

193. Gilbert GJ. Hypoxia and bedcovers. *JAMA*. 1972;221(10):1165–1166.

194. Kallweit U, Hidalgo H, Uhl V, Sandor PS. Continuous positive airway pressure therapy is effective for migraines in sleep apnea syndrome. *Neurology*. 2011;76(13):1189–1191.

195. Johnson KG, Ziemba AM, Garb JL. Improvement in headaches with continuous positive airway pressure for obstructive sleep apnea: a retrospective analysis. *Headache*. 2013;53(2):333–343.

196. Evers S, Barth B, Frese A, Husstedt IW, Happe S. Sleep apnea in patients with cluster headache: a case-control study. *Cephalalgia*. 2014;34(10):828–832.

197. Nath Zallek S, Chervin RD. Improvement in cluster headache after treatment for obstructive sleep apnea. *Sleep Med*. 2000;1(2):135–138.

198. Rains JC, Poceta JS. Sleep and headache. *Curr Treat Options Neurol*. 2010;12(1):1–15.

199. Stepanski EJ, Wyatt JK. Use of sleep hygiene in the treatment of insomnia. *Sleep Med Rev*. 2003;7(3):215–225.

200. Spierings EL, McAllister PJ, Bilchik TR. Efficacy of treatment of insomnia in migraineurs with eszopiclone (Lunesta(R)) and its effect on total sleep time, headache frequency, and daytime functioning: a randomized, double-blind, placebo-controlled, parallel-group, pilot study. *Cranio*. 2015;33(2):115–121.

Sleep and the Autonomic Nervous System

M.G. Miglis

Neurology Department, Stanford University School of Medicine, Stanford, CA, USA

O U T L I N E

INTRODUCTION

The autonomic nervous system is a diffuse network that regulates virtually all of the unconscious homeostatic mechanisms of the human body. In this sense, sleep can be thought of as a highly complex autonomic function. When sleep is disrupted, many of these homeostatic functions, such as sweating and temperature regulation, blood pressure and heart rate variability, and bowel and bladder function, can also be affected. In addition, many of the

Sleep and Neurologic Disease. http://dx.doi.org/10.1016/B978-0-12-804074-4.00018-2

primary disorders of autonomic dysfunction also greatly affect sleep. Healthy sleep is thus paramount to healthy autonomic functioning, and the treatment of any autonomic disorder should include treatment of comorbid sleep disorders. While virtually all of the primary sleep disorders can cause some degree of autonomic impairment,[1] this chapter will focus primarily on the common sleep complaints seen in patients with autonomic disease.

ANATOMY OF THE AUTONOMIC AND SLEEP/WAKE SYSTEMS

The cell populations that regulate autonomic function and sleep utilize some of the same neurotransmitters—namely norepinephrine and acetylcholine. They are situated in close proximity to one another in the brainstem and hypothalamus (Fig. 12.1), and in some cases directly connect via complex neural pathways. The lateral zone of the hypothalamus, for instance, contains several important cell populations that regulate sleep and wake, hunger, and reward responses. Hypocretin (otherwise known as orexin) neurons located in this zone send projections to both sleep and autonomic nuclei. These pathways extend to all monoaminergic

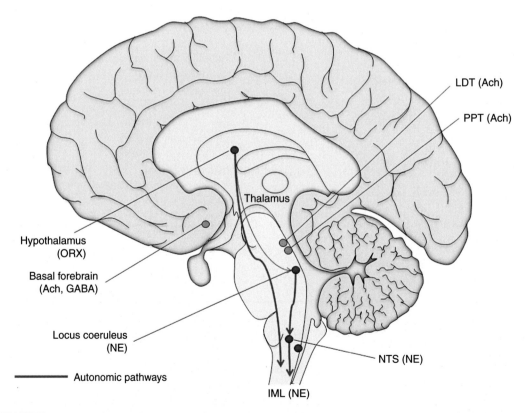

FIGURE 12.1 **Neuroanatomical locations of important autonomic and sleep nuclei.** *Red circles* indicate autonomic nuclei; *red arrows* indicate efferent autonomic projections. *ORX,* orexin; *Ach,* acetylcholine; *GABA,* gamma-aminobutyric acid; *NE,* norepinephrine; *LDT,* laterodorsal tegmentum; *PPT,* pedunculopontine nucleus; *NTS,* nucleus tractus solitarius; *IML,* intermediolateral cell column.

nuclei and stimulate the release of their alerting neurotransmitters to help promote wakeful-ness. Hypocretin neurons also send projections to many autonomic regulatory centers includ-ing the periaqueductal gray, nucleus of the solitary tract, nucleus ambiguus, and dorsal motor nucleus of the vagus nerve.[2] There are many animal studies to suggest that orexin exerts influ-ence on several autonomic functions including heart rate and blood pressure regulation,[3] ener-gy metabolism,[4] and gastrointestinal motility,[5] and is implicated in the autonomic changes seen in narcolepsy patients.[6] The lateral zone of the hypothalamus also contains autonomic neurons that send their efferent projections to the lateral medulla and the intermediolateral cell column of the spinal cord. These efferent pathways help regulate vascular tone and maintain our abil-ity to undergo postural change without significant alterations in blood pressure and heart rate.

The lateral medulla is another location of interest, and contains many important afferent and efferent autonomic pathways. The baroreceptors, chemoreceptors, cardiac receptors, and respiratory receptors all send their projections to the nucleus solitarius in the lateral medulla, via the glossopharyngeal and vagus nerves. These pathways are integral to the regulation of breathing during sleep, and accomplish this task via central pattern generators in the dorso-lateral pons, the dorsal respiratory group in the nucleus tractus solitarius, and the ventral re-spiratory group in the lateral medulla.[7] They also rely on chemoreceptors in the carotid body that send afferent projections to the nucleus solitarius.

While not all of these autonomic circuits have direct connections to sleep circuits, it is important to note their neuroanatomical proximity. For this reason, any disease that prefer-entially affects these areas of the rostral brainstem or hypothalamus can potentially disrupt both sleep and autonomic function. One classic example is the neurodegenerative disorders of α-synuclein deposition, discussed later in this chapter.

NORMAL AUTONOMIC FUNCTION DURING SLEEP

As humans transition from the waking state to drowsiness and into sleep, parasympathetic vagal tone increases, and sympathetic tone decreases. The Boetzinger complex in the lateral medulla initiates a regular firing rate that stimulates a slowing of the respiratory rate and induces even, regular breathing, helping to promote gas exchange.[8]

As sleep progresses from stage 1 nonrapid eye movement (NREM) sleep to the deeper stages of 2 and 3 NREM sleep, parasympathetic vagal tone increases even more. The net result is a reduction in heart rate, blood pressure, and cardiac output.[9] At the same time, sympathetic tone continues to decrease, leading to a reduction in peripheral vascular resis-tance and arterial blood pressure. The baroreceptors—stretch receptors in the aortic arch and carotid sinuses that help regulate arterial blood pressure—become more sensitive to changes in blood pressure, further promoting a state of regular respiration and gas exchange. Blood pressure in this stage is typically 10–20% lower than blood pressure during wake, a phenom-enon referred to as "dipping."[9] Heart rate also reaches its nadir in stage 3 NREM sleep. For these reasons, NREM sleep can be thought of as a state of parasympathetic dominance, auto-nomic stability and metabolic recovery.

During REM sleep, cholinergic discharges in the pedunculopontine nucleus and lat-erodorsal tegmental nucleus of the pons result in muscle atonia that inhibits body movement and dream enactment. REM can be divided into tonic REM, without rapid eye movements, and phasic REM, with rapid eye movements. Parasympathetic tone dominates during tonic REM;

however, during phasic REM the sympathovagal balance reverses and sympathetic tone increases significantly. In this stage of sleep, blood pressure and heart rate may fluctuate dramatically, and blood pressure can reach levels much higher than those of the waking state. Sympathetic nerve activity has been demonstrated to be higher in phasic REM than during wake.[10] In this sense, phasic REM can be thought of as a state of heightened sympathetic tone and relative autonomic instability.

The pressure to sleep is driven by the homeostatic sleep drive and the circadian cycle, and both likely affect autonomic balance in relation to sleep onset and sleep maintenance. The effect of circadian phase on autonomic tone has been questioned for some time, and it has long been observed that many disease states with some relation to sympathetic hyperactivity occur in the early morning hours or upon awakening, such as myocardial infarction, ischemic and hemorrhagic stroke, and congestive heart failure exacerbations. Some research indicates that parasympathetic tone is influenced by the circadian system, whereas sympathetic tone is influenced by the homeostatic sleep system;[11] however, this has not been firmly established.

Arousals from any stage of sleep can trigger bursts of sympathetic activity and exacerbate any preexisting elevations in blood pressure and heart rate. If frequent enough, these sympathetic surges may carry over into the waking state, leading to increased diurnal sympathetic tone and hypertension. This is one theory for the relationship between sleep apnea and hypertension, as well as one explanation for the heightened sympathetic tone seen in some postural tachycardia syndrome (POTS) patients.

MEASURING AUTONOMIC TONE DURING SLEEP

Peripheral Arterial Tone

Due to the diffuse and relatively inaccessible nature of the autonomic nerves, measuring autonomic tone during sleep can be difficult. Beat-to-beat blood pressure monitors are typically used in the autonomic laboratory during autonomic tilt table testing, and can also be worn during sleep (Fig. 12.2). These devices measure peripheral arterial tone (PAT) by

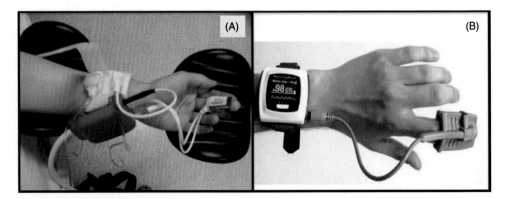

FIGURE 12.2 **Extremity devices used to measure autonomic tone during sleep.** (A) Pneumatic finger cuff used in the autonomic laboratory to measure peripheral arterial tone during tilt table testing (Nexfin, Edwards Instruments). (B) Pulse oximeter, from which photoplethysmogram and pulse transit time are derived.

inflating a pneumatic finger cuff on the distal phalanx of the patient's finger, thereby measuring small changes in blood volume with each heartbeat. The monitor connects to a small device worn on the wrist. When sympathetic tone increases, peripheral vasoconstriction results in attenuation of the PAT signal.[12] The advantage of these devices is that they can provide accurate, real-time measurements of changes in arterial blood pressure during sleep. The disadvantage is that they can be uncomfortable and disruptive to sleep, thus limiting their use outside of the research setting.

Pulse Transit Time

Pulse transit time (PTT) is a measurement of the time it takes for an arterial pulse wave to reach the periphery.[13] PTT can be calculated from the finger photoplethysmograph (PPG) of the oxygen saturation monitor and the R-wave of the electrocardiogram (ECG) during a polysomnogram (PSG, Fig. 12.3). Blood pressure can be indirectly calculated by assuming that the speed of the PTT wave is inversely proportional to systolic blood pressure, for example an increase in blood pressure results in a shorter PTT and vice versa. The ECG R-wave is used to estimate the opening of the aortic valve and contraction of the left ventricle.[12] The PPG is used to estimate the arrival of the pulse in the periphery, and from these two values the PTT is calculated. While not as accurate as the PAT, there is no inflation of the PPG finger probe, and thus the device can be worn with relative comfort during sleep. Since this method relies on the RR interval for its calculations, any arrhythmia, such as atrial fibrillation, can lead to significant artifact and invalidate PTT results.

Spectral Analysis

Spectral analysis of the RR interval is an indirect, noninvasive measurement tool. Spectral analysis of heart rate variability is often referenced in the literature as an estimate of sympathetic and parasympathetic tone during sleep, otherwise termed the sympathovagal balance. High-frequency RR signal (greater than 0.15 Hz) is associated with increased parasympathetic tone, and low-frequency RR signal (0.04–0.15 Hz) is associated with increased sympathetic tone.[14] The high frequency signal is most influenced by the vagal-mediated respiratory sinus arrhythmia of deep breathing, while the low frequency signal is most likely influenced by a baroreflex-mediated heart rate response to blood pressure.[15] In short, a greater LF/HF ratio suggests greater sympathetic drive, and a lower LF/HF ratio suggests greater parasympathetic drive. As in PPG analysis, arrhythmias can invalidate results.

Microneurography

Microneurography is a more invasive tool that has been utilized in the research setting as way to directly measure sympathetic output (Fig. 12.4). In this highly specialized technique, a small electrode is inserted into the sympathetic nerves, most commonly in the peroneal nerve as it courses through the anterior tibialis muscle.[16] Researchers have used microneurography to measure sympathetic nerve activity during wake and the various sleep stages to measure sympathetic tone during sleep.[10] It has been demonstrated that sympathetic burst activity is high during wake but even higher during REM sleep, and lowest during stage 3 NREM

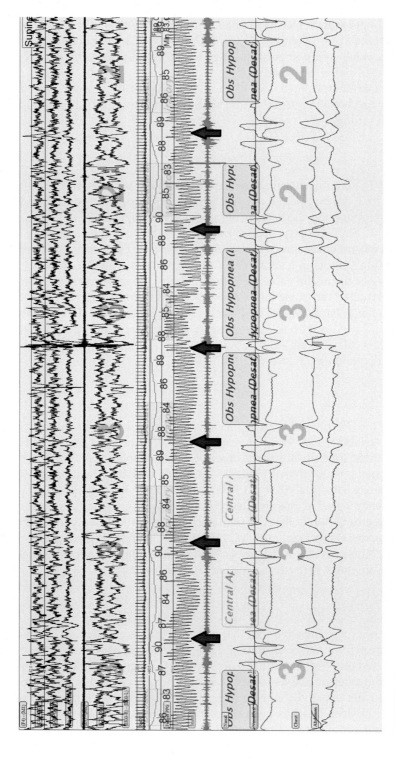

FIGURE 12.3   **Overnight polysomnogram demonstrating changes in pulse transit time, measured by photoplethysmography (PPG) from the pulse oximeter.** This particular patient has runs of cyclical apneas, which are both obstructive and central in nature. The corresponding oxygen desaturations that follow both obstructive and central events result in sympathetic activation, vasoconstriction, and elevation in PPG tone, most pronounced toward the end of the apnea (*arrows*). This is a good example of the fact that both central and obstructive apneas can result in sympathetic activation, suggesting that it is likely the oxygen desaturation, and not the type of apnea, that triggers the autonomic response.

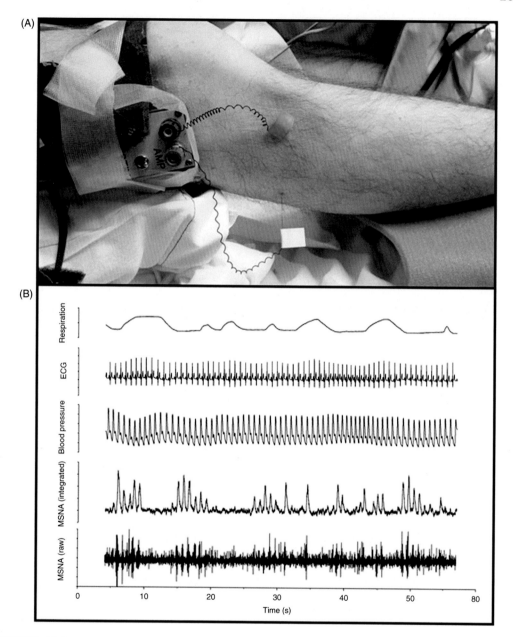

FIGURE 12.4 **Microneurography technique and measurement of sympathetic nerve activity.** (A) The micro-neurography needle is a fine tungsten microfilament that is inserted into unmyelinated sympathetic branches of a peripheral nerve, in this case the peroneal nerve. (B) Microneurography baseline recording demonstrating spontane-ous bursts of muscle sympathetic nerve activity *(MSNA)*. *Source: Istvan Bonyhay, Harvard Medical School.*

sleep. In addition, sympathetic bursts also occur during sleep stage transitions, and with the appearance of sleep spindles and K-complexes in stage 2 NREM sleep.

AUTONOMIC DISORDERS WITH PROMINENT SLEEP DISRUPTION

Postural Tachycardia Syndrome

POTS is an autonomic syndrome of inappropriate tachycardia triggered by postural change. It is defined as a sustained heart rate increase of $\geq$30 beats per minute within the first 10 min of standing or head-up tilt, with a normal blood pressure response, along with symptoms of orthostatic intolerance.[17] These symptoms typically include lightheadedness, dizziness, nausea, visual blurring, and presyncope; however, patients may also report palpitations, chest pain, exercise intolerance, mental clouding or "brain fog," fatigue and lack of refreshing sleep.[18] Of all these symptoms, fatigue, and lack of refreshing sleep can be the most debilitating. Although it can be difficult for patients to distinguish the sensation of sleepiness from that of fatigue, it is important for the clinician to attempt to tease this out in the history. Does the patient describe symptoms of sleepiness, such as the overwhelming desire to sleep or nap? Do they find themselves dosing off in meetings or while driving? Or do they describe more of a sensation of fatigue, of body or muscle "tiredness?" This can be helpful in attempting to understand what is driving the symptoms, and where to focus the workup and treatment. The Epworth sleepiness scale (ESS) was designed to measure the former,[19] and various fatigue scales, such as the fatigue severity scale and fatigue visual analog scale the later.[20] If framed in this context, patients will often be able to label their symptoms as either sleepiness or fatigue, though it is not uncommon for them to describe both.

Patients with POTS tend to describe more fatigue than sleepiness,[21] similar to patients with CFS or fibromyalgia. In one study, up to 50% of patients reported symptoms of insomnia, and up to 90% reported fatigue.[22] POTS patients often have difficulty falling and staying asleep, and their ESS scores tend to be low. Other patients may feel that even if they do sleep they never wake feeling refreshed, no matter how many hours they think they might have slept.

The literature on sleep in POTS is growing, and at the time of this publication there have been five studies evaluating the nature of the sleep complaints in this patient population. Thus far it has been demonstrated that patients report more sleep complaints than controls,[23] have greater subjective symptoms than objective findings and some element of sleep-state misperception,[24] and demonstrate no consistent abnormalities on PSG to objectively explain the severity of their symptoms.[22,25,26] Two studies demonstrated prolonged REM latency on PSG.[22,25] Patients in one study had mild OSA,[22] with a mean AHI of 6.6, however a third of these patients were also diagnosed with Ehlers–Danlos hypermobility subtype (EDS-HT), which may predispose patients to upper airway collapse.[27]

Researchers in another study used spectral analysis to evaluate heart rate variability in various sleep stages.[25] While they did report a reduction in LF/HF variability during sleep stage transitions in patients, there was no significant difference in stage-specific LF, HF, or LF/HF ratio when compared to controls. Nevertheless, it has been suggested that POTS patients might have altered heart rate variability during sleep, indicative of a hyperadrenergic state. Twenty-four-hour blood pressure monitoring has demonstrated a nondipping pattern in 55% of POTS patients, which is suggestive of either increased sympathetic tone or diminished parasympathetic tone during sleep.[28]

Patients with POTS tend to have a predisposition to an evening chronotype,[22] often finding it difficult to fall asleep until late into the evening. The reason for this is unclear; however, many patients note symptoms of hyperarousal, such as palpitations, racing thoughts, increased sweat output, and other symptoms of a hyperadrenergic state. The subtleties of circadian rhythm disorders in these patients have not been systematically evaluated, though there may be a connection between melatonin secretion and the autonomic system. Melatonin acts on two receptors, MT1 and MT2. MT1 receptors, when stimulated, lead to vasoconstriction of peripheral arterioles,[29] and MT2 receptor stimulation leads to vasodilation.[30] In addition, the stimulation of MT1 receptors in the suprachiasmatic nucleus may reduce central sympathetic output by stimulating GABA-ergic signaling to the paraventricular nucleus, which in turn inhibits adrenergic nuclei in the lateral medulla.[31] Many POTS patients are treated with nonselective beta-blockers, which can theoretically impair melatonin secretion via blockade of pineal beta-1-receptors,[32] though these medications may also provide relief from nocturnal palpitations and are sometimes dosed at bedtime for this purpose. Interestingly, in one small study, 3 mg of melatonin dosed in the morning produced a moderate decrease in standing tachycardia in patients with POTS, although it did not improve symptoms of orthostatic intolerance.[31]

POTS patients frequently describe a sensation of generalized hyperarousal that prevents them from falling asleep even in the setting of extreme exhaustion, a presentation often referred to by sleep psychologists as "tired but wired." This sensation is experienced by many insomnia patients and is supported physiologically by several studies that have examined functional imaging and EEG analysis. PET imaging in chronic insomnia patients during sleep has demonstrated increased activation and hypermetabolism in the arousal networks of the hypothalamus and brainstem, as well as their efferent projections in the medial prefrontal cortex and amygdala.[33] EEG frequency analysis has demonstrated that these patients have increased beta (14–35 Hz) and gamma (35–45 Hz) activity, frequencies that are typically associated with waking cortical activity.[34] Many patients with sleep-state misperception, or "paradoxical insomnia" may in fact have increased beta and gamma frequencies during sleep. While these studies have not been performed in POTS patients, it would not be surprising if they too had evidence of hyperactive arousal networks, as Bagai et al. had suggested in their POTS actigraphy study. The hyperarousal model, with its focus on heightened hypothalamic–pituitary–adrenal tone, increased catecholamine secretion, and excessive cortical activity during wake and sleep, may provide a window into the understanding of the autonomic complaints of patients with POTS.

Why do some patients with POTS exhibit symptoms of hyperarousal? One often overlooked fact is that arousals from any source, if frequent enough, can result in increased sympathetic tone. This has been demonstrated in the literature on insomnia, obstructive sleep apnea, periodic leg movements and restless leg syndrome (RLS).[1] Prior to an electrocortical arousal, there is a typical cardiac response that occurs: initial tachycardia, which often precedes the arousal by several seconds, followed by bradycardia. With every arousal there is a small elevation in sympathetic tone. If the arousals become frequent enough, this elevation can persist after the patient has returned to sleep,[35] and even into the next day.

Based on these results, it is difficult to say with any confidence that the sleep-related complaints of POTS patients are the result of a primary sleep disorder unique to POTS. A more plausible explanation is that these patients experience disrupted sleep from a combination of systemic factors, such as body fatigue, chronic pain, hyperactivity of arousal networks and, in some cases, sleep-state misperception. There are unfortunately many questions that remain to be answered about this condition. Like systemic exertion intolerance disease (previously chronic fatigue syndrome), fibromyalgia, and Ehlers–Danlos syndrome (EDS), POTS is a multifactorial syndrome with many possible etiologies, and thus there is unlikely to be one single associated sleep disorder or treatment algorithm that applies to all patients. In the interim we can only hope that with more detailed research over the coming decades, the mechanisms of POTS and other conditions of chronic fatigue will be understood in greater detail, allowing us to develop targeted treatments that address the underlying disease and not just symptom management.

Ehlers–Danlos Syndrome

EDS is a connective tissue disorder of abnormal collagen production characterized by joint hypermobility, vascular fragility, and skin hyperextensibility.[36] Of the six subtypes of EDS—classical, hypermobility, vascular, kyphoscoliosis, arthrochalasia, and dermatosparaxix—EDS-HT is the most common, with an estimated prevalence of 1 in 5000.[37] This disorder has been identified in increasing amounts of patients with autonomic impairment, especially those with POTS.[38]

Patients with EDS-HT may report a wide array of neurological, autonomic, and sleep-related symptoms including orthostatic intolerance,[38] migraine,[39] gastrointestinal pain and nausea,[40] insomnia, fatigue, and nonrestorative sleep.[41] Many of the same sleep-related complaints seen in patients with POTS are also seen in this patient population. An association with obstructive sleep apnea has been suggested,[27] and while the prevalence of OSA in EDS is unknown, it is likely more common than OSA in the general population. The underlying pathophysiology of collagen deficiency in EDS-HT may lead not only to hyperelastic joints and hyperextensible skin but also a hyperelastic upper airway, increasing the likelihood of collapsibility during sleep. This same mechanism may be involved in the pathogenesis of autonomic impairment, with decreased arteriolar contractility leading to vasomotor instability and orthostatic intolerance, though this has not been proven. In short, even though this condition is being recognized with increasing regularity in the autonomic clinic, the genetic abnormalities and pathophysiology of EDS-HT have yet to be elucidated, and there are many clinical questions that warrant further exploration.

There is currently no accepted treatment for the sleep complaints of those with either EDS-HT or POTS, although anecdotal evidence exists for the use of various traditional and nontraditional hypnotics, such as the benzodiazepine receptor agonists and α-2-δ ligands for sleep initiation insomnia, and long-acting stimulants, such as modafinil for fatigue and excessive daytime sleepiness. It is preferable to select a medication with a dual purpose, for example gabapentin if there is a strong pain component or nortriptyline if the patient has a history of migraine headaches. Some patients with nocturnal palpitations may benefit from bedtime dosing of a low dose, short acting beta-blocker, such as propranolol. Stimulants should only be considered when all other primary sleep disorders—such as OSA and RLS—have been ruled out or appropriately treated, and if used should be done so with caution, noting that these medications may worsen tachycardia and thus lead to some anxiety for patients.

Fatal Familial Insomnia

Fatal familial insomnia (FFI) is a rare but illustrative disorder of sleep and autonomic dysfunction. FFI is an autosomal dominant prion disease resulting from a missense mutation in the prion protein gene. Patients with FFI develop spongiform degeneration of the medio-dorsal and anterior thalamic nuclei, areas that regulate both sleep and autonomic control.[42] Insomnia occurs early in the disease, followed by apathy and confusion.

Autonomic instability is prominent and manifests as hypertension, tachycardia, elevated body-core temperature and hyperhidrosis, all features of a hyperadrenergic state. Patients become "nondippers"[43] and sympathetic tone is elevated during both wake and sleep.[44] PSG demonstrates a progressive reduction and disappearance of spindles, K complexes, and slow wave sleep, indicating a disruption in thalamocortical circuits.[42] Plasma melatonin concentrations diminish as the disease progresses, leading to complete disruption of circadian rhythms.[45] The mean age at onset is 51 years of age, and death typically occurs within 8–72 months.

Migraine

Autonomic hyperactivation occurs frequently in migraine and other autonomic cephalgias, and many of the pathways involved in pain generation also project to nuclei of the autonomic and sleep/wake systems. It is therefore not surprising that patients with chronic or refractory migraine report both sleep disruption and autonomic complaints, which can be difficult to differentiate from those of a primary autonomic disorder. Patients with migraine, for instance, may find it difficult to fall asleep because of headache, wake in the middle of the night or in the early morning hours with headache, and develop a strong component of insomnia that only serves to exacerbate their headaches. There are also circadian components to many headache syndromes that are often overlooked. Treatment should focus primarily on migraine control; once the headaches are treated, the autonomic symptoms and sleep disruption typically improves as well, though psychophysiological insomnia may persist. For a complete review of this topic, the reader is directed to the chapter in this text on sleep and headache.

Disorders of α-Synuclien Deposition

The α-synucleinopathies are neurodegenerative disorders resulting from the deposition of the protein α-synuclein in the central and/or peripheral nervous system. When these protein deposits occur exclusively in the peripheral autonomic nerves, the disease is termed pure autonomic failure (PAF). Parkinson's disease (PD), multiple system atrophy (MSA), and dementia with Lewy bodies (DLB) are diseases of both central and peripheral α-synuclein deposition (Table 12.1). With the exception of PAF, REM behavior disorder (RBD) is extremely common in these disorders and, along with symptoms of autonomic impairment, may precede the development of motor symptoms by several decades. It is estimated that up to 80% of patients with idiopathic RBD (iRBD) may eventually develop some form of α-synucleinopathy,[46–48] with an estimated median latency between RBD and motor symptoms of 12–14 years.[49] Within the α-synucleinopathies, RBD is estimated to occur in 30–50% of patients with PD, 50–80% of patients with DLB, and 80–95% of patients with MSA.[46–48,50] In this sense, RBD can be thought of as the strongest nonmotor predictor of future clinical disease.

TABLE 12.1   Sites and Aggregate Type of α-Synuclein Deposition

α-Synucleinopathy	PD	DLB	PAF	MSA
Deposition site	Cortex, substantia nigra, brainstem, postganglionic autonomic neurons	Cortex, brainstem, postganglionic autonomic neurons	Postganglionic autonomic neurons	Basal ganglia, brainstem, cerebellum
Aggregates	Lewy bodies	Lewy bodies	Lewy bodies	Glial cytoplasmic inclusions

Autonomic impairment is another early nonmotor manifestation, also occurring during the prodromal period, often in tandem with dream-enacting behavior.[50] This association is likely due to the proximity of the cholinergic REM nuclei and the autonomic nuclei in the brainstem. In the Braak model of neurodegeneration, these nuclei become impaired before the motor nuclei are affected, as the deposition of α-synuclein progresses in a rostral–caudal fashion from the lower brainstem to the cortex.[51] Many patients with RBD have some degree of autonomic impairment, and many patients with α-synucleinopathies and autonomic impairment also have RBD. This is a good example of disease states cosegregating due to the neuroanatomical proximity of their control nuclei. The topic of RBD, including workup and treatment, is also covered in the movement disorders section of this text.

Pure Autonomic Failure

Patients with PAF typically present in their fifth or sixth decade and may note symptoms of orthostatic hypotension, diminished sweat output, and gastrointestinal or genitourinary dysfunction, but without cognitive impairment or behavioral changes.[52] If followed for long enough, some of these patients may eventually convert to PD, MSA, or DLB, as predicted in the Braak model. If patients are eventually diagnosed with one of the central α-synucleinopathies, it is most likely to be MSA. In one small study, 4 out of 10 patients originally diagnosed with PAF were eventually diagnosed with MSA at follow-up exams 5–7 years later.[53] Therefore, it is typically recommended to follow patients for a minimum of 5 years before a definite diagnosis of PAF can be reached.[54]

Sleep data on patients with PAF is limited. In one study, both PAF and MSA patients demonstrated reduced sleep efficiency on PSG; however, RBD was seen in all patients in the MSA group, and in no patients in the PAF group.[53] This observation supports the concept of PAF as a purely peripheral disorder without brainstem involvement. However, the results of other studies have disputed this notion. One small case-series described RBD in three patients with PAF.[55] Another small survey study described frequent nocturnal vocalizations in 9 of 13 PAF patients (69%), suggestive of possible RBD.[56] PSGs were not performed in these patients and RBD could not be confirmed. This is a very interesting topic and a potential area for future research, though a challenging one owing to the relative rarity of PAF and the clinical overlap with disorders of central autonomic impairment, such as MSA.

Multiple System Atrophy

Of all the α-synucleinopathies, RBD is most commonly seen in MSA, affecting 80–95% of patients by some estimates.[50,57,58] Sleep-disordered breathing is also common in these patients, and affects patients with both parkinsonian and cerebellar subtypes. Both obstructive and central sleep apnea (CSA) have been reported with relative frequency. OSA is more common, and has been reported in up to 37% of patients.[59] CPAP may be considered as a potential treatment, however it should be noted that MSA patients can have a floppy epiglottis,[60] making CPAP a relative contraindication. It is not clear how to screen for these patients effectively and there is currently no consensus statement in this regard; therefore, PAP therapy is still generally recommended as first line therapy. CSA and Cheyne–Stokes respirations may occur due to degeneration of brainstem chemoreceptors.[61] This may be the presenting sleep related breathing pattern; however, at other times CSA emerges once the patient is started on CPAP, termed treatment-emergent CSA.[62] Adaptive servo-ventilation devices may be considered in these cases.

Stridor during sleep occurs with some regularity, and is thought to be secondary to selective deterioration of the vocal cord abductors.[63] It manifests as a harsh, high-pitched, inspiratory tone, and is often more disturbing to the bed partner than snoring. The pathophysiology of the vocal muscle weakness is unknown, and examination of motor neurons in the nucleus ambiguous has failed to confirm selective deterioration of neurons serving the vocal cord musculature. Patients with stridor have a greater incidence of OSA, as well as a greater risk of sleep-related death.[64] CPAP has been demonstrated to reduce the severity of stridor.[65] In severe cases, symptoms may persist during the waking hours. In these cases, tracheostomy may be the only effective treatment.

RLS is not as common as it is in PD patients; however, some studies estimate prevalence rates as high as 23%.[66] Patients with MSA also report a high degree of excessive daytime sleepiness, sleep fragmentation, unrefreshing sleep and insomnia due to rigidity, pain, and frequent nocturnal urination secondary to autonomic detrusor hyperactivity. Patients with PD report many of the same sleep complaints, though MSA patients tend to have more severe urinary dysfunction.

Parkinson's Disease

Sleep disorders are extremely common in PD, reported in up to two thirds of patients.[67] The sleep complaints are myriad and span the entire spectrum of sleep medicine, largely owing to the neuroanatomical deposition of Lewy bodies throughout the brainstem and cortex. The severity of these sleep complaints tends to correlate with the severity of the disease itself.[68] For a detailed review of the sleep complaints in these patients, the reader is directed to the movement disorders chapter of this text. Parkinson's patients with RBD are more likely to develop axial rigidity, have less response to levodopa therapy, and have an increased risk of falls, suggesting a more diffuse disease process.[69] Patients with RBD are also more likely to develop autonomic impairment, and symptoms of autonomic impairment predict the conversion to clinically apparent neurodegenerative disease.[70]

Dementia with Lewy Bodies

Patients with DLB have many of the same sleep issues as those with PD, though the extent of nocturnal rigidity depends on the degree of Parkinsonism in each individual patient. In addition, DLB patients have varying levels cognitive impairment, which is in turn associated with its own set of sleep problems, such as sundowning and circadian rhythm abnormalities.[71]

Autonomic complaints are common, and many patients are diagnosed with orthostatic hypotension.[72] Thus it is not surprising that there is a high prevalence of RBD, again indicating involvement of brainstem structures integral to both sleep and autonomic control.[73]

Idiopathic RBD

Symptoms of autonomic impairment are common in patients with iRBD. In a large multi-center case–control study, 318 patients with iRBD were administered the scale for outcomes in PD-autonomic (SCOPA-AUT), a standardized 25-item autonomic questionnaire that addresses several domains including gastrointestinal, urinary, cardiovascular, thermoregulatory, pupillomotor, and sexual function.[74] Patients with iRBD reported greater impairment than did controls, especially in relation to gastrointestinal, urinary, and cardiovascular function. There is also evidence that these patients have a higher conversion rate to clinically apparent neurodegenerative disease.[75] In addition, within the α-synucleinopathies, the presence of RBD seems to be correlated with a greater degree of autonomic dysfunction.[49]

Patients with iRBD have reduced heart rate variability during REM sleep.[76] They also have more significant systolic blood pressure falls on active standing. Postuma and coworkers demonstrated an average systolic blood pressure reduction of 15.2 mmHg in these patients on active stand tests, compared to 3.7 mmHg in age-matched controls.[48] Most patients reported symptoms of orthostatic intolerance, though did not meet criteria for orthostatic hypotension.

This data was corroborated by Frauscher and coworkers, who performed autonomic testing on iRBD patients and compared them to both controls and patients with PD.[77] While patients with iRBD did not demonstrate orthostatic hypotension on tilt testing, they had slightly greater blood pressure falls on active stand than controls did. In addition, the Valsalva ratio, a measurement of cardiovagal function, was reduced in iRBD patients.

Patients with iRBD have evidence of postganglionic cardiac sympathetic denervation, as measured by cardiac scintigraphy,[78] further supporting the theory of prodromal autonomic impairment. Thus autonomic impairment, like RBD itself, can be thought of as a feature that may help to identify at-risk patients, providing a potential window for disease modifying therapies should such therapies become available to clinicians in the future.

CONCLUSIONS

Sleep is a complex and highly orchestrated autonomic function, and when disrupted can lead to diurnal symptoms of both sleepiness and autonomic impairment. Healthy sleep is critical to homeostasis, and should be one of the primary goals in the treatment plan for all patients with autonomic disorders. Each disorder has its unique set of sleep complaints, thus a detailed history should always include questions about the patient's sleep. While some associations have been firmly established, such as the association between RBD and the neurodegenerative disorders of α-synuclein deposition, others remain elusive, such the symptoms of fatigue and sleepiness in patients with POTS or EDS-HT. With greater awareness of these conditions and associations, the practicing neurologist can query his patients for these symptoms, and if present and successfully treated, lead to significant improvement in their quality of life.

References

1. Miglis MG. Autonomic dysfunction in primary sleep disorders. *Sleep Med*. 2016;19:40–49.
2. Grimaldi D, Silvani A, Benarroch EE, Cortelli P. Orexin/hypocretin system and autonomic control: new insights and clinical correlations. *Neurology*. 2014;82:271–278.
3. Shirasaka T, Nakazato M, Matsukura S, Takasaki M, Kannan H. Sympathetic and cardiovascular actions of orexins in conscious rats. *Am J Physiol*. 1999;277:R1780–R1785.
4. Schuld A, Hebebrand J, Geller F, Pollmächer T. Increased body-mass index in patients with narcolepsy. *Lancet*. 2000;355:1274–1275.
5. Nozu T, Kumei S, Takakusaki K, Ataka K, Fujimiya M, Okumura T. Central orexin-A increases colonic motility in conscious rats. *Neurosci Lett*. 2011;498:143–146.
6. Donadio V, Liguori R, Vandi S, Giannoccaro MP, Pizza F, Leta V, Plazzi G. Sympathetic and cardiovascular changes during sleep in narcolepsy with cataplexy patients. *Sleep Med*. 2014;15:315–321.
7. Feldman JL, Del Negro CA. Looking for inspiration: new perspectives on respiratory rhythm. *Nat Rev Neurosci*. 2006;7:232–241.
8. Somers VK, Dyken ME, Skinner JL. Autonomic and hemodynamic responses and interactions during the Mueller maneuver in humans. *J Auton Nerv Syst*. 1993;44:253–259.
9. Van de Borne P, Nguyen H, Biston P, Linkowski P, Degaute JP. Effects of wake and sleep stages on the 24-h autonomic control of blood pressure and heart rate in recumbent men. *Am J Physiol*. 1994;266:H548–H554.
10. Somers VK, Dyken ME, Mark AL, Abboud FM. Sympathetic-nerve activity during sleep in normal subjects. *N Engl J Med*. 1993;328:303–307.
11. Burgess HJ, Trinder J, Kim Y, Luke D. Sleep and circadian influences on cardiac autonomic nervous system activity. *Am J Physiol*. 1997;273:H1761–H1768.
12. Pépin JL, Tamisier R, Borel JC, Baguet JP, Lévy P. A critical review of peripheral arterial tone and pulse transit time as indirect diagnostic methods for detecting sleep disordered breathing and characterizing sleep structure. *Curr Opin Pulm Med*. 2009;15:550–558.
13. Smith RP, Argod J, Pepin JL, Levy PA. Pulse transit time: an appraisal of potential clinical applications. *Thorax*. 1999;54:452–457.
14. Malliani A, Pagani M, Lombardi F, Cerutti S. Cardiovascular neural regulation explored in the frequency domain. *Circulation*. 1991;84:482–492.
15. deBoer RW, Karemaker JM, Strackee J. Hemodynamic fluctuations and baroreflex sensitivity in humans: a beat-to-beat model. *Am J Physiol*. 1987;253:H680–H689.
16. White DW, Shoemaker JK, Raven PB. Methods and considerations for the analysis and standardization of assessing muscle sympathetic nerve activity in humans. *Auton Neurosci*. 2015;193:12–21.
17. Freeman R, Wieling W, Axelrod FB, Benditt DG, Benarroch E, Biaggioni I, Cheshire WP, Chelimsky T, Cortelli P, Gibbons CH, Goldstein DS, Hainsworth R, Hilz MJ, Jacob G, Kaufmann H, Jordan J, Lipsitz LA, Levine BD, Low PA, Mathias C, Raj SR, Robertson D, Sandroni P, Schatz I, Schondorff R, Stewart JM, van Dijk JG. Consensus statement on the definition of orthostatic hypotension, neurally mediated syncope and the postural tachycardia syndrome. *Clin Auton Res*. 2011;21:69–72.
18. Raj SR. Postural tachycardia syndrome (POTS). *Circulation*. 2013;127:2336–2342.
19. Johns MW. A new method for measuring daytime sleepiness: the Epworth sleepiness scale. *Sleep*. 1991;14:540–545.
20. Krupp LB, LaRocca NG, Muir-Nash J, Steinberg AD. The fatigue severity scale: application to patients with multiple sclerosis and systemic lupus erythematosus. *Arch Neurol*. 1989;46:1121–1123.
21. Miglis MG, Muppidi S, Feakins C, Fong L, Prieto T, Jaradeh S. Sleep disorders in patients with postural tachycardia syndrome. *Clin Auton Res*. 2016;26:67–73.
22. Deb A, Morgenshtern K, Culbertson CJ, Wang LB, Hohler AD. A survey-based analysis of symptoms in patients with postural orthostatic tachycardia syndrome. *Proc Bayl Univ Med Cent*. 2015;28:157–159.
23. Bagai K, Song Y, Ling JF, Malow B, Black BK, Biaggioni I, Robertson D, Raj SR. Sleep disturbances and diminished quality of life in postural tachycardia Syndrome. *J Clin Sleep Med*. 2011;7:204–210.
24. Bagai K, Wakwe CI, Malow B, Black BK, Biaggioni I, Paranjape SY, Orozco C, Raj SR. Estimation of sleep disturbances using wrist actigraphy in patients with postural tachycardia syndrome. *Auton Neurosci*. 2013;177:260–265.
25. Mallien J, Isenmann S, Mrazek A, Haensch CA. Sleep disturbances and autonomic dysfunction in patients with postural orthostatic tachycardia syndrome. *Front Neurol*. 2014;5:118.

26. Pengo MF, Higgins S, Drakatos P, Martin K, Gall N, Rossi GP, Leschziner G. Characterisation of sleep disturbances in postural orthostatic tachycardia syndrome: a polysomnography-based study. *Sleep Med*. 2015;16:1457–1461.

27. Guilleminault C, Primeau M, Chiu HY, Yuen KM, Leger D, Metlaine A. Sleep-disordered breathing in Ehlers–Danlos syndrome: a genetic model of OSA. *Chest*. 2013;144:1503–1511.

28. Figueroa JJ, Bott-Kitslaar DM, Mercado JA, Basford JR, Sandroni P, Shen WK, Sletten DM, Gehrking TL, Gehrking JA, Low PA, Singer W. Decreased orthostatic adrenergic reactivity in non-dipping postural tachycardia syndrome. *Auton Neurosci*. 2014;185:107–111.

29. Masana MI, Doolen S, Ersahin C, et al. MT(2) melatonin receptors are present and functional in rat caudal artery. *J Pharmacol Exp Ther*. 2002;302:1295–1302.

30. Jin X, von GC, Pieschl RL, et al. Targeted disruption of the mouse Mel(1b) melatonin receptor. *Mol Cell Biol*. 2003;23:1054–1060.

31. Green EA, Black BK, Biaggioni I, Paranjape SY, Bagai K, Shibao C, Okoye MC, Dupont WD, Robertson D, Raj SR. Melatonin reduces tachycardia in postural tachycardia syndrome: a randomized, crossover trial. *Cardiovasc Ther*. 2014;32:105–112.

32. Stoschitzky K, Sakotnik A, Lercher P, Zweiker R, Maier R, Liebmann P, Lindner W. Influence of beta-blockers on melatonin release. *Eur J Clin Pharmacol*. 1999;55:111–115.

33. Nofzinger EA, Buysse DJ, Germain A, Price JC, Miewald JM, Kupfer DJ. Functional neuroimaging evidence for hyperarousal in insomnia. *Am J Psychiatry*. 2004;161:2126–2128.

34. Perlis ML, Smith MT, Andrews PJ, Orff H, Giles DE. Beta/Gamma EEG activity in patients with primary and secondary insomnia and good sleeper controls. *Sleep*. 2001;24:110–117.

35. Blasi A, Jo J, Valladares E, Morgan BJ, Skatrud JB, Khoo MC. Cardiovascular variability after arousal from sleep: time-varying spectral analysis. *J Appl Physiol*. 2003;95:1394–1404.

36. Beighton P, DePaepe A, Steinmann B, Tsipouras P, Wenstrup RJ. Ehlers–Danlos syndromes: revised nosology, Villefranche, 1997. Ehlers–Danlos National Foundation (USA)and Ehlers–Danlos Support Group (UK). *Am J Med Genet*. 1998;77:31–37.

37. Hakim AJ, Sahota A. Joint hypermobility and skin elasticity: the hereditary disorders of connective tissue. *Clin Dermatol*. 2006;24:521–533.

38. Wallman D, Weinberg J, Hohler AD. Ehlers-Danlos Syndrome and Postural Tachycardia Syndrome: a relationship study. *J Neurol Sci*. 2014;340:99–102.

39. Jacome DE. Headache in Ehlers–Danlos syndrome. *Cephalalgia*. 1999;19:791–796.

40. Nelson AD, Mouchli MA, Valentin N, Deyle D, Pichurin P, Acosta A, Camilleri M. Ehlers Danlos syndrome and gastrointestinal manifestations: a 20-year experience at Mayo Clinic. *Neurogastroenterol Motil*. 2015;27:1657–1666.

41. Voermans NC, Knoop H, van de Kamp N, Hamel BC, Bleijenberg G, van Engelen BG. Fatigue is a frequent and clinically relevant problem in Ehlers–Danlos syndrome. *Semin Arthritis Rheum*. 2010;40:267–274.

42. Lugaresi E, Provini F, Cortelli P. Agrypnia excitata. *Sleep Med*. 2011;12(suppl 2):S3–S10.

43. Donadio V, Montagna P, Pennisi M, Rinaldi R, Di Stasi V, Avoni P, Bugiardini E, Giannoccaro MP, Cortelli P, Plazzi G, Baruzzi A, Liguori R. Agrypnia Excitata: a microneurographic study of muscle sympathetic nerve activity. *Clin Neurophysiol*. 2009;120:1139–1142.

44. Portaluppi F, Cortelli P, Avoni P, Vergnani L, Contin M, Maltoni P, Pavani A, Sforza E, degli Uberti EC, Gambetti P, et al. Diurnal blood pressure variation and hormonal correlates in fatal familial insomnia. *Hypertension*. 1994;23:569–576.

45. Montagna P, Gambetti P, Cortelli P, Lugaresi E. Familial and sporadic fatal insomnia. *Lancet Neurol*. 2003;2:167–176.

46. Iranzo A, Tolosa E, Gelpi E, Molinuevo JL, Valldeoriola F, Serradell M, Sanchez-Valle R, Vilaseca I, Lomeña F, Vilas D, Lladó A, Gaig C, Santamaria J. Neurodegenerative disease status and post-mortem pathology in idiopathic rapid-eye-movement sleep behavior disorder: an observational cohort study. *Lancet Neurol*. 2013;12:443–453.

47. Postuma RB, Gagnon JF, Vendette M, Montplaisir JY. Markers of neurodegeneration in idiopathic rapid eye movement sleep behaviour disorder and Parkinson's disease. *Brain*. 2009;132:3298–3307.

48. Schenck CH, Boeve BF, Mahowald MW. Delayed emergence of a parkinsonian disorder or dementia in 81% of older men initially diagnosed with idiopathic rapid eye movement sleep behavior disorder: a 16-year update on a previously reported series. *Sleep Med*. 2013;14:744–748.

49. Postuma RB, Aarsland D, Barone P, Burn DJ, Hawkes CH, Oertel W, Ziemssen T. Identifying prodromal Parkinson's disease: pre-motor disorders in Parkinson's disease. *Mov Disord*. 2012;27:617–626.

50. Boeve BF, Silber MH, Ferman TJ, Lin SC, Benarroch EE, Schmeichel AM, Ahlskog JE, Caselli RJ, Jacobson S, Sabbagh M, Adler C, Woodruff B, Beach TG, Iranzo A, Gelpi E, Santamaria J, Tolosa E, Singer C, Mash DC,

Luca C, Arnulf I, Duyckaerts C, Schenck CH, Mahowald MW, Dauvilliers Y, Graff-Radford NR, Wszolek ZK, Parisi JE, Dugger B, Murray ME, Dickson DW. Clinicopathologic correlations in 172 cases of rapid eye movement sleep behavior disorder with or without a coexistingneurologic disorder. *Sleep Med*. 2013;14:754–762.

51. Braak H, Del Tredici K, Rüb U, de Vos RA, Jansen Steur EN, Braak E. Staging of brain pathology related to sporadic Parkinson's disease. *Neurobiol Aging*. 2003;24:197–211.

52. Kaufmann H. Consensus statement on the definition of orthostatic hypotension, pure autonomic failure and multiple system atrophy. *Clin Auton Res*. 1996;6:125–126.

53. Plazzi G, Cortelli P, Montagna P, De Monte A, Corsini R, Contin M, Provini F, Pierangeli G, Lugaresi E. REM sleep behaviour disorder differentiates pure autonomic failure from multiple system atrophy with autonomic failure. *J Neurol Neurosurg Psychiatry*. 1998;64:683–685.

54. Bannister R, Mathias CJ. Clinical features and investigation of the primary autonomic failure syndromes. 3rd ed. Autonomic failure, a textbook of clinical disorders of the nervous system. London: Oxford University Press; 1992:531–547.

55. Weyer A, Minnerop M, Abele M, Klockgether T. REM sleep behavioral disorder in pure autonomic failure (PAF). *Neurology*. 2006;66:608–609.

56. Kashihara K, Ohno M, Kawada S, Imamura T. Frequent nocturnal vocalization in pure autonomic failure. *J Int Med Res*. 2008;36:489–495.

57. Plazzi G, Corsini R, Provini F, Pierangeli G, Martinelli P, Montagna P, Lugaresi E, Cortelli P. REM sleep behavior disorders in multiple system atrophy. *Neurology*. 1997;48:1094–1097.

58. Palma JA, Fernandez-Cordon C, Coon EA, Low PA, Miglis MG, Jaradeh S, Bhaumik AK, Dayalu P, Urrestarazu E, Iriarte J, Biaggioni I, Kaufmann H. Prevalence of REM sleep behavior disorder in multiple system atrophy: a multicenter study and meta-analysis. *Clin Auton Res*. 2015;25:69–75.

59. Gaig C, Iranzo A. Sleep-disordered breathing in neurodegenerative diseases. *Curr Neurol Neurosci Rep*. 2012;12:205–217.

60. Shimohata T, Tomita M, Nakayama H, et al. Floppy epiglottis as a contraindication of CPAP in patients with multiple system atrophy. *Neurology*. 2011;76:1841–1842.

61. Iranzo A. Sleep and breathing in multiple system atrophy. *Curr Treat Options Neurol*. 2007;9:347–353.

62. Suzuki M, Saigusa H, Shibasaki K. Multiple system atrophy manifesting as complex sleep-disordered breathing. *Auris Nasus Larynx*. 2010;37:110–113.

63. Ghorayeb I, Yekhlef F, Chrysostome Y, et al. Sleep disorders and their determinants in multiple system atrophy. *J Neurol Neurosurg Psychiatry*. 2002;72:798–800.

64. Munschauer FE, Loh L, Bannister R, Newsom-Davis J. Abnormal respiration and sudden death during sleep in multiple system atrophy with autonomic failure. *Neurology*. 1990;40:677–679.

65. Iranzo A, Santamaria J, Tolosa E, et al. Long-term effect of CPAP in the treatment of nocturnal stridor in multiple system atrophy. *Neurology*. 2004;63:930–932.

66. Gama RL, Távora DG, Bonfim RC, et al. Sleep disturbances and brain MRI morphometry in Parkinson's disease, multiple system atrophy and progressive supranuclear palsy: a comparative study. *Parkinsonism Relat Disord*. 2010;16:275–279.

67. Tandberg E, Larsen JP, Karlsen K. A community-based study of sleep disorders in patients with Parkinson's disease. *Mov Disord*. 1998;13:895–899.

68. Happe S, Lüdemann P, Berger K. FAQT study investigators. The association between disease severity and sleep-related problems in patients with Parkinson's disease. *Neuropsychobiology*. 2002;46:90–96.

69. Postuma RB, Gagnon JF, Vendette M, Charland K, Montplaisir J. REM sleep behaviour disorder in Parkinson's disease is associated with specific motor features. *J Neurol Neurosurg Psychiatry*. 2008;79:1117–1121.

70. Postuma RB, Gagnon JF, Vendette M, Charland K, Montplaisir J. Manifestations of Parkinson disease differ in association with REM sleep behavior disorder. *Mov Disord*. 2008;23:1665–1672.

71. Raggi A, Neri W, Ferri R. Sleep-related behaviors in Alzheimer's disease and dementia with Lewy bodies. *Rev Neurosci*. 2015;26:31–38.

72. Thaisetthawatkul P, Boeve BF, Benarroch EE, Sandroni P, Ferman TJ, Petersen R, Low PA. Autonomic dysfunction in dementia with Lewy bodies. *Neurology*. 2004;62:1804–1809.

73. Ferman TJ, Boeve BF, Smith GE, Lin SC, Silber MH, Pedraza O, Wszolek Z, Graff-Radford NR, Uitti R, Van Gerpen J, Pao W, Knopman D, Pankratz VS, Kantarci K, Boot B, Parisi JE, Dugger BN, Fujishiro H, Petersen RC, Dickson DW. Inclusion of RBD improves the diagnostic classification of dementia with Lewy bodies. *Neurology*. 2011;77:875–882.

74. Ferini-Strambi L, Oertel W, Dauvilliers Y, Postuma RB, Marelli S, Iranzo A, Arnulf I, Högl B, Manni R, Miyamoto T, Fantini ML, Puligheddu M, Jennum P, Sonka K, Santamaria J, Zucconi M, Rancoita PM, Leu-Semenescu S, Frauscher B, Terzaghi M, Miyamoto M, Unger M, Stiasny-Kolster K, Desautels A, Wolfson C, Pelletier A, Montplaisir J. Autonomic symptoms in idiopathic REM behavior disorder: a multicenter case-control study. *J Neurol.* 2014;261:1112–1118.

75. Postuma RB, Iranzo A, Hogl B, Arnulf I, Ferini-Strambi L, Manni R, Miyamoto T, Oertel W, Dauvilliers Y, Ju YE, Puligheddu M, Sonka K, Pelletier A, Santamaria J, Frauscher B, Leu-Semenescu S, Zucconi M, Terzaghi M, Miyamoto M, Unger MM, Carlander B, Fantini ML, Montplaisir JY. Risk factors for neurodegeneration in idiopathic rapid eye movement sleep behavior disorder: a multicenter study. *Ann Neurol.* 2015;77:830–839.

76. Lanfranchi PA, Fradette L, Gagnon JF, Colombo R, Montplaisir J. Cardiac autonomic regulation during sleep in idiopathic REM sleep behavior disorder. *Sleep.* 2007;30:1019–1025.

77. Frauscher B, Nomura T, Duerr S, Ehrmann L, Gschliesser V, Wenning GK, Wolf E, Inoue Y, Högl B, Poewe W. Investigation of autonomic function in idiopathic REM sleep behavior disorder. *J Neurol.* 2012;259:1056–1061.

78. Miyamoto T, Miyamoto M, Inoue Y, Usui Y, Suzuki K, Hirata K. Reduced cardiac 123I-MIBG scintigraphy in idiopathic REM sleep behavior disorder. *Neurology.* 2006;67:2236–2238.

Index